MW00845591

Neuroimaging in Dementia

Frederik Barkhof · Nick C. Fox
António J. Bastos-Leite · Philip Scheltens

Neuroimaging in Dementia

Authors

Frederik Barkhof
Professor of Neuroradiology
Department of Radiology and Image
Analysis Center (IAC)
VU University Medical Center
De Boelelaan 1117
1081 HV Amsterdam
The Netherlands
f.barkhof@vumc.nl

Nick C. Fox
Professor of Neurology
Dementia Research Centre
UCL Institute of Neurology
London WC1N 3BG
United Kingdom
and
Professor of Neurology
VU University Medical Center
De Boelelaan 1117
1081 HV Amsterdam
The Netherlands
n.fox@dementia.ion.ucl.ac.uk

António J. Bastos-Leite
Assistant Professor of Radiology
and Medical Imaging
University of Porto
Faculty of Medicine
Department of Medical Imaging
Alameda do Professor Hernâni Monteiro
4200-319 Porto
Portugal
abastosleite@med.up.pt

Philip Scheltens
Professor of Cognitive Neurology
Department of Neurology
and Alzheimer Center
VU University Medical Center
De Boelelaan 1117
1081 HV Amsterdam
The Netherlands
p.scheltens@vumc.nl

ISBN 978-3-642-00817-7   ISBN 978-3-642-00818-4 (eBook)
DOI 10.1007/978-3-642-00818-4
Springer Heidelberg Dordrecht London New York

Library of Congress Control Number: 2011921695

© Springer-Verlag Berlin Heidelberg 2011

This work is subject to copyright. All rights are reserved, whether the whole or part of the material is concerned, specifically the rights of translation, reprinting, reuse of illustrations, recitation, broadcasting, reproduction on microfilm or in any other way, and storage in data banks. Duplication of this publication or parts thereof is permitted only under the provisions of the German Copyright Law of September 9, 1965, in its current version, and permission for use must always be obtained from Springer. Violations are liable to prosecution under the German Copyright Law.

The use of general descriptive names, registered names, trademarks, etc. in this publication does not imply, even in the absence of a specific statement, that such names are exempt from the relevant protective laws and regulations and therefore free for general use.

Product liability: The publishers cannot guarantee the accuracy of any information about dosage and application contained in this book. In every individual case the user must check such information by consulting the relevant literature.

Cover design: eStudioCalamar, Figueres/Berlin

Printed on acid-free paper

Springer is part of Springer Science+Business Media (www.springer.com)

Preface

This book is inspired by the previous work entitled 'Magnetic Resonance in Dementia' published in 2002 with our dear colleague and friend, Jaap Valk. We were encouraged by many positive reactions from colleagues in different disciplines including radiology, neurology, psychiatry and geriatrics. Since then, so many developments have taken place that a completely new title was needed. First of all, there are new diseases, or new insights into existing disorders presenting with dementia. Secondly, a vast amount of new imaging studies on dementia disorders have become available. Thirdly, image processing techniques have made their entrance into clinical practice and have now been fully integrated. To reflect the incorporation of these developments and the more extensive coverage of other imaging modalities such as PET, we have chosen a new title, 'Neuroimaging in Dementia'.

Preparing a completely new book with a new team of authors also provides an opportunity to reorganise the material presented in our previous title. "Neuroimaging in Dementia" provides a consistent focus on MRI appearance as the guiding principle. In this vein, the classification of dementia has also been revised to follow the MRI appearance as strictly as possible. To enhance legibility and to be as clinically useful as possible, many tables and boxes are included. Last, but not least, a considerable number of new MRI and PET images has been introduced to the backbone of this book to create a true imaging atlas of dementia alongside the text.

We trust you will find this title informative and hope that it will find a place in your daily practice of managing patients with dementia.

Amsterdam	Frederik Barkhof and Philip Scheltens
London	Nick C. Fox
Porto	António Bastos-Leite

Acknowledgements

This book was only possible because of the generosity of many colleagues and friends in sharing images and ideas. We are particularly grateful to the following contributors and collaborators:

Bart van Berckel
Ed Wild, Nicola Hobbs and Susie Henley
Giorgos Karas
John O'Brien and Michael Firbank
Dominic Paviour
Jonathan D. Rohrer
Annemiek Rozemuller
Jonathan Schott
Wiesje van der Flier
Mike Wattjes

Contents

List of Abbreviations

ACA	Anterior cerebral artery
AChEI	Acetylcholinesterase inhibitors
AD	Alzheimer's disease
ADC	Apparent diffusion coefficient
ADEM	Acute disseminated encephalomyelitis
AGD	Argyrophilic grain disease
ALD	Adrenoleukodystrophy
AMN	Adrenomyeloneuropathy
ANA	Anti-nuclear antibodies
APBD	Adult polyglucosan body disease
APOE	Apolipoprotein E
APOEe4	Apolipoprotein E4 allele
APP	Amyloid precursor protein
ARWMC	Age-related white matter changes
ASA	Arylsulfatase A
ASL	Arterial spin labelling
AVM	Arteriovenous malformations
BBSI	Brain-boundary shift integral
BOLD	Blood oxygen-level dependent
BSE	Bovine spongiform encephalopathy
CA	Cornu ammonis
CAA	Congophilic amyloid angiopathy
CACH	Childhood ataxia with central hypomyelination
CADASIL	Cerebral autosomal dominant arteriopathy with subcortical infarcts and leukoencephalopathy
CBD	Corticobasal degeneration
CBF	Cerebral blood flow
CBS	Corticobasal syndrome
CHMP2B	Charged multivesicular body protein 2B
CJD	Creutzfeldt-Jakob disease
CKD	Chronic kidney disease
CNS	Central nervous system
CO	Carbon monoxide
COPD	Chronic obstructive pulmonary disease
Cr	Creatine
CRP	C-reactive protein
CSF	Cerebrospinal fluid

CT	Computed tomography
CTX	Cerebrotendinous xanthomatosis
DAI	Diffuse axonal injury
DAVF	Dural arteriovenous fistula
DIR	Double-inversion recovery
DIS	Dissemination in space
DIT	Dissemination in time
DLB	Dementia with Lewy bodies
DMN	Default-mode network
DNTC	Diffuse neurofibrillary tangles with calcification
DPHL	Delayed posthypoxic leukoencephalopathy
DRPLA	Dentatorubral-pallidoluysian atrophy
DSA	Digital subtraction angiography
DSC	Dynamic susceptibility contrast
DTI	Diffusion tensor imaging
DWI	Diffusion-weighted imaging
EEG	Electro-encephalography
ELISA	Enzyme-linked immunosorbent assay
EMG	Electromyograph
EPI	Echo-planar imaging
ESR	Erythrocyte sedimentation rate
FA	Fractional anisotropy
FBD	Familial British dementia
FD	Fabry's disease
FDD	Familial Danish dementia
FDG	Fluorodeoxyglucose
FDG-PET	Fluorodeoxyglucose-positron emission tomography
FFI	Fatal familial insomnia
FLAIR	Fluid attenuation inversion recovery
FMR1	Fragile X mental retardation gene-1
fMRI	Functional magnetic resonance imaging
FSE	Fast spin echo
FTD	Frontotemporal dementia
FTDP-17	Frontotemporal dementia with parkinsonism linked to chromosome 17
FTLD	Frontotemporal lobar degeneration
FUS	Fused-in-sarcoma
FXTAS	Fragile X-associated tremor/ataxia syndrome
GALC	Galactocerebrosidase
GCA	Global cortical atrophy
GE	Gradient echo
GFAP	Glial fibrillary acid protein
GFR	Glomerular filtration rate
GLD	Globoid leukodystrophy
GM	Grey matter
GRN	Progranulin
GSD	Glycogen storage disorder
GSS	Gerstmann-Sträussler-Scheinker
HAART	Highly active anti-retroviral therapy

HAND	HIV-associated neurocognitive dysfunction
HCHWA	Hereditary cerebral hemorrhage with amyloidosis
HCV	Hippocampal volume
HD	Huntington's disease
HDL	Huntington's disease-like
HIV	Human immunodeficiency virus
HIVE	Human immunodeficiency virus encephalitis
HLA	Human leukocyte antigen
HMPAO	Hexamethylpropylene amine oxime
HSE	Herpes simplex encephalitis
HSV	Herpes simplex virus
HSV-1	Herpes simplex virus type-1
IBMPFD	Inclusion body myopathy associated with Paget's disease and frontotemporal dementia
ICA	Independent component analysis
^{123}I-MIBG	Iodine-123 metaiodobenzylguanidine
IRIS	Immune reconstitution inflammation syndrome
IVL	Intravascular lymphomatosis
KSS	Kearns-Sayre syndrome
LB	Lewy bodies
LBD	Lewy body dementia
LE	Limbic encephalitis
LPA	Logopenic aphasia
MAO	Monoamine oxidase
MAPT	Microtubule-associated protein tau
MBs	Microbleeds
MCA	Middle cerebral artery
MCI	Mild cognitive impairment
MCP	Middle cerebellar peduncles
MD	Mean diffusivity
MELAS	Mitochondrial encephalomyopathy with lactic acidosis and stroke-like episodes
MGC	Multinucleated giant cells
MID	Multi-infarct dementia
MLD	Metachromatic leukodystrophy
MNGIE	Mitochondrial neuro-gastrointestinal encephalopathy
MPR	Multi-planar reconstruction
MRA	Magnetic resonance angiography
MRI	Magnetic resonance imaging
MRS	Magnetic resonance spectroscopy
MS	Multiple sclerosis
MSA	Multiple system atrophy
MT	Magnetization transfer
MTA	Medial temporal lobe atrophy
MTR	Magnetization transfer ratio
NAA	N-acetyl aspartate
NAL	Neuroaxonal leukodystrophy
NBIA	Neurodegeneration with brain iron accumulation
NFT	Neurofibrillary tangles

NIFID	Neuronal intermediate filament inclusion disease
NINDS-AIREN	National Institute of Neurological Disorders and Stroke-Association Internationale pour la Recherche et l'Enseignement en Neurosciences
NMDA	N-methyl d-aspartate
NMO	Neuromyelitis optica
NPH	Normal pressure hydrocephalus
OPCA	Olivopontocerebellar atrophy
PAS	Periodic acid-Schiff
PC	Phase-contrast
PCA	Posterior cerebral artery
PCR	Polymerase chain reaction
PD	Parkinson's disease
PDD	Parkinson disease dementia
PET	Positron emission tomography
PIB	Pittsburgh compound B
PKAN	Pantothenate kinase-associated neurodegeneration
PML	Progressive multifocal leukoencephalopathy
PNFA	Progressive nonfluent aphasia
PPA	Primary progressive aphasia
PRES	Posterior reversible encephalopathy syndrome
PSEN	Presenilin
PSP	Progressive supranuclear palsy
PWI	Perfusion-weighted imaging
REM	Rapid eye movement
ROI	Region-of-interest
RPLS	Reversible posterior leukoencephalopathy syndrome
RRMS	Relapsing/remitting multiple sclerosis
SAE	Subcortical arteriosclerotic encephalopathy
SCA	Spinocerebellar ataxia
SCP	Superior cerebellar peduncle
SD	Semantic dementia
SIVD	Subcortical ischemic vascular dementia
SLE	Systemic lupus erythematosus
SPECT	Single-photon emission computed tomography
SPM	Statistical parametric mapping
SSPE	Subacute sclerosing panencephalitis
SVD	Small vessel disease
SVD	Subcortical vascular dementia
SWI	Susceptibility-weighted imaging
TBI	Traumatic brain injury
TEA	Transient epileptic attacks
TGA	Transient global amnesia
THC	Tetrahydrocannabinol
TSE	Transmissible spongiform encephalopathy
TSE	Turbo spin echo
VaD	Vascular dementia
VBM	Voxel-based morphometry
VCP	Valosin-containing protein

VGKC	Voltage-gated potassium channel
VLCFA	Very-long-chain fatty acids
VRS	Virchow-Robin spaces
VWM	Vanishing white matter
WE	Wernicke's encephalopathy
WK	Wernicke-Korsakoff's syndrome
WM	White matter
WMC	White matter changes
WMH	White matter hyperintensity
WML	White matter lesions

How to Use This Book?

1

Contents

1.1 General Background

The focus of this book is on the practical use of neuroimaging in dementia in a clinical diagnostic setting. We felt that there are many publications that describe the imaging findings of a particular disease, but you first need to know what the diagnosis is in order to look up articles describing those findings. As well as offering a summary of the findings in the most relevant conditions causing cognitive decline we wished to provide a guide to interpreting a particular imaging finding. The organization of the book therefore takes as departure point the dominant imaging findings and incorporates the clinical features along the way. The topic of vascular dementia does not easily fit with this approach due to its heterogeneous appearance; however, it does form a natural bridge between primary white and grey matter disorders. The 'route map' aims to direct the reader towards additional tests (imaging and non-imaging) and clinical features in a practical way.

Etiological, pathogenetic and clinical information are given as a reference, mainly as a background to understand and interpret imaging findings, not to provide an encyclopaedic text on all aspects on dementia – the interested reader will easily find her/his way to dedicated textbooks on genetics, biochemistry, histopathology and others.

Structural MR imaging is the lead theme largely because of its central position in clinical practice in many countries. Each chapter contains suggestion about the imaging strategy (e.g. which sequences to apply) and interpretation (e.g. salient features to look for) within a given clinical context (e.g. young age at onset). When appropriate, suggestion are provided for nonconventional MR techniques, such as diffusion-weighted

F. Barkhof et al., *Neuroimaging in Dementia*,
DOI: 10.1007/978-3-642-00818-4_1, © Springer-Verlag Berlin Heidelberg 2011

MR, indications for nuclear medicine techniques (e.g. PET), or other diagnostic tests, such as CSF analysis.

1.2 Main Classification System

There are many ways to classify dementing disorders, e.g. sporadic/inherited, cortical/subcortical, all of which have their limitations. Classification according to histopathology (e.g. with or without certain type of inclusion bodies) is conceptually attractive, but clinically not very useful. By contrast, structural (MR) imaging is often performed in the work-up of a patient presenting with cognitive decline (even if only to exclude surgical pathology) and provides an increasingly useful angle of thought – or point of departure.

Our classification system is based roughly on four dominant imaging patterns:

• Primary grey matter loss – neurodegenerative diseases such as Alzheimer
• Vascular dementia – combined white and grey matter damage

• Primary white matter disorders – e.g. HIV encephalitis and metabolic disorders
• Disorders associated with brain swelling

While the lead theme of dominant neuroimaging finding may be useful in many circumstances, there are many patients in whom a clinical clue (e.g. visual hallucinations) can be more relevant than the non-specific imaging findings (diffuse cortical atrophy in case of Lewy body dementia). Other clinical settings, e.g. rapidly progressive cognitive decline, may lead to a differential diagnosis that may run across the disease clusters as reflected by the main chapters. Such alternative slicing patterns are presented throughout the book.

1.3 A Route-Map or Classification Tree

A key goal of imaging is to exclude a neurosurgically treatable cause of dementia (see e.g. Practice Parameter AAN); an MR scan performed for such an indication will include a T2-weighted sequence (e.g. FLAIR) which provides a useful starting point for our purpose.

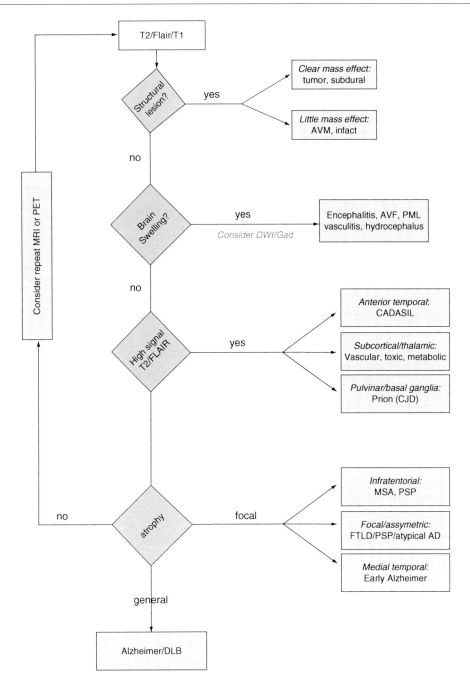

The flow diagram above provides an example of how a series of assessments can be used to lead into the main diagnostic groups as represented in this book. It should be noted that although a 'main finding' may be a key pointer there can be considerable overlap of findings. Incidental white matter lesions for example will present in many elderly subjects, and occur with increased frequency in patients with Alzheimer's disease. Additionally, combinations of pathology are the rule rather than the exception in the very old, especially Alzheimer's and Vascular dementia.

Dementia: Clinical Background

Contents

2.1 What Is Dementia?

Dementia refers to a clinical syndrome rather than a disease. Dementia is usually defined as an acquired condition involving multiple cognitive impairments that are sufficient to interfere with activities of daily living. It is usually but not necessarily progressive. Memory impairment is one of the most common deficits, but other domains such as language, praxis, visual-perceptive and most notably executive functions are often involved. With increasing loss of function due to these cognitive problems, there is progressive difficulty with activities of daily living. Many of the diseases that cause dementia have a relentlessly progressive course with an insidious onset; many have long durations (e.g. 5–10 years from diagnosis) and relatively prolonged end stage period where all self-care and -independence is lost. Dementia places tremendous burdens on patients, their families and carers and on health and social care systems. The most important causes of dementia have an age-related incidence. As a result, the prevalence and societal costs of dementia are predicted to rise dramatically over the coming decades.

2.2 Prevalence and Incidence

Of all diseases associated with age, dementia is the fastest growing entity (Fig. 2.1).

2.2.1 Prevalence

In 2000, prevalence data of 11 European population-based studies were pooled to obtain stable estimates of

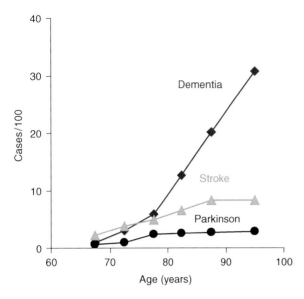

Fig. 2.1 Prevalence of three age-associated syndromes. Dementia shows the highest increase in numbers with advancing age (*Eurodem*)

prevalence of dementia in the elderly (>65 years). Age-standardized prevalence was 6.4% for dementia (all causes), 4.4% for AD and 1.6% for VaD. Prevalence of dementia was higher in women than in men and nearly doubled with every 5 years increase of age: from 0.8% in the age group 65–69 years to 28.5% over the age of 90 years (Fig. 2.2).

Prevalence rates for dementia have been compared among 12 population-based European studies.

Crude prevalence rates varied between 5.9% (Italy, the Counselice study) and 9.4% (the Netherlands, Rotterdam study). Again, an almost exponential increase with age and a female excess – mostly after age 75 – was described, independent of country. As the age distribution of the Western population shifts, the rapid increase of the prevalence of dementia with increasing age means that both the number of affected individuals and the affected proportion of the total population are increasing. This will be most prominent in Europe, where the median age of the population is higher than in any other part of the world.

A consensus conference in 2005 under the auspices of Alzheimer Disease International estimated that 24.3 million people worldwide suffer from dementia, with 4.6 million new cases of dementia every year (one new case every 7 s) (Ferri et al. 2005). A recent update by ADI in 2009 estimated that 35.6 million people worldwide will be living with dementia in 2010. This number was estimated to nearly double every 20 years, to 65.7 million in 2030, and 115.4 million in 2050 (www.alz.co.uk). Much of the increase is clearly attributable to increases in the numbers of people with dementia in low and middle income countries. Rates of increase are not uniform and are driven by the population structure and life-expectancy changes; numbers in developed countries are forecasted to increase by 100% between 2001 and 2040, but by more than 300% in India, China and their south Asian and western Pacific neighbours.

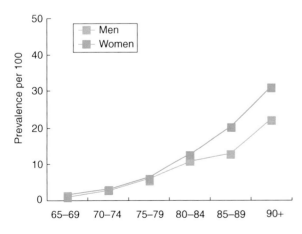

Fig. 2.2 Prevalence rates of dementia among men and women after the age of 65. (After Lobo et al. (2000) Neurology 54(11 Suppl 5):S4–S9)

2.2.2 Incidence

In the same collaborative effort that pooled prevalence data of European studies, data on incidence of dementia of eight population-based European studies were compared and pooled. In total, there were 42,996 person-years of follow-up with 835 new dementia cases. Of these, 60–70% were diagnosed with AD and 15–20% with VaD. Incidence rates of dementia increased exponentially with age from 2.4 per 1,000 person-years in the 65–69 age group to 70.2 per 1,000 person-years in the 90+ age group. Rates among women were higher, especially above the age of 80 (Fig. 2.3). The rates continue to increase with age in women, whereas the increase plateaus in men at age 85.

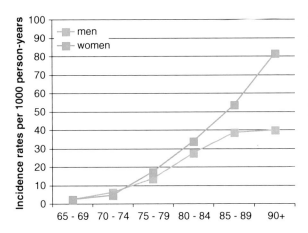

Fig. 2.3 Pooled incidence rates of dementia by sex. (Data from Fratiglioni et al. (2000) Neurology 54(11 Suppl 5):S10–S15)

2.3 Nosological Approach

As mentioned above, dementia is a syndrome, not a disease, and has many and varied causes. The diagnostic workup is meant to identify the underlying cause with a particular emphasis on picking up treatable conditions. Diagnosis is critically dependent on careful history taking from patient and informant followed by clinical and cognitive examination supported by ancillary investigations, of which neuroimaging is one of the most important. The a priori chance of a particular disease being present is dependent on age. The younger the patient, the greater the chance that one of a wide range of underlying pathologies is the cause of the cognitive problems. Diseases like FTD and HD tend to occur more often before the age of 70; genetic forms of AD almost exclusively occur at young ages and rare metabolic causes are more likely in early adulthood (see Table 2.1). In the older patient, AD, DLB and vascular disease are by far the most common pathologies. Mixed disease is very common: notably, AD with vascular disease has been shown to be the most prevalent in post-mortem series of older individuals (>85 years).

The nosological approach is facilitated by the use of clinical criteria, which are detailed in the remaining chapters of this book, where the diseases

Table 2.1 Differential diagnostic considerations in a patient presenting with dementia at young age (arbitrarily defined as onset before age 65). Note the wide variety of diseases in this age group and the particular emphasis on the use of imaging

Disease	MRI findings	Clinical clues	Additional tests
AD	Posterior cingulate atrophy, medial temporal atrophy	Family history, visuospatial and apraxia > memory	CSF (abeta and tau); FDG-PET; amyloid PET
FTLD	Frontotemporal atrophy Temporal atrophy (asymmetrical or symmetrical)	Family history, language, behaviour	FDG-PET
CBD	Frontoparietal atrophy; may be asymmetrical	Asymmetrical Parkinsonism, dyspraxia and myoclonus; alien limb	CSF; Dopamine imaging
SVD	Strategic infarcts, lacunes, WMH	TIA; stroke	Vascular risk factors
Vasculitis	WMH, patchy enhancement, multifocal diffusion restriction	TIA, multifocal	ESR and CRP elevation; CSF, DSA, serology
MS	Disseminated WM lesions, black holes; Gad-enhancement	Relapses; other neurological findings	CSF oligoclonal bands
CJD	Abnormal DWI basal ganglia or neocortex	Myoclonus; cerebellar ataxia	EEG, CSF tau and 14-3-3-protein
Paraneoplastic or limbic encephalitis	Temporal lobe lesions; thalamic swelling	Subacute onset; other neurological findings	CSF antibodies
Infectious	WM lesions, enhancement	Fever, HIV, Lues	Serology, CSF, culture
Metabolic	WM lesions, GM lesions, lactate in spectroscopy, diffusion restriction	Stroke-like episode	CSF, serology, muscle biopsy, genetics

Source: Modified from Ridha B, Josephs KA (2006) Neurologist 12:2–13

Note the wide variety of diseases and the particular emphasis on the use of imaging. For abbreviations see list on page XV

Table 2.2 Listing of the clinical criteria for the various dementia syndromes

Dementia type	Presenting symptom	Criteria	Year published	Imaging included
AD	Memory	NINCDS-ADRDA	1984	No
		DSM IV	1994	No
		Dubois	2007	Yes, MRI/PET
VaD	Memory	NINDS-AIREN	1993	Yes, CT/MRI
	Memory	DSM IV	1994	No
	Unspec	SCADDTC	2002	Yes, CT/MRI
	Dysexecutive	SIVD	2000	Yes, MRI
DLB	Fluctuating, Executive Dysfunction	McKeith	2005	SPECT and MRI
FT(L)D	Behaviour,	Neary	1998	Supportive
	Language	McKhann	2001	No
CJD sporadic	Various	Masters	1979	No
CJD variant	Psychiatric	Will	2000	Yes, MRI
PSP	Falls,	Litvan	1996	No
	Parkinsonism	Williams	2005	No
CBD	Limb Dyspraxia	Boxer	2006	No
NPH	Gait	Vanneste	2000	Yes, CT/MRI
Huntington	Chorea	CAG repeats	1993	No
MSA	Parkinsonism	Gilman	2008	MRI, PET supportive

The table illustrates that for some diseases in time neuroimaging features have been added to the strictly clinical features. For abbreviations see list on page XV

presenting with dementia are discussed. In Table 2.2, the main disease categories and their published clinical criteria are listed with the use of imaging highlighted. From the table, it may be inferred that for the majority of diseases, no specific imaging criteria have been formulated; however, it is also notable that more recent revisions of criteria are increasingly including imaging (for positive as well as negative predictive value).

2.3.1 Genetic/Protein Classification

Several genes have been implicated in the origin of dementia syndromes. Some diseases are almost exclusively genetic, like HD, while in AD, genetic forms account for <5% of all cases. While the gene product is known for many of the genes, effective therapy has not

evolved. In Table 2.3 the known genes and location are listed.

2.3.2 Clinical and Pathological Uncertainty

Using clinical criteria various levels of diagnostic certainty may be reached. For instance, the NINCDS-ADRDA criteria for probable AD have a diagnostic sensitivity and specificity compared to the pathological diagnosis, ranging between 50% and 90%, mainly depending on the setting (clinical expertise) and the age of the patients studied. This diagnostic uncertainty applies to other clinical criteria as well. Of note is that when imaging is included in the criteria such as in the NINDS-AIREN a higher degree of specificity (>90%) is reached. In general, the use of imaging has shifted from excluding disorders that may mimic a dementia

Table 2.3 Genetic causes of dementia

Disease/phenotype	Gene	Gene product	Chromosome	Age at onset (typical)
AD	PSEN 1 PSEN 2 APP	Amyloid Amyloid Amyloid	14 1 21	30–55 variable 45–65
HCHWA	APP Cystatin C BRI2	Amyloid Variant cystatin C ABri and ADan	21 20 13	<65 variable variable
CJD, FFI, GSS	PRNP	Prion protein	20	variable
FTD (esp bvFTD), CBS, PSP	MAPT	Tau	17	25–65
FTD, PNFA	Progranulin, GRN	TDP43 + ve intranuclear inclusions in neurons	17	35–90
BvFTD, FTD-MND	TARDBP (TDP-43); CHMP2B; VCP	Idem; Ubiquitin	1, 3, 9	Very variable and rare
CADASIL	Notch3	Notch protein	19	25–65
Huntington's disease	IT-15	Huntingtin	20	Variable (CAG repeat length)

For abbreviations see list on page XV

syndrome or may be (surgically) treatable to using it to identify specific abnormalities that may aid the clinician to diagnose underlying disease, i.e. to increase specificity over sensitivity. One has to bear in mind that the ultimate 'gold standard' for diagnosis does not exist. In many criteria, a definite diagnosis is often designated as either being made post-mortem or on the basis of genetic information. The former is obviously too late to be helpful in the clinical situation and usually becomes available many years after the first clinical manifestation. The latter may be available during the clinical workup, and probably better serves to inform the clinician about the underlying pathology than anything else. In this respect, certain tau mutations leading to an unexpected clinical diagnosis of AD or vice versa presenilin mutations with unexpected clinical FTD presentation are particularly informative. However, one has to be careful about generalising from familial to sporadic cases. The future of clinical diagnosis making lies within the realm of making a diagnosis at protein level, regardless of the clinical presentation. Possibly, molecular imaging (e.g. demonstrating amyloid deposition rather than a given clinical presentation) will allow a more rational approach towards disease modifying treatment; other imaging of specific pathological markers (e.g. tau) would be very valuable in differential diagnosis. Until that is possible, clinical and radiological information has to be pooled to make the best possible judgement to enable treatment and management of the patient.

2.4 Differential Diagnosis

In the twentieth century the perspective on dementia evolved tremendously. Before 1900, there was very little in the way of specific diagnoses, but with much effort from clinicians to recognize subtypes and help from pathologist, geneticists, neuro-imagers and others, it is now possible to make a list of differential diagnoses and to have a fair chance of predicting pathology in a number of conditions.

Memory deficits, a key feature of the DSM IIIR definition of dementia is no longer essential for dementia and a number of criteria for different diseases causing dementia incorporate the different cognitive profiles expected in the different disorders. This shift in conceptual thinking is illustrated in Table 2.2.

2.4.1 Diagnostic Evaluation

A full diagnostic evaluation is warranted in every patient who present with cognitive or behavioural complaints. Current EFNS and AAN guidelines stipulate what tests are evidence based and need to be done. Below, the main ancillary investigations are summarised. Note that in general the tendency is to move away from excluding other (brain) diseases, towards finding specific clues to make a diagnosis. Imaging has taken the lead in this, followed closely by CSF examinations and to a lesser extent EEG.

2.4.1.1 Laboratory Tests

These should be used to explore whether the patient has co-morbidity, risk factors for dementia, and risk for delirium or has a primary cause for dementia. For this matter, the following tests are generally proposed as mandatory: full blood count and erythrocyte sedimentation rate (ESR), electrolytes, calcium, glucose, renal, liver and thyroid function tests. More extensive tests will be required in individual cases (and places), like serological test for syphilis and vitamin B12 levels, HIV and Borrelia. Patients should be treated for co-morbidity, especially thyroid and vitamin B12 deficiency.

2.4.1.2 Cerebrospinal Fluid (CSF)

Like imaging, CSF provides a 'window on the brain' as biochemical changes, such as extracellular aggregation of beta amyloid in plaques and formation of tau tangles, are reflected in it. A 50% decrease of CSF $A\beta42$ is seen in patients with AD or MCI in comparison to age-matched controls. The decrease has been associated with enhanced $A\beta42$ deposition in the brain. With specificity set at 90% the mean sensitivity is 86% in comparison to normal aging. In the differential diagnosis between AD and other dementias, CSF $A\beta42$ is only moderately specific, with reduced levels also seen in DLB and to a lesser extent also in FTLD and VaD.

CSF tau levels are on average increased 2–3 times in AD and MCI in comparison to controls. Tau is thought to reflect the amount of neuronal degeneration in chronic neurodegenerative disorders. With specificity set at 90%, mean sensitivity is 81% for AD. Elevation of CSF tau is also observed in CJD and after acute stroke; in VaD and FTLD it may also be elevated. The concentration of CSF phosphorylated tau (e.g. p-tau181 or p-tau231) reflects the phosphorylated state of tau protein, and thus the formation of tangles. CSF levels of p-tau in AD patients can be increased by an order of magnitude compared to controls. Increased levels of p-tau are considered to be more specific for AD (Box 2.1).

Assessment of 14-3-3 protein in the sporadic form of CJD has a sensitivity and specificity well above 90%. False positive test results have been noted in patients with encephalitis, cerebral infarcts, metastases, paraneoplastic syndromes and rapidly progressive AD, making it likely that the protein is a marker of brain cell death rather than for CJD.

2.4.1.3 Electro-Encephalography (EEG)

Generalised slowing of background rhythm on EEG is a frequent finding in AD and DLB. These changes are not specific for AD and can also be found in other diffuse encephalopathies (Box 2.2). In FTD patients, the EEG is generally normal. Typical sharp wave complexes are relatively specific for CJD, particularly for the sporadic form. Another possible important finding is temporal epileptic activity which can cause transient epileptic amnesia, a rare cause of memory deficits. In the Box 2.2 the main EEG findings in various dementias are listed.

Box 2.1 Levels of CSF markers in some dementias

	$A\beta1-42$	Total tau	Ptau-181
AD	↓↓	↑↑	↑/↑↑
DLB	↓	=/↑	=
VAD	=/↓	=/↑	=
FTLD	=/↓	=/↑	=
CJD	↓	↑↑↑	=
Normal aging	=	=	=

↑ mildly elevated, ↑↑ elevated, ↑↑↑ strongly elevated, = normal CSF markers in the most prevalent dementia syndromes. (Courtesy of Dr. N.S.M. Schoonenboom). For abbreviations see list on page XV

Box 2.2 EEG characterisctics in the most prevalent dementia syndromes. Note that for FTLD in particular EEG is normal, while in CJD it has the highest diagnostic yield

	AD	FTLD	VaD	Delirium	DLB	CJD
Decreased alpha rhythm	+	-	0	++	+	+
Decreased alpha reactivity	+	-	0	+	+	+
Asymmetric alpha rhythm	-	--	+	-	-	-
Theta waves	+	-	0	++	+	+
Temporal slow waves	0	0	++	0	++	0
Focal abnormalities	-	-	+	-	0	0
Sharp waves	-	--	+	+	0	0
Periodic discharges	--	--	0	+	0	++
FIRDA	--	--	0	++	+	0

-- never occurring, - incidental, 0 sometimes, + frequent, ++ almost always, *FIRDA* frontal intermittent delta activity. (Courtesy Prof C.J. Stam)

2.5 Multidisciplinary Integration

The diagnosis of a patient with (suspicion of) dementia may require input from many disciplines: clinicians (e.g. geriatricians, neurologists, psychiatrists), psychologists, neurophysiologists as well as radiologists and nuclear medicine specialists. The increasing need for an earlier and more specific diagnosis to guide management and treatment poses a burden on the diagnostic skills of the team involved.

The appropriate use of neuroimaging covered in this book may help to advance diagnosis and alleviate burden for the patient and carer. This book aims to take an approach that fits with clinical reality – that one is presented with patients with clinical and imaging features rather than diagnoses. The arrangement of the book aims to help the reader to move from imaging and clinical features to a diagnosis; and to be accessible to all those seeking to improve diagnosis in dementia.

Suggested Reading

Dubois B, Feldman HH, Jacova C, DeKosky ST, Barberger-Gateau P, Cummings J, Delacourte A, Galasko D, Gauthier S, Jicha G, Meguro K, óBrien J, Pasquier F, Robert P, Rossor

M, Salloway S, Stern Y, Visser PJ, Scheltens P (2007) Research criteria for the diagnosis of Alzheimer's disease: revising the NINCDS–ADRDA criteria. Lancet Neurol 6:734–746

Ferri CP, Prince M, Brayne C et al (2005) Global prevalence of dementia: a Delphi consensus study. Lancet 366:2112–2117

Harvey RJ, Skelton-Robinson M, Rossor MN (2003) The prevalence and causes of dementia in people under the age of 65 years. J Neurol Neurosurg Psychiatry 74(9):1206–1209

Hort J, O´Brien JT, Gainotti G, Pirttila T, Popescu BO, Rektorova I, Sorbi S, Scheltens P, on behalf of the EFNS scientist panel on dementia (2010) EFNS guidelines for the diagnosis and management of Alzheimer's disease. Eur J Neurol 17(10):1236–1248

Knopman DS, DeKosky ST, Cummings JL et al (2001) Practice parameter: diagnosis of dementia (an evidence-based review). Report of the Quality Standards Subcommittee of the American Academy of Neurology. Neurology 56: 1143–1153

Scheltens P, Fox N, Barkhof F, De Carli C (2002) Structural magnetic resonance imaging in the practical assessment of dementia: beyond exclusion. Lancet Neurol 1:13–21

Schott JM, Fox NC, Rossor MN (2002) Genetics of the dementias. J Neurol Neurosurg Psychiatry 73:ii27–ii31

The Toolbox

3

Contents

3.1 Introduction: Choosing the Right Tools

The focus of imaging in patients suspected of having dementia has shifted from an exclusionary to an inclusionary approach over the past decades. Exclusion of a (surgically) treatable cause of dementia (e.g. tumour or subdural haematoma) can be ascertained by using CT, but demonstration of positive disease markers (e.g. hippocampal atrophy for Alzheimer's disease [AD]) becomes increasingly more relevant in an era where disease-specific pharmacological treatment is becoming available. It is in this regard – adding positive predictive value to the diagnosis in dementia – that MRI offers clear benefits over CT. Catheter angiography (DSA) is hardly ever indicated, except perhaps for suspicion of vasculitis that cannot be confirmed otherwise.

Although MRI is the neuroimaging modality of choice for the investigation of dementia, the possibilities to evaluate the pattern of brain atrophy have considerably increased with the advent of multi-slice CT, due to the availability of high resolution coronally reformatted images to examine the medial temporal lobe. However, for certain indications, CT is still clearly inferior to MRI. For example, CT is clearly less informative in subjects suspected of having some rare disorders causing dementia, such as encephalitis or Creutzfeldt–Jakob disease (CJD).

When structural imaging is equivocal or does not lead to the diagnosis, functional imaging may add diagnostic value (Fig. 3.1). Second-line neuroimaging investigation includes metabolic information obtained by using single-photon emission computed tomography (SPECT) or positron emission tomography (PET), or physiological information obtained by using diffusion or perfusion MRI. For example, in the early stages

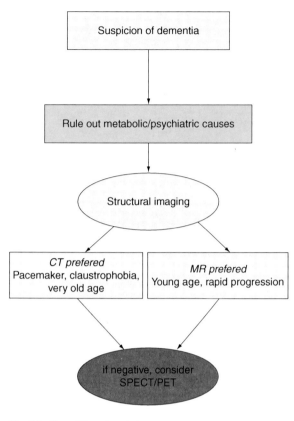

Fig. 3.1 Overall imaging strategy

Table 3.1 CT acquisition parameters

Single slice detector	Multi-detector row CT
Negative angle	Transverse orientation
3 mm collimation	0.6 mm collimation spiral
Consecutive slices	Pitch 1.15
	0.5 mm reconstruction interval
	Coronal/oblique MPR 3/3 mm
KvP 120	KvP 120
mAs 380	mAs 380

See Fig. 3.2 for planning of slices

Examples of sequential and spiral CT acquisition protocols are given in Table 3.1. Beyond exclusion of a surgical lesion, it is possible to evaluate the presence and extent of cerebrovascular disease by means of CT and, with appropriate techniques, it is also possible to assess atrophy of the medial temporal lobe. For the latter, it is important to realise that the long axis of the medial temporal lobe is somewhat tilted with respect to the transverse plane. In order to obtain slices that are either parallel or perpendicular to the hippocampus, the acquisition protocols need to be adjusted (see Fig. 3.2).

Slices obtained parallel to the hippocampus will suffer from more partial volume averaging in that region. Coronal multi-planar reconstructions (MPR) from a multi-detector row CT, perpendicular to the long axis of the hippocampus, are therefore advantageous for the evaluation of medial temporal lobe structures (Fig. 3.2), whereas sagittal reconstructions are more appropriate to evaluate the paramedian cerebral cortex, such as the precuneus.

of fronto-temporal lobar degeneration (FTLD), there may not exist any discernible atrophy. FDG-PET or HMPAO-SPECT might demonstrate decreased metabolism or hypoperfusion preceding tissue loss on structural imaging. In the future, molecular imaging may provide even more disease-specific information. PET tracers binding to amyloid are a good example of such developments, although their role in the diagnostic algorithm still remains to be established.

3.2 Structural Imaging

3.2.1 Data Acquisition

3.2.1.1 CT Protocol

A CT scan without contrast may yield significant information in patients suspected of having dementia.

3.2.1.2 Standard Structural MR Imaging Protocol

The prevalence of AD and vascular pathology means that the suggested first aims in the imaging evaluation of a patient suspected of having dementia – beyond exclusion of a surgically treatable disorder – are:

1. To assess the extent and pattern of brain atrophy, in particular medial temporal lobe atrophy (for evidence of Alzheimer's pathology)
2. To determine the degree of vascular damage, including the occurrence of strategic vascular lesions

Fig. 3.2 Planning slices perpendicular to hippocampus. Since the long axis of the hippocampus (outlined approximately on the *right*) is slightly angulated relative to the transverse plane, coronal images should be tilted as illustrated to get a true perpendicular cross section of the medial temporal lobe

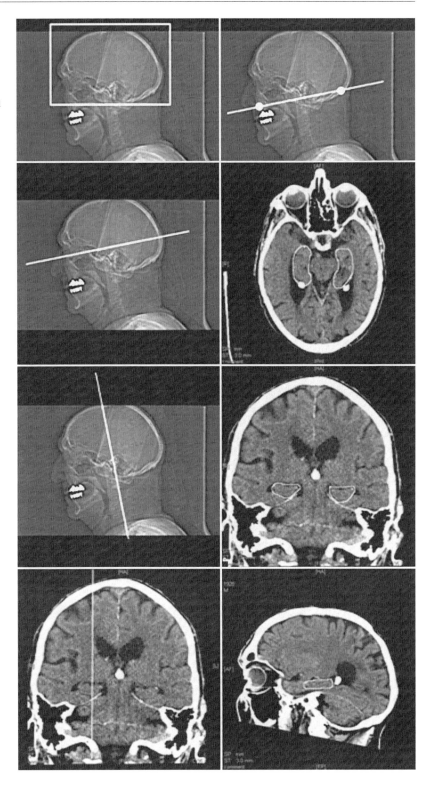

Table 3.2 Standard structural MRI protocol

Sequence name	Typical duration (minutes)
I. Coronal 3D T1-weighted gradient echo 1 mm isotropic voxels	8'
II. Transverse T2-weighted TSE/FSE 3–5 mm slices	4'
III. Transverse FLAIR TSE/FSE 3–5 mm slices	4'
IV. Transverse T2* gradient echo 3–5 mm slices	4'

Table 3.2 provides the essential sequences that will provide the important minimum set of information required (Fig. 3.3) to address those two fundamental questions. Additional sequences (see below) may need to be added in certain circumstances.

3D T1-Weighted Gradient Echo

The main features of this sequence include both high spatial resolution and high contrast between grey matter and white matter. Within reasonable scanning times, this can only be achieved by using a 3D gradient-echo T1-weighted sequence with preparation of the contrast by means of an additional inversion pulse – MPRAGE (Siemens), IR-SPGR (General Electric) or 3D-TFE (Philips). If 3D techniques are unavailable, coronal-oblique 2D T1-weighted images can serve as an alternative.

3D T1-weighted gradient-echo coronal images can be acquired with 140 partitions covering a slab of 210 mm, which is typically sufficient to cover the head without infolding and to obtain 1.5-mm thick slices with a 1×1 mm^2 in-plane resolution. Multi-planar reformatting (MPR) tools can then be applied to such images, in order to reslice the data into 3-mm thick slices perpendicular to the long axis of the hippocampus, as depicted in Fig. 3.4.

In addition, the 3D T1-weighted volume can also be reformatted into sagittal and transverse T1-weighted

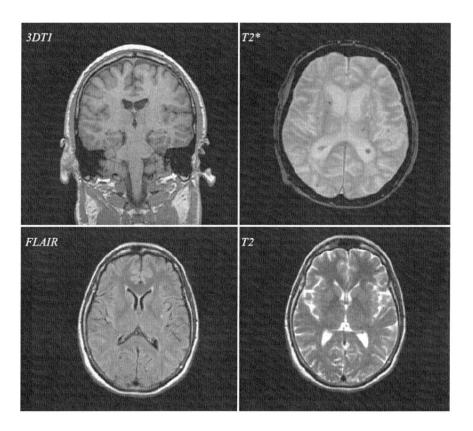

Fig. 3.3 Four critical MRI contrasts in dementia evaluation

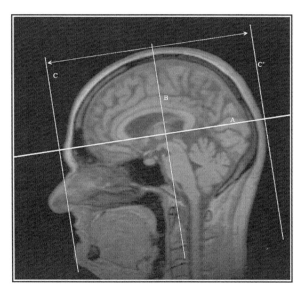

Fig. 3.4 Planning 3D slab perpendicular to the long axis of the hippocampus

images. Sagittal T1-weighted images are particularly useful for the detection of atrophy in the posterior fossa.

Fluid Attenuation Inversion Recovery (FLAIR)

FLAIR is a special inversion recovery sequence, in which the inversion time chosen is approximately the T1 relaxation time of water. By using such an inversion time, the signal of water is suppressed. A strongly T2-weighted FLAIR sequence is generally used. The advantage of this sequence is that hyperintense lesions close to the cerebrospinal fluid (CSF) spaces (e.g. ventricles) are easily seen. The contrast between grey and white matter is low (especially in the elderly population), which facilitates the detection of lesions in the cortical and subcortical regions. The sequence is highly valued in many centres and its use is standard, certainly in combination with a TSE T2-weighted sequence. FLAIR performs less well than SE techniques in the posterior fossa; also its use in the spinal cord has been disappointing. Sensitivity to detect thalamic lacunes is inferior to T2-weighted TSE/FSE (Fig. 3.5), although FLAIR may readily depict the subtle thalamic changes in prion diseases like CJD (Sect. 5.6). Another disadvantage of FLAIR is the loss of distinction of pathology with extremely prolonged T1 relaxation times (comparable to free water), and the

inability to separate structures with low signal (such as vessels or calcification) from CSF (Fig. 3.6). Taken together, FLAIR should not be used as the only T2-weighted sequence, but should always be combined with a T2-weighted TSE/FSE (see below) – the alternative is simply to use a dual-echo TSE/FSE.

T2-Weighted Turbo or Fast Spin Echo Sequences (TSE/FSE)

In conventional SE sequences, each RF excitation pulse is followed by a single phase encoding step. In TSE/FSE sequences each RF excitation pulse is followed by multiple (typically between 5 and 25) phase encoding steps, saving a considerable amount of time. The repetitive 180° pulses reduce the susceptibility to magnetic field inhomogeneities, and thus lead to a reduced sensitivity to detect microbleeds, for example (Fig. 3.6). Also, because of the difference between echo time in conventional SE and the so-called effective echo time in TSE/FSE (the echo in the middle of k-space) fatty tissue will not have lost its signal at long echo times. Both factors have to be taken into account when reading the images. As a result of increased incidental magnetisation transfer effects, brain tissue will be darker, and CSF relatively brighter.

T2*-Gradient Echo (GE) Sequences

Instead of using a 180° RF pulse, an echo can also be produced by reversal of gradients, which again saves time. Since no refocusing RF pulse is used, inhomogeneities in the magnetic field will not be cancelled, introducing susceptibility to local disturbances of the magnetic field. Whereas this leads to artefacts at bone and air interfaces, susceptibility weighting can also be used to advantage and allow detection of calcification and blood pigment. By using a sufficiently long TE (>20 ms) and a low flip angle (to avoid T1-weighting), microbleeds (MBs) are very well depicted by means of this sequence (Fig. 3.6). Alternatives to T2*-GE images for the detection of MBs include echo-planar imaging (EPI) and susceptibility-weighted imaging (SWI). The latter uses the phase information of the MR signal to enhance susceptibility effects; phase shift information also allows differentiation of paramagnetic effects (e.g. iron in MBs) from diamagnetic effects (e.g. in calcification).

Fig. 3.5 FLAIR misses thalamic lesions. Axial T2-weighted images (*left side*) of three different patients (aged 68, 52 and 79 years) showing a left tuberothalamic artery infarct, and two right-sided paramedian thalamic infarcts (*red arrows*). Corresponding FLAIR images (*right side*) do not reveal discernable thalamic abnormalities

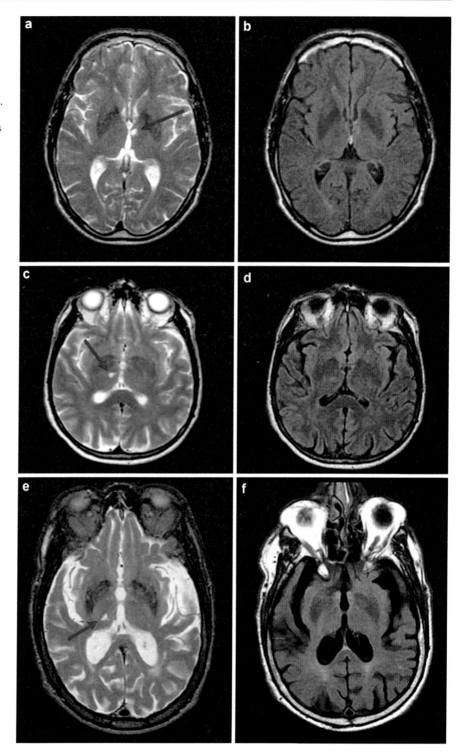

Fig. 3.6 Microbleed (MB) detection: impact of pulse-sequence. The most sensitive sequences are susceptibility weighted. Both T2*GE and SWI detect many MBs (*dark spots*) that go undetected on T2-TSE and FLAIR. SWI is slightly more sensitive than T2*GE and detects a few additional MBs (*red arrows*)

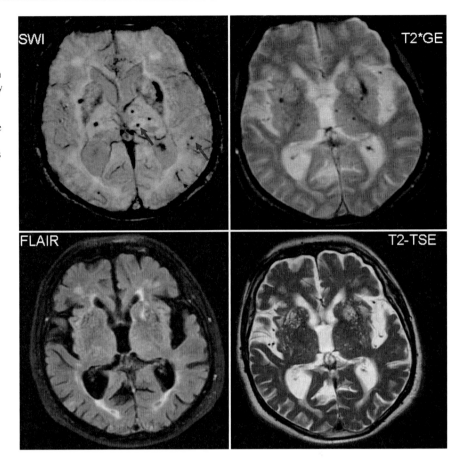

3.2.1.3 Indications for Gadolinium Administration and Diffusion-Weighted Imaging (DWI)

MR imaging offers a variety of non-conventional imaging techniques that can be used to expand the examination and obtain additional information on BBB-integrity, vessel patency, cerebral blood flow, cortical metabolism and function. The discussion of the various pulse sequences used for MR angiography (MRA) is beyond the scope of this book, but non-contrast-enhanced (e.g. time-of-flight) MRA can be useful in the work-up of patients suspected of having vascular dementia, especially those with large-vessel disease. Two other MR techniques that are very frequently used in a clinical setting (Table 3.3) include diffusion-weighted imaging (DWI) and gadolinium-enhanced – best depicted on (2D) T1-weighted spin-echo images.

Table 3.3 Indications for DWI and gadolinium administration

- Rapidly progressive dementia
  - Especially in a young patient
- Suspicion of CJD
  - Restricted diffusion on DWI (basal ganglia or cortex)
    - No enhancement with contrast
- Suspicion of infection
  - Enhancement of inflamed areas
  - Restricted diffusion on DWI, e.g. in HSV
- Vasculitis
  - Vascular and leptomeningeal enhancement
  - Areas of infarction on DWI
- Recent ischaemia
  - Restricted diffusion on DWI
  - Cave luxury perfusion with enhancement in subacute phase

Diffusion-Weighted Imaging (DWI)

DWI uses the Brownian microscopic movement and transport of water in biological tissues over distances comparable to the cell's dimensions as a source of contrast. In the presence of magnetic field gradients, protons carried by free moving water molecules undergo a phase shift of their transverse magnetisation. Normal diffusion is characterised by the random Brownian displacements of molecules, resulting in phase shifts that are widely dispersed. The phase dispersion results in attenuation of the MR signal. The amount of attenuation depends directly on the amplitude of the molecular displacement and on the intensity of the magnetic field gradient. In conventional MRI, diffusion effects are negligible. By adding a strong magnetic field gradient to an (arbitrary) imaging pulse sequence, diffusion sensitivity becomes stronger and measurable. Strength and deviation of the gradient pulse determine the degree of diffusion weighting: the 'diffusion sensitivity' or 'b-value'.

On DWI, structures with unrestricted diffusion (such as CSF) are dark, whereas structures with restricted diffusion are bright (Fig. 3.7). By using a simple formula, the apparent diffusion coefficient (ADC) can be estimated from the slope of the diffusion measurements with increasing b-values. On ADC maps, low signal intensity represents restricted diffusion, and high signal intensity represents high diffusion values. The superiority of DWI to conventional MRI in the detection of recent infarctions (probably due to cytotoxic oedema) is well established, but is limited to the first few days after symptom onset. DWI can be useful to identify recent infarcts in patients with (multi-infarct) vascular dementia, transient global amnesia (hippocampal lesions), or in the context of vasculitis (Table 3.3). Other pathologies, such as vacuolating myelinopathy and spongiform encephalopathies (including CJD), can also be demonstrated by

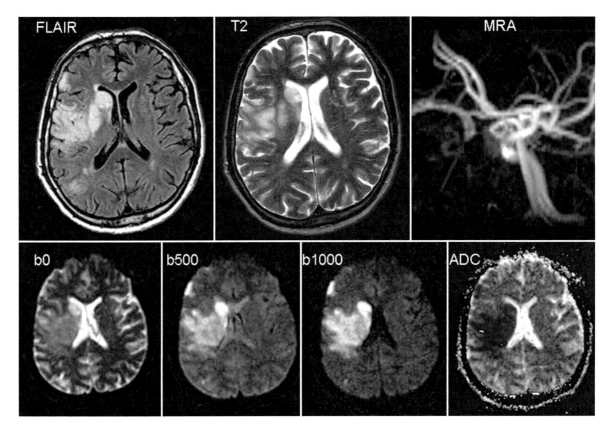

Fig. 3.7 DWI in recent infarction. This 67-year-old man incurred a partial medial cerebral artery infarction seen on FLAIR/T2 following occlusion of the right internal carotid (*red arrow* on MRA image). Trace diffusion-weighted images in the *bottom row* with b0/500/1,000 s/mm^2 show persistently bright signal in the infarcted area, leading to a low apparent diffusion coefficient (ADC)

using DWI. In a clinical setting, the so-called trace maps can be calculated by averaging the effects of gradients in three or more directions. Trace maps display results of averaged measurements, avoiding the effects of diffusion anisotropy (vide infra).

Gadolinium-Enhanced T1-Weighted Imaging

Small paramagnetic molecules, like gadolinium-DTPA, can be used to demonstrate leakage of the BBB. Extravasation of such a paramagnetic substance leads to T1-shortening, which is best appreciated on (2D) T1-weighted spin-echo images. Post-contrast FLAIR images can be useful to detect leptomeningeal enhancement.

In many primary neurodegenerative disorders (e.g. AD), the BBB is intact, and therefore routine administration of contrast material is not indicated for the imaging work-up. By contrast, disruption of the BBB may occur in certain rare disorders causing dementia, such as infectious (e.g. herpse simpex virus encephalitis) and inflammatory disorders (vasculitis, sarcoid, multiple sclerosis; see Table 3.3). Typically, these conditions will occur relatively more commonly in younger patients (see also Table 2.1) when compared to neurodegenerative diseases. In addition, the clinical presentation is more often atypical; there may be systemic involvement, and/or rapid progression.

3.3 Feature Extraction

3.3.1 Introduction: Structured Reporting

For the interpretation of structural images (especially MRI), we suggest a six-step approach as depicted in Table 3.4.

When all steps fail to reveal any abnormalities, (beyond what is expected for age) the scan has to be considered/reported as normal (for age). If clinical suspicion persists, consider repeat scanning after a suitable interval depending on the speed of clinical

Table 3.4 Structured evaluation of MRI in work-up of dementia

1. Exclude a structural lesion which may be amenable to neurosurgical intervention. Consider both lesions with significant mass effect (e.g. subdural haematoma, meningioma) and lesions with minor mass effect (e.g. AVM)

2. Exclude brain swelling, either generalised swelling (e.g. associated with dural AVF or hydrocephalus) or focal swelling (e.g. medial temporal lobe swelling in HSV encephalitis). Whenever swelling is present, consider adding DWI and Gd-enhanced images

3. Assess signal increase on T2/FLAIR, both in white matter (e.g. white matter changes [WMC] in vascular disease) and in grey matter (e.g. thalamic infarction in vascular disease or pulvinar sign in CJD). Consider applying a white matter rating scale, such as the Fazekas scale or the ARWMC scale. When describing vascular changes, assess/report specifically lacunes, état criblé and (bilateral) thalamic lesions and whether changes involve the brain stem or basal ganglia

4. Assess microbleeds (MBs) on T2*GE, especially in subjects with white matter changes

5. Determine the degree and pattern of general cortical atrophy (GCA), and specifically report whether the atrophy
   (a) Is abnormal for age or not
   (b) Is symmetric of asymmetric
   (c) Has a regional pattern
   (d) Has a posterior or anterior gradient

6. Assess focal atrophy, especially in the following regions
   (a) Medial temporal lobe (e.g. as seen in AD)
   (b) Temporal pole and/or frontal lobes (e.g. consistent with FTLD)
   (c) Biparietal atrophy (posterior cortical atrophy, mostly AD)
   (d) Occipital atrophy (posterior cortical atrophy or Balint syndrome, usually AD but there is a considerable overlap with other neurodegenerative pathologies, e.g. DLB)
   (e) Posterior cingulate and precuneus (e.g. presenile or posterior AD)
   (f) Mesencephalic atrophy (e.g. PSP)
   (g) Pontine (and cerebellar) atrophy (e.g. MSA)
   (h) Cerebellar atrophy (e.g. alcohol abuse, prion etc.)

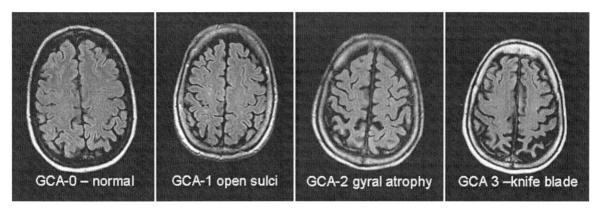

Fig. 3.8 Examples of the four-step rating of global cortical atrophy (GCA)

progression and the possible differential diagnoses (but typically at least 6 months in possible degenerative causes). In addition, consider the possibility of a nuclear medicine examination (see Fig. 3.1).

We recommend the use of standard rating scales (vide infra for detailed description) for white matter changes, regional and global atrophy and microbleeds in the radiological report. These scores will be discussed now.

3.3.2 Visual Rating

3.3.2.1 Global Cortical Atrophy (GCA)

When determining the degree of global cortical atrophy (GCA), it is important to always use the same sequences (e.g. FLAIR or 3D T1-weighted images), because the size of the CSF spaces appears differently on each type of sequence, tending to be overestimated on (heavily) T2-weighted sequences. Simple rating scales can be used to quantify the degree of GCA, such

as the four-step scale proposed by Pasquier, ranging from no atrophy (score 0) to knife-blade atrophy (score 3), as illustrated in Fig. 3.8. Care should be taken not to influence the GCA score by focal (exaggerated) changes, which are better captured separately (e.g. using the MTA score). The GCA score should give an overall estimate of atrophy, without regional bias.

3.3.2.2 Medial Temporal Lobe Atrophy (MTA)

In addition to GCA, focal areas of atrophy (i.e. disproportionate for the overall picture) should be identified. For the medial temporal lobe, this can be done on coronal MR images (or coronally reformatted images from multi-slice CT) by using the medial temporal lobe atrophy (MTA) score, based on the scheme presented in Table 3.5, and illustrated in Fig. 3.9. On axial transverse CT scans, even when using a negative angulation parallel to the long axis of the hippocampus, the rating of hippocampal atrophy is more difficult, even when attempting to measure the minimal diameter of the hippocampus.

Table 3.5 Visual assessment of medial temporal lobe atrophy (MTA)

Score	Width of choroid fissure	Width of temporal horn	Height of hippocampus
0	N	N	N
1	↑	N	N
2	↑↑	↑	↓
3	↑↑↑	↑↑	↓↓
4	↑↑↑	↑↑↑	↓↓↓

Source: According to Scheltens et al. (1992)

↑ increase, ↓ decrease, N normal

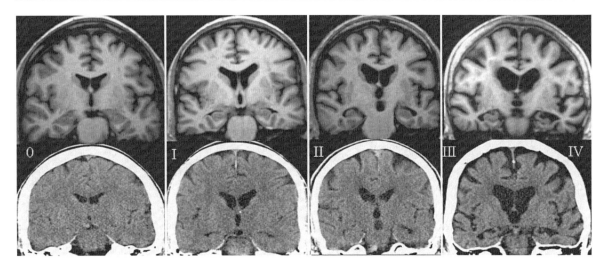

Fig. 3.9 Visual rating of MTA. In these same day scans, a perfect similarity between MRI and CT is noted for assessment of the medial temporal lobe for visual rating of MTA (Modified with permission from Radiology 2009; 253:174–183)

As it occurs with GCA, the MTA scores increase as a function of age, and scores of 1 may be normal under the age of 75, whereas a score of 2 may even be considered normal over the age of 75. This issue will be explored in more detail in Chap. 4. Rating of MTA has been shown to be very sensitive to the occurrence of AD, but is not specific for this disease, given that MTA occurs in other dementias as well, especially FTLD (see Table 5.6). For the latter, a specific rating scale has been proposed by Galton and colleagues, which includes more detailed scoring of the temporal pole and lateral temporal lobe. Additionally, the frontal lobe (especially the orbitofrontal and mesial frontal surfaces) should be scrutinised in patients suspected of having (the frontal variant of) FTLD. Other more comprehensive ratings scales exist, for example the scale for MTA in AD developed by Duara, assesses not only the hippocampus, but also the entorhinal cortex and inferior temporal gyrus (see Suggested Reading).

3.3.2.3 Posterior/Parietal Atrophy Rating

While MTA is the hallmark finding in senile-onset AD, especially in APOE4 positive patients with an amnestic presentation, it can be relatively mild in subjects with presenile-onset, without APOE4 and non-amnestic presentations. In such patients, the pattern of atrophy will be dominated by posterior/parietal atrophy, much more in keeping with the spatial distribution of reduced FDG uptake on SPECT. To appreciate this using structural imaging, a specific rating scale has been designed, evaluating the posterior cingulate, precuneus and superior parietal regions, as detailed in Table 3.6 and Fig. 3.10.

Overall score is based on the presence of atrophy in the three different planes for left and right separately. In case of difference in the scores on different orientations (e.g. score 1 on the sagittal slide and a score 2 on the axial slide), the highest score should be considered.

Table 3.6 Visual rating of posterior/parietal atrophy from multi-planar MRI

	Sagittal	Axial	Coronal
Rating	Width of posterior cingulate sulcus and parieto-occipital sulcus, atrophy of the precuneus	Widening of the posterior cingulate sulcus and sulcal dilatation in the parietal lobes	Widening of the posterior cingulate sulcus and sulcal dilatation in the parietal lobes
0	Normal	Normal	Normal
1	↑	↑	↑
2	↑↑	↑↑	↑↑
3	↑↑↑	↑↑↑	↑↑↑

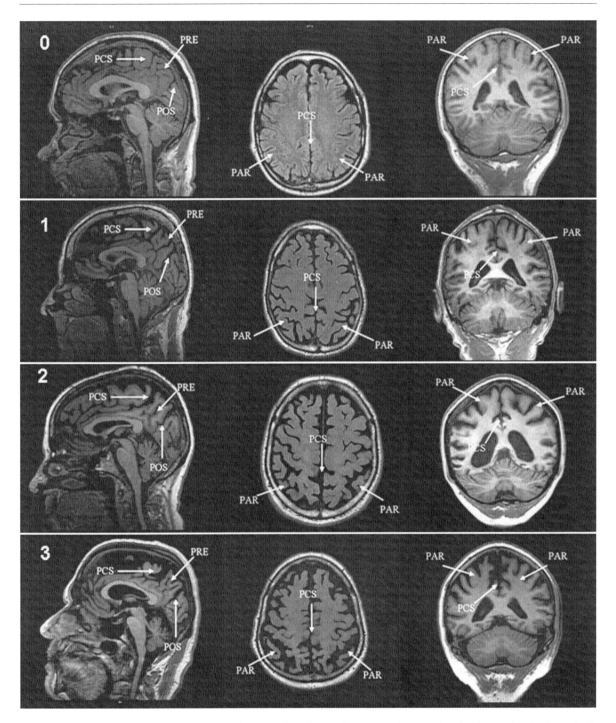

Fig. 3.10 Posterior/parietal atrophy rating scale (See Table 3.6). On this four-step scale, posterior cingulate and parietal atrophy is core visually by examining widening of sulci and gyral shrinkage. (Figure prepared by Esther Koeman)

3.3.2.4 White Matter Changes (WMC) Rating Scales

Many scales have been proposed to examine the amount, size and distribution of vascular WMC in ageing and dementia. The majority has been designed specifically for MRI or CT; some are applicable to both (e.g. van Swieten and ARWMC scales)

The choice of a rating scale will depend on the set-ting. In routine patient care, a simple scale, such as the

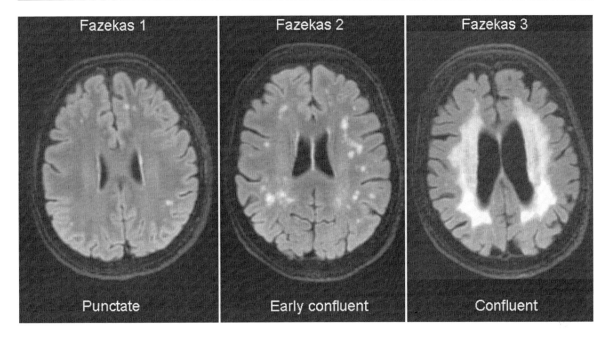

Fig. 3.11 Examples of WMC scoring using Fazekas' rating scale

four-step Fazekas scale (Fig. 3.11) may suffice, whereas for certain purposes (e.g. diagnosis of vascular dementia), more complex scales, such as the Age-Related White Matter Changes (ARWMC) scale, may be used to provide regional scores of WMC severity (Table 3.7). Alternatives include the Scheltens WMC scale (with more steps per region), and template-based scoring by means of the Manolio's scale. For a more complete review of possible rating scales and their advantages/disadvantages, the interested reader is referred to the reviews of Fazekas and Mäntylä. A dedicated scale for the assessment of WMC along the cholinergic projections, from the basal nucleus of Meynert to the neocortex, has been proposed by Black and colleagues.

The simplest WMC scale is the Fazekas scale. A score of 1 (punctate small WMC only) can be considered normal even below the age of 65, but certainly above 65 years of age. A score of 2 (early confluent WMC) is abnormal in subjects under the age of approximately 70 years, while a score of 3 (confluent WMC) is always abnormal and conveys a poor outcome in terms of cognition and survival.

3.3.2.5 Lacunes, Perivascular Virchow–Robin Spaces, Microbleeds

Visual assessments can also identify lacunes, enlarged perivascular spaces (also known as Virchow–Robin spaces [VRS]), and microbleeds (MBs). Lacunes are usually defined as infarcts of deep small vessels areas with CSF-like signal on all sequences, measuring between 3 and 15 mm. On FLAIR, they are often surrounded by a hyperintense rim (Fig. 3.12). VRS are typically smaller than 3 mm in cross-sectional diameter, and follow the orientation of perforating small arteries. When occurring in the white matter of the

Table 3.7 ARWMC rating scale for WMC on CT or MRI

Rating score WML

0 No lesions (including symmetrical, well-defined caps or bands)

1 Focal lesions

2 Beginning confluence of lesions

3 Diffuse involvement of the entire region/with or without involvement of U fibres

Rating score basal ganglia lesions

0 No lesions

1 One focal lesion (>5 mm)

2 More than one focal lesion

3 Confluent lesions

WMC on MRI are defined as bright lesions 5 mm on T2, PD or FLAIR images. Lesions on CT are defined as hypodense areas of 5 mm; left and right hemispheres are rated separately. The following brain areas are used for rating: frontal, parieto-occipital, temporal, infratentorial/cerebellum and basal ganglia (striatum, globus pallidus, thalamus, internal/external capsule and insula).

Source: Wahlund et al. Stroke 2001; 32:1318–1322

Fig. 3.12 Lacunes. This patient with confluent white matter changes developed a new lacune at follow-up. The lacune (*arrow*) is defined as a sharply demarcated lesion with CSF-like signal (liquefaction after complete infarction) and is surrounded by a rim of high signal, probably representing partial ischaemic changes

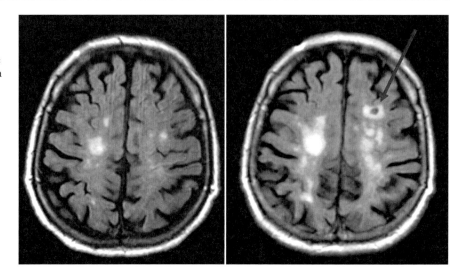

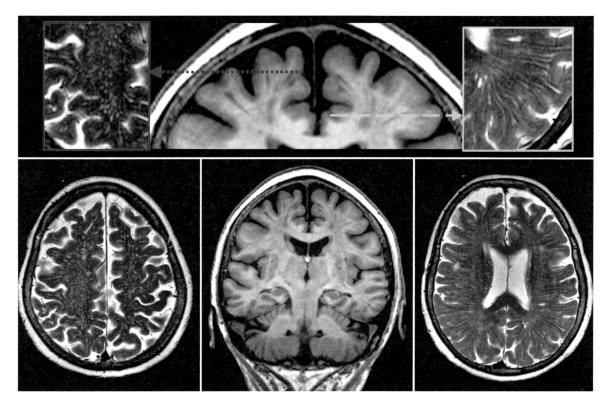

Fig. 3.13 Virchow–Robin spaces (VRS). This 71-year-old woman presented with subjective memory complaints only. The coronal T1-weighted image in the middle shows normal hippocampi and ventricles. There is diffuse widening of the VRS which are seen as sharply demarcated structures with signal intensity close to CSF. Note that their appearance on axial T2-weighted images depends on their orientation relative to the imaging plane, leading to a dot-like appearance when cut cross-sectionally (*left*) or stripe-like when cut tangentially (*right*)

cerebral hemispheres, they are perpendicular to the ventricular surface (Fig. 3.13). VRS have a predilection for the basal ganglia, leading (in extreme cases) to the appearance of 'état criblé' (Fig. 3.14). In general,

widened VRS can be regarded as a sign of focal atrophy, caused by volume loss in the surrounding tissue. When restricted to the perforated substance around the anterior commissure, VRS can be considered fully

Fig. 3.14 État criblé. Multiple small CSF-like structures are seen in the basal ganglia bilaterally. Note their monotonic size and regular distribution within the affected grey matter structures that differentiates them from lacunes (although impossible to determine on an individual level). In contrast to VRS in the white matter, état criblé is invariably an abnormal finding and often associated with severe ischaemic white matter changes

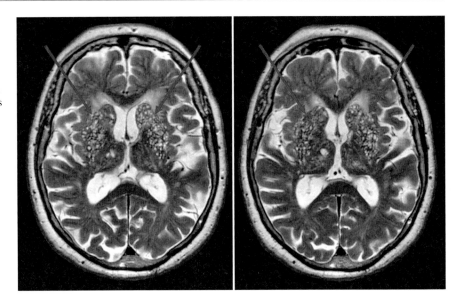

Fig. 3.15 Cerebral microbleeds (MBs) on T2* gradient echo. Cerebral MBs are defined as dot-like areas of low signal on susceptibility-weighted MR images with a size between 3 and 10 mm. They are located within the parenchyma (*red arrows*) in contrast to vessels caught in cross section, which are located in the sulci (*green arrows*). Scrolling up and down through images often helps in differentiating MBs from vessels. Note also low signal in the basal ganglia due to iron deposition and/or calcification (asterisk)

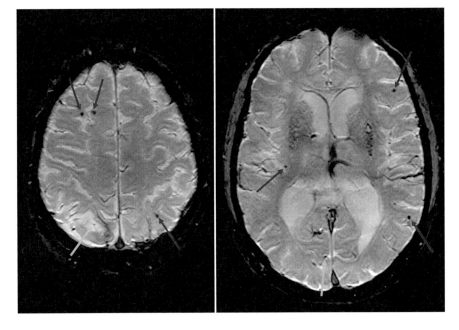

benign (see also Chap. 4). By contrast 'état criblé' is often accompanied by diffuse confluent WMC. Giant VRS will be discussed in Chap. 8.

Cerebral microbleeds (MBs) are defined as small areas (<10 mm) of low signal intensity on gradient-echo (GE) T2*-weighted images, not related to areas of calcification (e.g. in the globus pallidus) or flow voids due to vessels, which are typically located in sulci (Fig. 3.15). MBs are in fact local susceptibility effects caused by haemosiderin deposition in vessel walls, probably related to arteriosclerosis. They typically go undetected on routine T2-weighted and FLAIR images (Fig. 3.6), which is partially due to the so-called turbo/fast read-out phenomenon (where the multiple refocusing pulses annihilate T2*-effects).

Compared to the prevalence in normal ageing (<10%), MBs are found with increased frequency in AD (~20%), and vascular dementia (~65%), and are a very typical finding in congophilic amyloid angiopathy (CAA). Their clinical significance in terms of treatment with thrombolytic therapy is uncertain (see reviews Cordonnier and Greenberg). Several scales have been proposed (e.g. 'BOMBS' and 'MARS') to rate MBs by location. Lobar MBs are more common in CAA and AD, while MBs in a central location (basal ganglia, thalamus or brainstem) are more reflective of systemic vascular disease.

- Basal nucleus of Meynert thickness on coronal MRI
  - Diameter of tissue below anterior commisure and lentiform nucleus
  - Reportedly abnormal in AD
- Ventricular width – measured on transaxial images
  - Commonly between lateral borders of frontal horns
  - Non-specific finding related to neurodegeneration
- Intercaudate distance on transaxial CT/MRI
  - Reduced in Huntington disease

3.3.3 Linear Measurements for MRI and/or CT

In addition to the visual rating scales described above, a number of linear measurements for specific structures of interest have been proposed (Box 3.1)

An in-depth discussion of more detailed technical description, normative data and diagnostic performance of the above measurements is beyond the scope of this book. However, it is probably fair to conclude that none of these have been widely adopted.

Box 3.1 Linear atrophy measurements

- Radial temporal horn distance measured on transaxial CT/MRI
  - Largest diameter perpendicular to the long axis of the temporal horn, when it curves medially
  - Indirect measure of hippocampal atrophy in AD
- Interuncal distance measured on transaxial CT/MRI
  - Shortest distance between the uncus on both sites
  - Indirect measure of hippocampal atrophy in AD
- Fimbriosubicular distance
  - Measures the hippocampal sulcus (fissure) width on coronal slices

3.3.4 Quantitative Techniques

Contribution by Giorgos Karas

3.3.4.1 Introduction: Overview of Techniques

In many clinical situations, a qualitative visual assessment of images is sufficient for diagnostic purposes. With trained 'eyes', it is reproducible and differentiates reasonably well a normal from a diseased state. However, quantitative assessment gives more detailed information, provided that the technique of analysis is suited for the type of images available. Generally speaking, when quantification is pursued, the requirements on imaging quality are more strenuous.

In this subsection, we will discuss certain types of regional volumetric analyses that may be useful, especially hippocampal volumetry. Furthermore, we will provide some background information on more advanced image processing techniques that are mostly used in a research setting at present, but that may yield didactic information, as well as to evolve into clinically applicable tools in the near future.

3.3.4.2 Regional Volumes and Region-of-Interest Approaches

Beyond the determination of whole brain volume, the hippocampal volume (HCV) is probably the most

Fig. 3.16 Hippocampal volume outline. A selection of coronal T1-weighted images from a 3D dataset is shown. The hippocampal formation (including Ammon's horn (CA), subiculum, dentate gyrus and part of the alveus) are indicated by green regions of interest

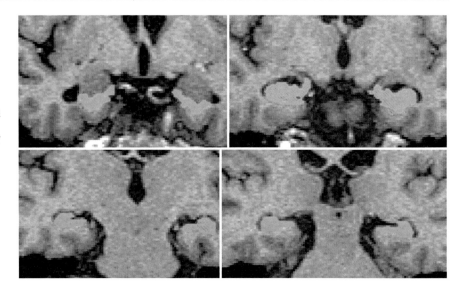

frequently determined parameter in patients with dementia, certainly in patients with Alzheimer's disease.

Regional volumes can be derived from segmentation of brain images (e.g. FreeSurfer software http://surfer.nmr.mgh.harvard.edu/), but such methods critically depend on the quality of the input image, as well as on availability of adequate computing power (see below for more detail). In addition, they often still require significant human interaction.

The most common approach is the manual outlining of a region-of-interest (ROI), such as the hippocampus, by a human operator. High-resolution 3D T1-weighted images, ideally reformatted to an oblique-coronal plane perpendicular to the long axis of the hippocampus, are usually needed to maximise reliability (Fig. 3.16). In order to reduce measurement errors, clear anatomic boundaries and strict landmark definitions are required, especially for the first (anterior) and the last (posterior) slices corresponding to the ends of the HCV delineation (see Box 3.2). The HCV is defined to include the cornu ammonis (CA fields 1–4), dentate gyrus, subiculum, alveus and the fimbria. For convenience, the posterior extent of the tail is usually excluded.

It is hard to give the exact reference data for HCV, since this depends on the exact protocol used and the software employed (especially how edge pixels are dealt with). In healthy volunteers, each hippocampus will measure around 4 mL, with a predictable age-related decline. This can be accommodated for by

transforming Z-scores (representing standard deviations from the mean) into W-scores.

Box 3.2 – Guidelines for hippocampal volume measurements

On coronal viewing, the most anterior slice is the slice on which the hippocampus is first visible. This is seen as a notch in the medial border of the temporal horn of the lateral ventricle: this notch is the border between the amygdala and the hippocampus. The structure of the hippocampus is usually visible, with the white matter of the alveus as a border between the hippocampus and the amygdala. If the alveus is not visible, a straight line is drawn between the notch between the hippocampus and amygdala on the medial border of the temporal horn of the lateral ventricle and the sulcus semiannulare. The border between the entorhinal cortex is in the edge that the subiculum makes with the entorhinal cortex. Here a straight line can be drawn between the most medio-dorsal point of this edge and the most medio-dorsal point of the white matter. The border between the subiculum and the white matter is formed by the grey-value difference of these structures.

In the following slices, the border with the amygdala usually becomes clearer. The borders are defined by the CSF in the temporal horn, the alveus or by a straight line drawn between the notch and the sulcus semiannulare, as described above. After some slices, the amygdala disappears. The uncal apex of the hippocampus is then visible; sometimes the hippocampus appears to be in two parts – if so, include both. Do not include choroid plexus. This is more homogeneous grey than the hippocampus. Most of the times, a thin white line is visible on top of the hippocampus: this line corresponds to the alveus and fimbria, which function as a border between the hippocampus and the CSF/choroid plexus.

When there is no atrophy or only minor atrophy, the hippocampus will border the temporal stem and midbrain. The tail of the caudate nucleus, the optical tract and the lateral geniculate nucleus as well as the more posterior part of the pulvinar are grey matter structures that should not be included. When there is atrophy, the hippocampus is mostly surrounded by CSF. The border with the entorhinal cortex/parahippocampal gyrus and the white matter of the temporal lobe is the one described above.

In the most posterior slices, the fimbria continues into the fornix. A straight horizontal line is drawn through the fimbria/fornix at the dorsal border (the more grey appearing part) of the hippocampus. The last, most posterior, slice to measure is the first slice on which the total length of the crus of the fornix is seen.

Modified from Jack CR. MRI-based hippocampal volume measurements in epilepsy. Epilepsia 1994;35(suppl.6):S21–S29.

3.3.4.3 Segmentation-Based Analysis

Most of the automated brain segmentation algorithms aim to classify separately (i.e. partition or segment) grey matter (GM), white matter (WM) and CSF voxels, mostly on the basis of their signal intensity on T1-weighted images, but also on the basis of the correspondence between their location on the input image and a given template, which is usually in so-called standard space. Therefore, such segmentation algorithms examine the intensity of each voxel and compare it with the intensity of the corresponding voxel in the target anatomical space. Furthermore, they examine the location of each voxel and compare it with the neighbouring voxels. After a series of iterations, a voxel is assigned a probability value of belonging to GM, WM or CSF. There are many segmentations strategies currently available, but the most frequently used so far is perhaps the one included in the Statistical Parametric Mapping (SPM) software package, downloadable from www.fil.ion.ucl.ac.uk/spm.

The outputs of this kind of segmentation process are the following three probability maps: GM, WM and CSF probability maps. Each voxel on each of the maps has a probability value between 0% and 100% of belonging to a particular tissue class. One can measure the number of voxels on a probability map above a certain probability threshold. Alternatively, one can proceed further and compare probability maps (e.g. GM probability maps) of various subjects with each other in the standard space by using a statistical model, a technique often referred to as voxel-based morphometry (VBM). Before entering the probability maps in such a model, one has to smooth the data in order to render them suitable for parametric statistics. VBM will be used in this book to illustrate patterns of atrophy on a group level for education purposes. Its application to scans on an individual basis is still under development and, therefore, the technique is not currently suited for routine clinical applications, although machine learning techniques are a hot research topic.

3.3.4.4 Registration-Based Analyses

The term registration refers to the process of bringing one MRI volume into anatomical correspondence with another. The volume being processed is usually termed the source volume and the other volume is called the target volume. The target volume can be a previous scan of the same patient (if one wishes to compare time points) or an average anatomical scan (if one wishes to fit the individual volume into a standard anatomical space). The reason why such a process is feasible is due to the anatomical characteristics of the brain, which has complex detailed structure but similar overall morphology and is enclosed by a rigid skull.

A registration process can be described according to the degrees of freedom it uses. If one only rotates an MRI volume in the x, y and z plane, then there are three degrees of freedom. If we also introduce translation (shifting) in three directions, then we will have six degrees of freedom, which is called a rigid body transformation. Zooming (in or out, also termed scaling) in three directions adds another three degrees of freedom. Finally, shearing (stretching of the volume or perspective manipulation in which line parallelism remains) makes the total degrees of freedom to be 12, a process often referred to as 'affine registration'.

Registration-based methods can be used to determine the rate of cerebral atrophy within patients over time, for example by using SIENA (http://www.fmrib. ox.ac.uk/fsl/siena/index.html) or the brain-boundary shift integral (BBSI). Registration-based methods are extremely robust and precise, and provide a direct measurement of atrophy over time, obviating the need to determine two individual brain volumes (Fig. 3.17). The computation time needed is relatively low, and application to individual patient data is possible, provided that there is a comparable acquisition (scanner, coil and pulse sequence).

3.3.4.5 Deformation-Based Analysis

Registration beyond 12 degrees of freedom gets more complicated. At an intermediate level, one can apply various mathematical models to more precisely match the anatomy between the two scans. These models can be for example polynomial models. When even higher accuracy of registration is required, it is possible to introduce millions of degrees of freedom with various advanced and computationally intensive methods. Examples include elastic warping, voxel compression, finite element analysis or surface-based cortical matching techniques.

A so-called FLUID analysis uses a viscous fluid model to determine the deformation field needed to match the source to the target volume. This transformation matrix can be displayed in colour on an anatomical overlay to show the changes over time

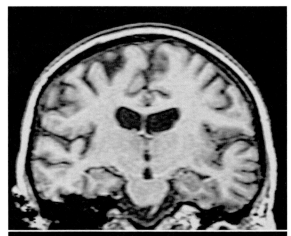
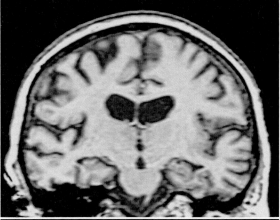
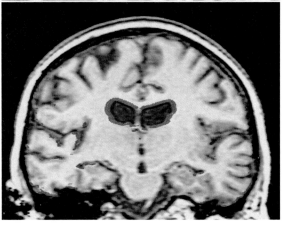

Fig. 3.17 Brain-boundary shift. Registration of scans taken at a 1-year interval in a 55-year-old woman at risk of familial AD. Notice the excellent alignment of the scans acquired at year 1 (*top panel*) and year 2 (*middle panel*) after registration. Simple visual inspection may fail to identify the progression of atrophy as shown in red (*bottom panel*) using image subtraction

Fig. 3.18 Non-linear
(FLUID) registration in AD.
MRI scans taken 2 years
apart in an AD patient were
subjected to a non-linear
registration with viscous
constraints. The deformation
field needed to register the
images is superimposed on
the baseline image, to
illustrate which brain areas
shrank (*green/blue*), and
which CSF spaces expanded
to accommodate the
intracranial volume loss
(*yellow/red*). (Image kindly
prepared by Jasper Sluimer)

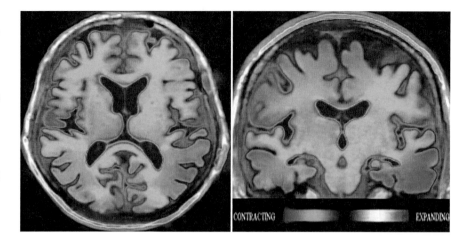

(Fig. 3.18). The Jacobian values in the matrix can be used to regionally or globally calculate the amount of compression. FLUID analysis is computationally demanding, but can be applied in a clinical setting, provided that the requirements in terms of acquisition stability over time are met.

3.3.4.6 Fusion and Automatic Diagnostic Systems

The ultimate goal of computational techniques may be the one of diagnostic assistants, much like the algorithms already employed in digital mammography, where the computer suggests a lesion and the human observer agrees or disagrees with the suggestion. There are various theoretical ways to design such an algorithm. They fall into two basic categories: single-modality systems and multi-modality systems. For example, a single-modality system could be simply an automated scorer of hippocampal atrophy. A more complex single-modality system could provide simultaneous scoring of atrophy for multiple regions. Alternatively, a multi-modality system could take into account structural and functional imaging data fused into a single image set. Attempts to obtain diagnostic classification algorithms for dementia scans are gradually beginning to appear in the literature, but the results are still very preliminary. Maybe computerised systems (including human interpretation and checking) will be robust enough to be used on a clinical basis in the next few years.

3.4 Advanced MR Techniques

The following four techniques are not currently used in routine clinical practice, but will be briefly discussed, given that they will be used throughout this book to illustrate salient features of certain types of dementia:

- Diffusion tensor imaging (DTI)
- Perfusion-weighted imaging (PWI)
- MR spectroscopy (MRS)
- Functional imaging (fMRI)

Their future role in the clinical work-up of dementia is currently uncertain.

3.4.1 Diffusion Tensor Imaging (DTI)

In the human brain, grey matter diffusion is more or less isotropic due to the random orientation of structures. White matter diffusion, however, is extremely variable and dependent on the relative orientation of the myelin sheaths along the axonal tracts, resulting in anisotropic diffusion. By applying a strong gradient magnetic field in a single direction, the so-called diffusion-sensitising gradient, the signal becomes sensitised for diffusion in that direction. Fibre tracts parallel to this gradient field will show maximal signal loss, whereas the effect is minimal if the gradient field is perpendicular to the fibre tracts.

By applying gradients in three or more different directions, one can display the anisotropy of tissue, especially in the white matter. By using more than six non-collinear diffusion gradients, it is possible to determine the full diffusion tensor, which is the starting point for techniques like fibre tracking and quantitative analyses of the principal eigenvectors of diffusion.

Studies of DTI in dementia have consistently shown altered diffusion (tract) properties in accordance with the pattern of neurodegenerative pathology. For example, widespread abnormalities in the temporal lobe (but also elsewhere in the brain) were found in AD. By using DTI and fibre tracking, tract-specific pathology can be demonstrated, for example in the cerebellar peduncles of patients with movement disorders, which may be specifically linked to the clinical syndrome at hand.

3.4.2 MR Perfusion Imaging

There are various ways to determine cerebral perfusion by using MRI. Perfusion-weighted imaging (PWI) can be performed invasively by injecting a paramagnetic contrast bolus, or non-invasively by applying a magnetic tag (arterial spin labelling or ASL) to the blood flowing towards the skull. Dynamic susceptibility contrast (DSC) imaging is a fairly well-developed technique that can be routinely applied and analysed by using standard MR machines (typically by means of EPI sequences). Advantages of DSC-PWI include both a short examination time and a high signal intensity-to-noise ratio. The more elegant and non-invasive alternative of ASL is less well developed and still not routinely available, but is an area of active research. ASL critically depends on certain assumptions regarding transit time between tagging and read-out (which is often done in an interleaved fashion), and suffers from long acquisition times (and/or limited signal intensity-to-noise ratio) at field strengths below 3T.

The type of information that can be obtained with MR perfusion imaging is comparable to that from nuclear medicine examinations, such as HMPAO-SPECT, and, to a lesser extent, FDG-PET, although ASL allows absolute quantification as well. An advantage of MR is that its spatial resolution is higher than SPECT or PET. Perfusion-weighted imaging in patients with dementia can be helpful in AD, where a temporo-parietal hypoperfusion pattern can be found (Fig. 3.19), and in fronto-temporal lobar degeneration (symmetrical or asymmetrical), where a fronto-temporal pattern of hypoperfusion can be found, even in early stages of the disease.

3.4.3 MR Spectroscopy

MR spectroscopy (MRS) measures the concentration of certain brain metabolites. This can be achieved by exploiting the minor differences in resonance frequency of these metabolites. Because of its relative simplicity, proton spectroscopy is most generally used, but MRS of other nuclei, such as phosphorous or fluorine is also possible. The signal of water protons is immense compared with that of the other proton-containing brain metabolites. To appreciate those millimolar concentrations (the metabolic peaks of interest), the water signal has to be suppressed, for example by applying a saturation pulse with the exact frequency of water.

Various metabolites can be routinely depicted by using proton MRS (Table 3.8). MRS can be performed as a single voxel study ($1 \times 1 \times 1$ cm^3 or larger), or as chemical shift imaging, measuring a number of voxels in one plane at the same time. In the latter technique, information is obtained over a whole slice, subdivided into voxels. The overview over a larger area is sometimes helpful. The spectra, however, are usually much less resolved, especially at short echo times.

The results obtained by means of MRS can be divided into two categories: specific abnormalities indicating a unique disease condition or changes in the ratios of the normal metabolites. In patients with neurodegenerative disorders, the usual finding is a change in the ratio between metabolites or a general decrease in metabolites. In AD, a decrease in N-acetyl aspartate (NAA) is found together with an increase in myo-inositol, respectively indicating neuronal pathology and gliosis or increased tissue osmolality (see Fig. 3.20). Most of the MRS studies have been done in the parieto-occipital cortex of patients with AD. Attempts have been made to obtain spectra in the medial temporal lobe, but these are hindered by the close approximation of the hippocampus to the skull base (with associated susceptibility effects). In fronto-temporal dementia, one may find MRS abnormalities similar to those described for AD, but in a different anatomical

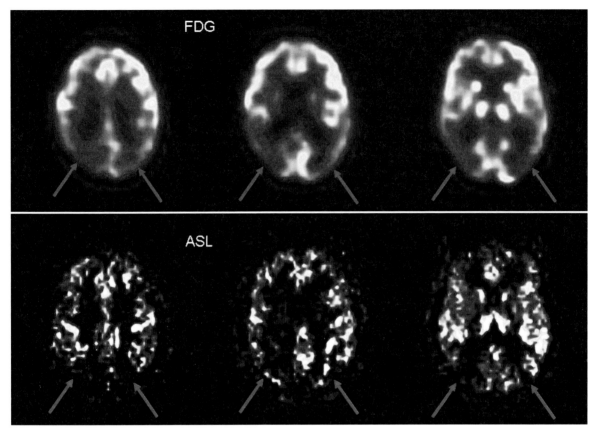

Fig. 3.19 MR perfusion imaging – comparison with PET. In a subject with AD, FDG-PET (*top row*) showed typical temporo-parietal hypometabolism. Arterial spin labelling (ASL) images (*bottom row*) were acquired non-invasively by MR revealing a similar pattern of hypoperfusion. The low signal-to-noise in this 1.5T example can be improved by the use of 3T scanners. (Thanks to Joost Kuijer)

Table 3.8 Metabolites detectable by proton-MRS

Metabolite	Chemical shift (ppm)	Representing
N-acetylaspartate (NAA)	2.02 (2.48, 2.60)	Neurons, axons
Creatine (Cr)	3.02, 3.94	Energy level, cellular density
Choline (Cho)	3.22	Membrane turnover, neurotransmission
Myo-inositol (m-Ino)	3.56	Gliosis, osmotic balance
Glutamine/glutamate (Glu)	2.1, 2.5, 3.8	Excitatory amino acids
Lactate (Lact)	1.33 (doublet)	Anaerobic glycolysis

location (as with atrophy). In vascular dementia, lactate may be present. In mitochondrial disorders, lactate is generally considered to be helpful for the diagnosis. Loss of NAA in the basal ganglia may differentiate multisystem atrophy (MSA) from idiopathic Parkinson's disease.

3.4.4 Brain Activation – Functional MRI (fMRI)

Although there are many techniques that convey 'functional' information (e.g. DWI, PWI, MRS), the abbreviation fMRI typically refers to images obtained by using

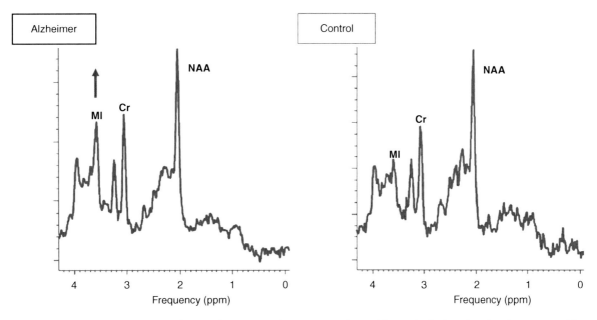

Fig. 3.20 MR spectroscopy in AD. Spectra from a posterior cingulate voxel in a presymptomatic familial AD subject (*left*) and a healthy control subject (*right*). Myo-inositol (MI) is increased (as usual in AD); N-acetyl aspartate (NAA) is in this case unchanged but typically is reduced – the most sensitive measure is usually a reduction in NAA/MI ratio which – as in this case; creatine/phosphocreatine = Cr. (Adapted from Godbolt et al. (2006) Neurology 66:718–722)

the blood oxygen-level dependent (BOLD) contrast. In BOLD fMRI, the difference in magnetic susceptibility between oxygenated and deoxygenated blood serves as an intrinsic contrast medium to compare conditions (e.g. an active task with rest). Given its non-invasive nature and lack of radioactive contrast media, fMRI has largely replaced nuclear medicine techniques for activation studies. PET is increasingly being reserved for the study of receptor density and function (vide infra).

A wide variety of paradigms for brain activation exist. Some are very basic, such as stroboscopic stimulation of the visual cortex, stimulation of the motor cortex by voluntary movements (either truly executed or imagined) or silent speech generation. Other more complicated paradigms were developed to study neuropsychological processes, such as language and memory functions, the later allowing separate analyses of encoding, retrieval and working memory. Such paradigms are of interest in the (differential) diagnosis of dementia, especially in the early stages; as well as in its follow-up, and, perhaps more importantly in the future, in monitoring the effect of therapy. Selective activation of the hippocampus is possible with a variety of stimuli (Fig. 3.21). Memory tasks, encoding

tasks and tasks evaluating the frontal lobe can be accomplished by using working memory tests (e.g. N-back task). In subjects at risk of developing Alzheimer's disease (e.g. APOE4-carriers and early MCI patients), increased activation can be found, perhaps best regarded as compensatory activity. In subjects with frank AD or FTLD, reduced hippocampal activation or frontal lobe activation have been reported, respectively.

A relatively novel field of fMRI is the analysis of so-called resting-state networks. Although the brain never 'rests' completely, in the absence of an active task (e.g. when lying still in the scanner with eyes closed and not thinking of anything in particular), multiple networks can be detected by fMRI using techniques like independent component analysis (ICA) that identify functional networks characterised by low-frequency temporally coherent oscillations. Among the resting state networks is the so-called default-mode network (DMN) that is the equivalent of consciousness and becomes less active during engagement of an active task; it is virtually absent in coma and vegetative state. Not by coincidence, the DMN encompasses the posterior cingulate and hippocampus, areas crucially involved in AD.

Fig. 3.21 Visual encoding fMRI in AD. Activation in the parahippocampal gyrus during a visual encoding task using BOLD (blood-oxygen-level dependent) contrast. On the left side a normal volunteer is shown with activation in the ventral stream extending well into the medial temporal lobe and on the right side a patient with early AD, with clearly diminished activation. (Figure kindly provided by S.A.R.B. Rombouts)

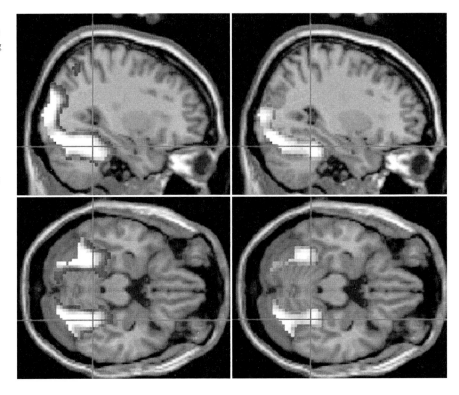

3.5 SPECT and PET

Contribution by Bart van Berckel

3.5.1 Data Acquisition – Introduction

Single-photon emission computed tomography (SPECT) and positron emission tomography (PET) both rely on the detection of radioactive signals from a labelled compound (tracer) that selectively binds in the brain. SPECT is technically less demanding and more widely available, whereas PET is more sensitive but technically more demanding. In the past, SPECT was mostly used to evaluate brain perfusion, and PET to evaluate the metabolism of glucose in the brain. However, due to the development of radioligands for in vivo detection of Alzheimer's pathology, SPECT and PET became more important in dementia (especially research).

Currently, most efforts are focused on the development of radioligands for in vivo detection of β-amyloid by using PET and SPECT. However, the development of these new agents involves major challenges, as the molecules clearly must be safe/non-toxic, need to be labelled by a radioactive tracer, must be able to cross the blood–brain barrier, and then must specifically bind to β–amyloid.

Especially for PET, the half-life of the tracer may be a limiting factor. For example, ^{11}C labelled PET-tracers have a half-life of 20 min and therefore require a cyclotron for on-site production, whereas ^{18}F labelled PET-tracers can be produced off-site and distributed regionally. SPECT tracers generally have sufficiently long half-lives to be produced off-site (e.g. [123]I) or can be prepared locally with simple procedures (e.g. [99m]Tc) (Table 3.9).

3.5.2 SPECT Technique and Tracer Generation

Tracers for SPECT are photon-emitting isotopes. The emitted gamma rays are detected by using a rotating camera containing two to three detectors that convert radioactive rays into electric signals used to compute a cross-sectional image. Due to the low energy of the photons, collimation of the signal and scatter in tissue, resolution is limited to ~1 cm. One of the advantages of SPECT is that commonly used tracers (e.g. the

Table 3.9 Common PET/SPECT tracers and their indications

Tracer	Modality	Pathophysiology	Indication
^{99m}Tc-HMPAO	SPECT	Hypoperfusion	AD
[123]I-ioflupane	SPECT	Dopamine-receptor loss	DLB
^{123}I-FP-CIT	SPECT	Dopamine-receptor loss	DLB
^{18}F-FDG	PET	Hypometabolism	AD, FTD
[11]C-PIB	PET	Amyloid deposition	AD, CAA
^{18}F-FDDNP	PET	Amyloid and NFT	AD

metastable nuclear isomer [99m]Technetium) can be readily produced from a local [99]Molybdenum generator. It can be delivered on a weekly basis to the hospital or can be bought from a central distributor.

3.5.2.1 Blood-Flow and Metabolism

The most commonly used tracer to examine cerebral perfusion is ^{99m}Tc-HMPAO (hexamethylpropylene amine oxime). [99m]Tc has a half-life of 6 h. HMPAO is a general flow tracer that can be used to determine areas of hypoperfusion in dementia, but ^{123}I-IMP and ^{99m}Tc ECD are also used for this purpose (Fig. 3.22).

3.5.2.2 Dopamine Transporter Tracers

Tracers binding to the pre-synaptic dopamine transporter are well established for the diagnosis of Parkinson's disease and related disorders. In the setting of dementia, they are useful for the distinction between Lewy body dementia (with reduced binding in basal ganglia) and AD (with normal binding). [123]I FP-CIT (I 123–radiolabelled 2-carbomethoxy-3-(4-iodophenyl)-N-(3-fluoropropyl) nortropane) and ^{123}I-ioflupane ([I-123] N-ω-fluoropropyl-2β- carbomethoxy-3β-(4-iodophenyl) nortropane, traded under the name DaT-SCAN are the most widely used tracers for this purpose. See Sect. 5.4.2.4 for indications and examples.

3.5.3 PET Technique and Scanners

Contrary to the random directions of photons emitted from a SPECT tracer, positron emission and annihilation lead to the emission of diametrically opposed photons, which can be detected by coincidence detectors. The latter improves sensitivity over SPECT, but comes at the cost of more complex detector systems and tracer production facilities. Recently, PET scanners have been combined with CT scanners. In addition to the advantage of having anatomical reference images, these combined PET-CT scanners also obviate the need to obtain a transmission scan for attenuation correction, and thus increase the speed of the examination.

3.5.3.1 Brain Metabolism (FDG)

[^{18}F]-2-fluoro-2-deoxy-D-glucose (FDG) is a glucose analogue that readily enters the brain and is taken up through the glucose transporter. Once metabolised, it becomes trapped in the cells and thus serves as a marker of cerebral (glucose) metabolism. Being labelled to ^{18}F, FDG has a half-life of 110 min., which obviates the need for local production. Reduced glucose metabolism occurs in many neurodegenerative diseases and reflects diminished metabolism of the tissue. While hypometabolism itself is non-specific, the pattern of abnormalities can be quite helpful and antedate tissue loss on structural imaging, thus allowing an earlier diagnosis.

3.5.3.2 Amyloid Tracers

Several amyloid-imaging tracers are being introduced into clinical practice recently, including ^{18}F-FDDNP (2-(1-{6-[(2-18F-fluoroethyl)(methyl)amino]-2-naphtyl}ethylidene)malo nitrile) and [11]C -PIB (N-methyl-11C-2-(4'-methylaminophenyl)-6-hydroxy benzothiazole). Both tracers bind with nanomolar affinity to amyloid and are being used to study amyloid deposition in the brain. Currently, ^{11}C-PIB

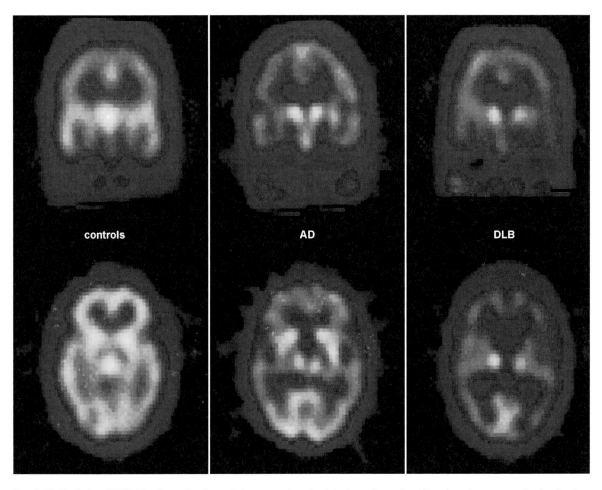

Fig. 3.22 Perfusion SPECT in dementia. Coronal (*upper row*) and transverse (*lower row*) projections of 99mTc-HMPAO SPECT images from a control, a patient with AD and a patient with dementia with Lewy bodies (DLB). Compared with the control there is reduced uptake of the tracer (indicating reduced blood flow) in the patients: the AD patient shows generalised reduction, including in the temporal lobes; in the DLB patient even further reductions are observed, including the occipital region that is relatively spared in the AD patient. (Figure kindly provided by S. Paling, S.J. Colloby and J. O'Brien, Newcastle, UK)

seems superior to ^{18}F-FDDNP for the identification of amyloid in AD on a case-to-case basis, but its widespread implementation is hampered by its short half-life (Fig. 3.23). Both[18]F-PIB and a number of new ^{18}F-labelled ligands for amyloid imaging are being evaluated, such as ^{18}F-BAY94-9172; these appear to have a good effect size and are likely to be increasingly available.

3.5.3.3 Research PET Tracers

Beyond clinical applications, there are many tracers that may be interesting for our understanding of dementia.

For example, PET tracers exist that bind to acetylcholine and nicotine receptors. Similarly, tracers exist (e.g. PK11195) that bind to the peripheral benzodiazepine receptor that is found in activated microglia.

3.5.4 Feature Extraction

Both PET and SPECT images are analysed for diagnostics purposes usually by means of visual inspection. (Semi)quantitative analyses of SPECT and PET may provide additional information, but they are more time consuming and often not necessary.

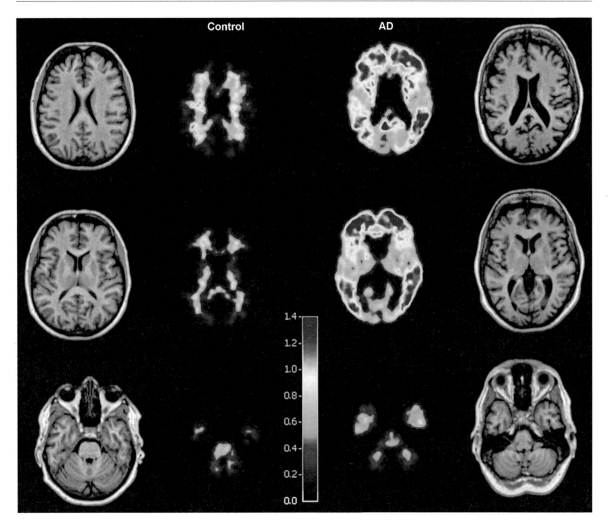

Fig. 3.23 Amyloid PET in AD. The binding-potential of the tracer ^{11}C-Pittsburgh Compound-B (PIB) is low in healthy controls and confined to the white matter. By contrast, abnormal uptake occurs in the cortex of AD patients, probably many years before the diagnosis due to abnormal amyloid deposition. (Reprinted with permission from J Nucl Med 2009; 50: 191–197)

3.5.4.1 Visual Interpretation

Images of regional cerebral blood flow and metabolism as respectively determined by HMPAO-SPECT and FDG-PET lend themselves fairly well to visual inspection. The cerebellum, basal ganglia and occipital cortex are typically used as reference regions. With normal ageing, minor decreases in prefrontal activity can be observed.

FDG-PET has good sensitivity (~90%) for Alzheimer's disease, and is reimbursed in the USA for the distinction between AD and FTLD. Findings in AD included reduced metabolism in the posterior cingulate, and temporo-parietal association cortices, which initially may be asymmetric. Frontal association cortices may become involved in advanced AD. The primary visual and sensorimotor cortex as well as the basal ganglia and the cerebellum remain largely unaffected and can serve as reference regions. The hippocampus has relatively little FDG uptake in normal subjects and is usually not a very useful area for visual diagnostic purposes. Similar patterns can be observed with HMPAO-SPECT, but the diagnostic accuracy is generally lower than for FDG-PET, mainly due to its lower resolution and lower specificity.

The specificity of an abnormal FDG-PET is lower than its sensitivity, probably in the order of 70%. In

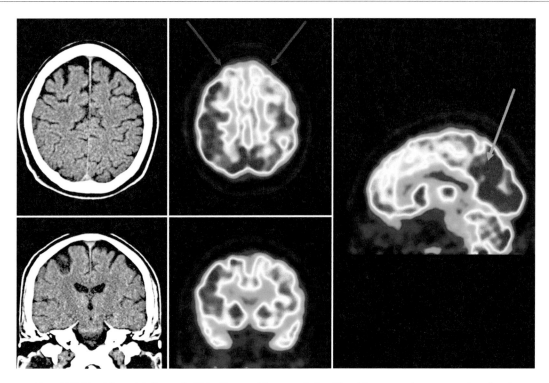

Fig. 3.24 Use of PET in differential diagnosis. This 60-year-old man presented with memory and speech impairment, but also with change in character. Symptoms progressed over 3 years and interfered with ability to work. His MMSE was 17/30. The initial clinical suspicion was AD, but structural imaging (CT on the left) revealed neither hippocampal, nor posterior cingulate atrophy. FDG-PET showed frontal hypometabolism (*purple arrows*), consistent with FTLD. Note preserved tracer uptake (*green arrow*) in the posterior cingulate; additional PIB-PET showed no evidence of abnormal amyloid deposition, thus making AD very unlikely

FTD, a different pattern is observed than in AD, with reduced uptake of FDG in the mesial frontal and anterior temporal lobes (Fig. 3.24). Whereas initially is often asymmetric, the picture becomes more symmetric in more advanced cases, and reduced metabolism is also found in the parietal lobes. In dementia with Lewy bodies (DLB), the FDG pattern may resemble AD, perhaps with a more pronounced involvement of the occipital lobe. In some cases, patients with AD may have an atypical and unexpected distribution of pathology with unexpected reductions in blood flow or metabolism in the frontal or the occipital lobes, which may lead to a false-positive diagnosis of FTLD or DLB.

3.5.4.2 Quantification

Nuclear medicine techniques, especially PET, lend themselves well to quantification purposes. However, sometimes an arterial input function is necessary, for instances to quantify cerebral perfusion and metabolism. For diagnostic purposes, they may not be needed. An alternative is the usage of reference tissue techniques in which specific binding in cortical brain areas is compared with a region devoid of specific binding, like the cerebellum or the pons.

Statistical parametric mapping (SPM) of regional glucose metabolism uses these reference tissue techniques. By using SPM, one can compare a subject to a reference group of normal age-matched subjects and compute a probability of abnormality per pixel (Fig. 3.25). This may be advantageous to determine whether there is hypometabolism in the posterior cingulate, normally one of the most metabolically active regions in the brain. By using commercially available software packages (http://www.pmod.com/technologies/products/alzheimer_tool/alzheimer_tool.php), one can calculate a sum of T-scores reflecting the amount of abnormal pixels within a mask.

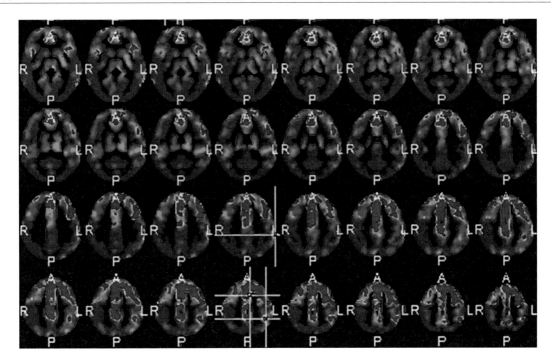

Fig. 3.25 Voxel-wise analysis of FDG-PET. The FDG images of the patient in Fig. 3.24 were analysed using software that analyses the pattern of tracer uptake voxel-wise by comparing the subject's scan to a reference data set of normal ageing and AD. T-maps are created that allow better recognition of the frontal pattern of hypometabolism (*in red*) and confirmed a low statistical likelihood for AD

The situation is slightly less complex for tracers like PIB that have little–if any–background activity in the normal brain, especially in the grey matter. Quantitative analysis has revealed that binding reaches a plateau 1–2 h after the injection, during which phase a steady-state image can be acquired. The white matter may serve as the reference tissue (with normal low uptake).

Being a technique that seeks to measure abnormal tissue metabolism, it would be desirable to correct parametric images for partial volume effects and atrophy, for example by using a voxel-wise approach with MRI-derived estimates of brain volume or grey matter density. This is certainly relevant for quantitative analysis. For visual analysis, the images can be evaluated side-by-side.

Suggested Reading

Bocti C, Swartz RH, Gao FQ, Sahlas DJ, Behl P, Black SE (2005) A new visual rating scale to assess strategic white matter hyperintensities within cholinergic pathways in dementia. Stroke 36:2126–2131

Cordonnier C, Al-Shahi Salman R, Wardlaw J (2007) Spontaneous brain microbleeds: systematic review, subgroup analyses and standards for study design and reporting. Brain 130:1988–2003

Duara R, Loewenstein DA, Potter E et al (2008) Medial temporal lobe atrophy on MRI scans and the diagnosis of Alzheimer disease. Neurology 9(71):1986–1992

Fazekas F, Barkhof F, Wahlund LO et al (2002) CT and MRI rating of white matter lesions. Cerebrovasc Dis 13(Suppl 2):31–36

Firbank MJ, Harrison RM, O'Brien JT (2002) A comprehensive review of proton magnetic resonance spectroscopy studies in dementia and Parkinson's disease. Dement Geriatr Cogn Disord 14(2):64–76, Review

Greenberg SM, Vernooij MW, Cordonnier C et al (2009) Microbleed Study Group. Cerebral microbleeds: a guide to detection and interpretation. Lancet Neurol 8:165–174

Herholz K, Salmon E, Perani D (2002 Sep) et a. Discrimination between Alzheimer dementia and controls by automated analysis of multicenter FDG PET Neuroimage 17(1):302–316

Klunk WE, Engler H, Nordberg A et al (2004 Mar) Imaging brain amyloid in Alzheimer's disease with Pittsburgh Compound-B. Ann Neurol 55(3):306–319

Mäntylä R, Erkinjuntti T, Salonen O et al (1997) Variable agreement between visual rating scales for white matter hyperintensities on MRI - comparison of 13 rating scales in a poststroke cohort. Stroke 28:1614–1623

Mori S, Crain BJ, Chacko VP, van Zijl PCM (1999) Three-dimensional tracking of axonal projections in the brain by magnetic resonance imaging. Ann Neurol 45:265–269

Silverman DH (2004 Apr) Brain 18F-FDG PET in the diagnosis of neurodegenerative dementias: comparison with perfusion

SPECT and with clinical evaluations lacking nuclear imaging. J Nucl Med 45(4):594–607

Small GW, Kepe V, Ercoli LM et al (21 Dec 2006) PET of brain amyloid and tau in mild cognitive impairment. N Engl J Med 355(25):2652–2663

Tedeschi G, Litvan I, Bonavita S, Bertolino A, Lundbom N, Patronas NJ, Hallett M (1997) Proton magnetic resonance spectroscopic imaging in progressive supranuclear palsy, Parkinson's disease and corticobasal degeneration. Brain 120:1541–1552

Normal Ageing

4

Contents

4.1 Introduction – Don't Worry, It is Normal?

Brain ageing can be classified into normal and pathological. The distinction may however be difficult with overlap in nearly all aspects seen on imaging. One of the most severe consequences of pathological brain ageing is dementia. Normal ageing may be subdivided into successful ageing (without any discernible deterioration) and the more commonly observed typical (usual) ageing. Typical ageing may encompass a variety of changes in the brain, including overall shrinkage, but also local alterations, such as white matter changes. Many of these 'normal' ageing phenomena have been linked to risk factors (e.g. vascular) and although cognitive function may appear preserved, subtle abnormalities may be detected on detailed neuropsychological testing. Such relationships are often only discernible on a group level, and inferences at the level of the individual subject are difficult to provide with any certainty.

4.1.1 Successful Ageing Versus Usual Ageing

In successful ageing, elderly subjects have apparently minimal morphological (and probably physiological) loss relative to younger individuals. Therefore, elderly subjects with a normal appearance of the brain on imaging may represent examples of successful brain ageing. In usual ageing, brain imaging abnormalities are present without overt clinical deficits or symptoms; subtle cognitive deficits may be detected with neuropsychological testing. Our view of both these models of ageing is clearly being continuously updated, as imaging

F. Barkhof et al., *Neuroimaging in Dementia*,
DOI: 10.1007/978-3-642-00818-4_4, © Springer-Verlag Berlin Heidelberg 2011

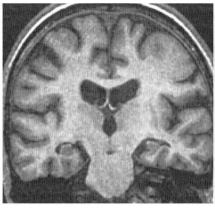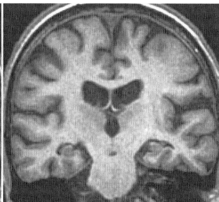

Fig. 4.1 Successful ageing. Serial (1 year apart) registered MRIs of a cognitively healthy 89-year-old man who was followed for 5 years after these images and who remained independent and clinically normal over that period. The baseline scan is on the left (acquired at the age of 89). On the right, the registered image acquired 14 months later is shown: note the lack of significant change – brain volume loss was 0.2% compared to baseline – similar to that seen in much younger adults

methods increasingly show us the effects of age on apparently healthy brains, and improved psychometric methods also show us the widespread nature of subtle cognitive decline. Nonetheless, within the spectrum of 'normal ageing', there are clearly individuals whose brains age rather better than others (Figs. 4.1 and 4.2).

Although normal ageing and pathological ageing may be difficult to distinguish, it is important to be familiar with changes that typically occur in subjects without a clinical correlate. Mild-to-moderate age-dependent brain volume loss, enlarged perivascular (Virchow–Robin) spaces, punctiform or minor white matter abnormalities, and iron accumulation in the basal ganglia are examples of such changes and will be the focus of this chapter (Table 4.1).

4.2 Brain Volume Loss

4.2.1 Global Brain Atrophy

Evidence on the effect of age on brain structures dates from autopsy studies of the nineteenth century. Such studies suggested that brain weight declines gradually, but only slowly (about 0.1% a year), from early adulthood until the age of 60 years. However, they were limited by the inclusion of few elderly people, a lack of clinical data, and were confounded by the secular changes that took place in the nineteenth century: as nutrition improved, 'normal' heights and weights increased. In the 1960s, Miller and Corsellis,

Table 4.1 Age-related brain imaging findings in usual ageing

Mild-to-moderate brain volume loss:

- Ventricular enlargement
  - Third ventricle dilatation
- Sulcal enlargement mostly due to cortical thinning
  - Equally affecting the frontal and parietal lobes
- Mild medial temporal lobe atrophy
  - Hippocampal sulcus cavities

Enlarged perivascular (Virchow–Robin) spaces in the:
- Basal ganglia region, near the anterior commissure
- White matter of centrum semiovale, near the vertex
- Mesencephalon

Changes of the vascular wall:
- Elongation and tortuosity (e.g. basilar artery)
- Wall-thickening and calcification (e.g. carotid syphons)

Vascular changes:
- Punctiform or early confluent ischemic white matter changes
- Lacunar infarcts and microbleeds

Iron accumulation in the:
- Globus pallidus, putamen, dentate nucleus

accounting for these changes, concluded that brain weights were stable between the ages of 20 and 50, progressively falling thereafter. Later studies of thousands of autopsies have found that brain weight peaked by the mid-to-late teens and declined slowly (0.1–0.2% a year) until the age of 60–70, after which losses accelerate. An understanding of cerebral atrophy in the individual during life had to wait until the advent of non-invasive imaging.

Imaging and autopsy studies suggest that cortical thinning, white matter loss and, in some cases, cortical

Fig. 4.2 Normal (usual) ageing. CT images in three orthogonal planes of an 81-year-old female with subjective memory complaints, but normal neuropsychological testing. There is minimal frontal atrophy (*red arrows*), and mild – but normal for age – hippocampal atrophy (grade 2 on the MTA scale, *pink circle*). Note that there is no atrophy in the precuneus or the posterior cingulate gyrus (*green circle*)

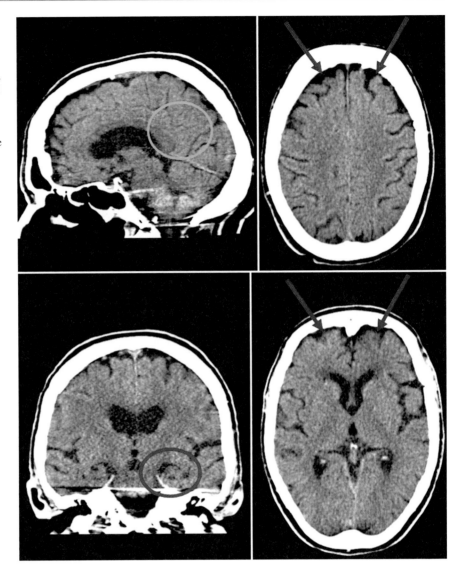

microinfarcts underlie cerebral atrophy. Cortical micro-infarcts represent relatively hidden pathology, because current imaging techniques have limited ability to detect these abnormalities. Several studies also suggest that shrinkage of cortical grey matter predominates over white matter loss. Although brain volume loss is measurable in young adults, there appears to be a gradual increase in rates of loss, which become more pronounced after the seventh decade of life (Fig. 4.2) with a suggestion that the parietal and frontal lobes are equally affected. It is clear, however, that there is a continuum, and that some brains remain almost normal until the ninth decade of life (Fig. 4.1) – it is unreasonable to suggest that pathological atrophy begins at a specific and generalisable age.

Voxel-based morphometry (VBM) studies of normal ageing suggest that global grey matter volume decreases linearly with age at least over the 30–60 year age range. By contrast, the white matter volume overall hardly declines with age, but local areas of relative accelerated loss or preservation can be seen. In fact, white matter volume might even increase (and show increased maturation in terms of diffusion characteristics) throughout middle adulthood, declining only after the age of 50 years. This quadratic pattern of change over time in the white matter volume suggests that the healthy adult brain, or at least its imaging characteristics, may continue to mature until the fifth decade of life. Longitudinal studies, using subjects as their own control, allow progression of atrophy to be quantified at the individual

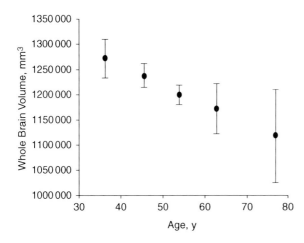

Fig. 4.3 Brain volume declines with age in typical ageing. (Adapted with permission from Archives of Neurology 2003; 60:989–994)

level. These studies show that rates of global atrophy in healthy people increase gradually with age from an annual rate of 0.2% a year at ages of 30–50 to 0.3–0.5% at ages of 70–80. Grey matter loss appears more prominent in the frontal and parietal cortices than in the temporal and occipital lobes. However, temporal lobe and hippocampal volumes also decline continuously, with gradually accelerating losses being most obvious (an inflection point) after the age of 70 years (Fig. 4.3).

4.2.2 Medial Temporal Lobe Atrophy Versus Hippocampal Cavities

Medial temporal lobe atrophy (MTA) can be assessed by using visual rating scales, linear measurements of temporal lobe structures, and volumetry of the hippocampus. The most well-known visual rating scale for MTA is based on the evaluation of the choroidal fissure width, the temporal horn width, and the hippocampal height by using coronal high-resolution T1-weighted images (T1-WI) perpendicular to the long axis of the temporal lobe (see Sect. 3.3.2). Given the relative preservation in volume of the temporal lobe structures with normal ageing, MTA scores below 2 are not frequently associated with dementia in subjects under the age of 75 years. However, an MTA score of 2 can be found in non-demented subjects over

this age, because the hippocampal volume decreases as a function of age (Fig. 4.4). Longitudinal studies of apparently healthy individuals vary considerably, but suggest that rates of hippocampal atrophy increase from around 0.2% a year in subjects aged 30–50 to 0.8% in those in their mid-70s, rising further, but very variably, to 1.5–2% a year at 80–90 years of age.

The fimbriosubicular distance is a linear measurement that enables to evaluate the hippocampal sulcus width, whose enlargement is associated with MTA in AD. The degree of widening of the hippocampal sulcus may therefore serve either as a measure to rate MTA severity or to complement its visual rating scale. By contrast, hippocampal cavities, which are generally considered to result from lack of obliteration of the hippocampal fissure, are not associated with considerable MTA and do not seem to have pathological value (Fig. 4.5).

In fact, hippocampal cavities are not a marker of AD; conversely, subjects with higher numbers

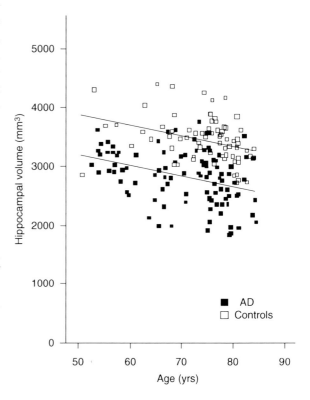

Fig. 4.4 Scatterplot displaying hippocampal volume (mm³) plotted against age. The parallel regression lines of patients with Alzheimer's disease (AD, *bottom line*) and control subjects (*top line*) indicate that the hippocampal volume is linearly affected by age in both groups, with an off-set introduced by disease. (Reproduced with permission from Neurology 2006; 66:236–238)

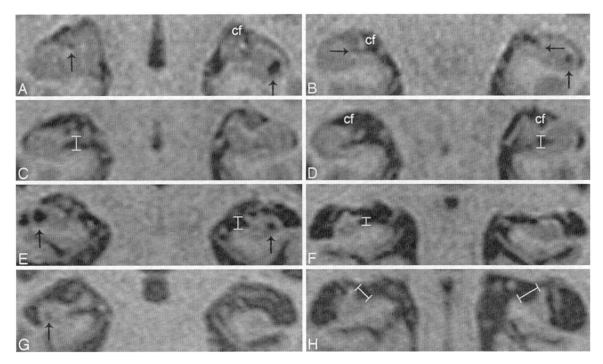

Fig. 4.5 Magnified coronal high-resolution T1-weighted images (T1-WI) of the hippocampal region. (**a** and **b**) A 68-year-old non-demented control showing discrete enlargement of the choroidal fissures (cf), minimal medial temporal lobe atrophy (MTA), grade 1, and bilateral hippocampal cavities (*vertical arrows*). Note that both hippocampal sulci are not enlarged (*horizontal arrows*). (**c** and **d**) A 54-year-old patient with AD showing enlargement of the hippocampal sulcus, measured between the fimbria and the subiculum (vertical measurement overlays), a MTA score of 1, and enlargement of the choroidal fissures (cf).

(**e** and **f**) A 76-year-old patient with AD showing enlargement of the hippocampal sulcus (vertical measurement overlays), moderate to severe MTA (grade 3) and bilateral hippocampal cavities (*vertical arrows*). (**g** and **h**) A 93-year-old patient with AD showing severe MTA (grade 4), and a small hippocampal cavity on the right side (*vertical arrow*). Note that the fimbria appears laterally displaced. This displacement contributes to an increase of the fimbriosubicular distance (oblique measurement overlays). (Reproduced with permission from Am J Neuroradiol 2006; 27:2141–2145)

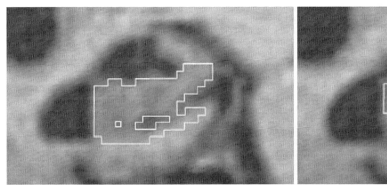

Fig. 4.6 Coronal baseline T1-weighted image (*left*) and registered-repeat image after approximately 1 year (*right*) of the right hippocampus from a patient with AD. The images demonstrate an enlargement of the hippocampal sulcus (*arrow*) as well as the

disappearance of a hippocampal cavity, which apparently merged with the medial part of the hippocampal sulcus. (Reproduced with permission from J Comput Assist Tomogr 2007; 31:581–587)

of hippocampal cavities have a lower probability of having AD. One explanation for this finding may be that some hippocampal cavities may disappear as the

hippocampus atrophy – cavities in effect merge (on imaging) with the hippocampal sulcus as the atrophy progresses and the sulcus enlarges (Fig. 4.6).

4.3 Enlarged Perivascular (Virchow–Robin) Spaces and Lacunes

Enlarged Virchow–Robin spaces (VRS) represent mild expansions (normally invisible) of the subarachnoid space around small vessels and are often considered to be a normal finding. Most of the enlarged VRS measure <2 mm in cross-sectional diameter and are often found (even in younger subjects) surrounding perforating arteries entering the striatum through the anterior perforated substance, around the anterior commissure. More extensive widening of VRS in the basal ganglia, a condition also referred to as 'état criblé', is a pathological finding, frequently associated with diffuse confluent white matter changes. Most likely, diffuse widening of the VRS in the basal ganglia is a sign of focal atrophy. In patients with 'état criblé', it can be difficult to rule out associated lacunar infarcts. Enlarged VRS are also frequently found in elderly subjects at the cortical-subcortical transition, near the vertex.

Although more difficult to discern on CT, enlarged VRS on MRI are characterised by signal intensity similar to CSF on all sequences. They are dark on T1 and on FLAIR, isointense on proton-density and bright on T2-weighted imaging and typically sharply demarcated (Fig. 4.7). Enlarged VRS may expand over time and incidentally become very large locally, causing mass effect. Giant VRS will be discussed in Chap. 8.

Lacunar infarcts are focal ischemic lesions of the deep small vessels that can be the result of minor strokes or transient ischemic attacks, but are often clinically silent. Contrary to focal incomplete infarcts, lacunar infarcts are hypointense on T1 and FLAIR (the latter may often reveal an irregular rim of hyperintensity around the lacune). Actually, lacunar infarcts may have signal intensity similar to CSF on all MR sequences. The distinction between lacunar infarcts and enlarged VRS is therefore not always easy. Apart from the size and location criteria, shape is an important factor to take into account. Enlarged VRS are usually round or linearly shaped, whereas lacunar infarcts are typically wedged or irregularly shaped. Occasionally, enlarged VRS are identifiable by their orientation around the course of perforating vessels on sagittal or coronal MR images, but this feature may be missed due to partial volume effects.

4.3.1 Microbleeds

Microbleeds, also known as haemorrhagic lacunes, are hypointense foci (<5 mm) on gradient-echo T2*-weighted images (T2*-WI). Routine T2*-WI with a long echo time or dedicated susceptibility-weighted imaging (SWI) are the preferred sequences for their detection (Fig. 4.8). Microbleeds are usually found in the basal ganglia or thalamus and posterior fossa in hypertensive patients, sometimes in combination with larger cerebral haemorrhages. More peripheral corticomedullary microbleeds are found in cerebral

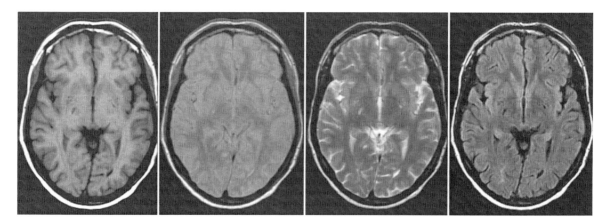

Fig. 4.7 Axial T1-weighted, proton-density weighted, T2-weighted, and FLAIR images (from *left* to *right*) showing sharply demarcated areas, around the anterior commissure, with a signal intensity similar to the cerebrospinal fluid – enlarged Virchow–Robin spaces

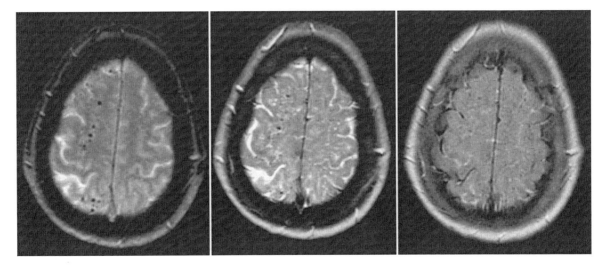

Fig. 4.8 Axial gradient-echo T2*-weighted image, spin-echo T2-weighted image, and turbo spin-echo fluid-attenuated inversion recovery (FLAIR) image (from *left* to *right*) showing hypointense foci consistent with microbleeds, which are better seen on the T2*-weighted image

autosomal dominant arteriopathy with subcortical infarcts and leukoencephalopathy (CADASIL), and are a mandatory finding in cerebral amyloid angiopathies (CAA). Cerebral microbleeds can also be found in asymptomatic individuals. In these circumstances, their clinical relevance is unclear, but relates to a possibility that they may predispose to macroscopic haemorrhages, especially when they are found in combination with ischemic diffuse confluent white matter abnormalities or in patients on anticoagulants.

4.4 Age-Related White Matter Hyperintensities

White and deep grey matter hyperintensities on T2-WI, FLAIR images, and proton-density weighted images are generally considered as a marker of ischemic small vessel disease. On CT, white matter changes appear as mildly hypodense areas, sometimes referred to as leukoaraïosis. White matter hyperintensities (WMH) progressively increase with age, and are sometimes referred to as age-related white matter changes (ARWMC). ARWMC are associated with vascular risk factors and with other cerebrovascular lesions. Several WMH visual rating scales have been proposed. In the commonly used Fazekas scale, WMH on MRI can be classified as: punctiform, early confluent, and diffuse confluent (Fig. 4.9). The ARWMC visual rating scale is applicable both to CT and MR imaging and involves the scoring of five different regions for each hemisphere: frontal, parietal, occipital and temporal lobes, basal ganglia, and infratentorial structures (range of scores for each region: 0–3).

WMH can be subdivided into deep and periventricular, but it is generally assumed that both types result from ischemia. The periventricular white matter is a terminal zone in terms of arterial supply, because it is perfused by long perforating arteries and, therefore, it is particularly vulnerable to decreases of blood flow, both at the microvascular level and at the level of large arteries. 'Caps' and 'bands' are examples of periventricular WMH attributable to breakdown of ependymal lining, increase of extracellular fluid, and subependymal gliosis. They generally occur before deep WMH and represent usual ageing phenomena (Fig. 4.10).

Deep white and grey matter hyperintensities are thought to represent ischemic lesions due to small vessel disease causing hypoperfusion and incomplete infarction. As periventricular and deep WMH progress, they may become confluent.

In spite of being considered as a surrogate marker of ischemic small vessel disease, punctiform and early confluent deep WMH often have surprisingly

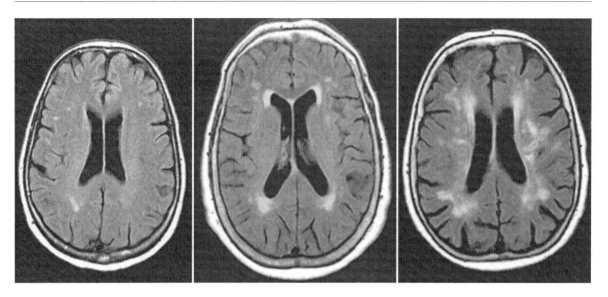

Fig. 4.9 Axial FLAIR images showing punctiform (Fazekas score 1), early confluent (score 2), and diffuse confluent (score 3) white matter hyperintensities. A score of 3 is always abnormal and carries an overall poor prognosis for survival

Fig. 4.10 Axial FLAIR images showing punctiform white matter hyperintensities, as well as periventricular caps (*right*) and bands (*left*), attributable to breakdown of ependymal lining, increase of extracellular fluid, and mild subependymal gliosis. Smooth periventricular caps and bands are normal ageing phenomena and clinically less relevant than ischemic deep WMH

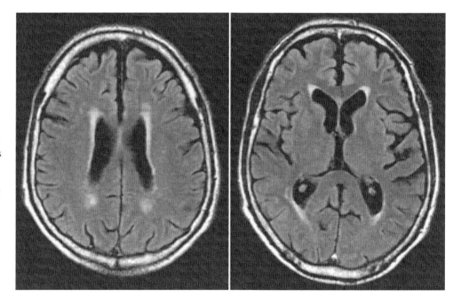

little discernible clinical consequences. By contrast, functionally independent elderly subjects with diffuse confluent WMH are at increased risk of loss of independence in activities of daily living secondary to motor and cognitive deterioration. The occurrence of hypoperfusion in subjects with diffuse confluent WMH (Fig. 4.11) perhaps explains the high risk of decline. Therefore, early confluent, and certainly diffuse con-

fluent WMH, are definitive signs of pathological ageing, whereas punctiform WMH are more benign.

A key prognostic factor is progression of the WMH. Prospective, longitudinal studies suggest, unsurprisingly, that the most important predictor of WMH progression is their severity at baseline. Influences on (and how to prevent) progression of WMH await well-conducted large scale trials.

| FLAIR | T1-WI | Cortical CBF | Subcortical CBF | Global CBF |

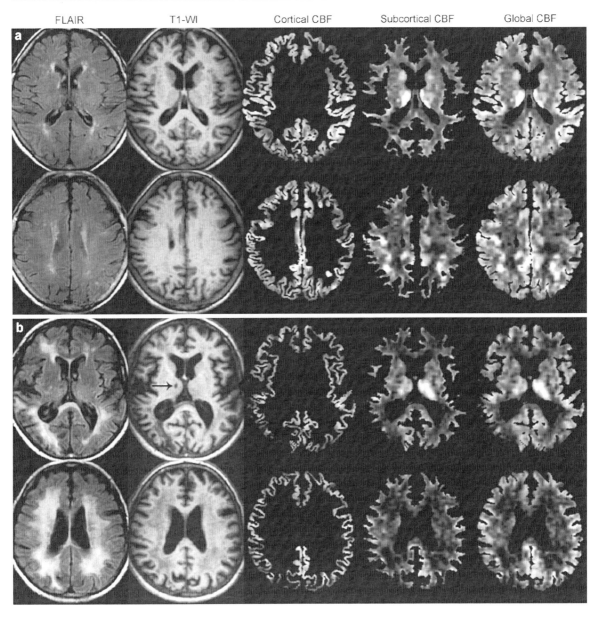

Fig. 4.11 (**a**) Axial fluid-attenuated inversion recovery (FLAIR) images, T1-weighted images (T1-WI), cortical grey matter (GM) cerebral blood flow (CBF) images, subcortical CBF images, and global CBF images combining all GM and white matter (WM) at the level of basal ganglia and thalamus (*above*), and at the level of cerebral WM (*below*) from a subject with punctiform WMH on FLAIR images; (**b**) Corresponding images from a subject with diffuse confluent WMH on FLAIR images. On the CBF images, note that the cortex, the cortical-subcortical transition, and the thalamus are highly perfused. Note also the perfusion differences between the subject with diffuse confluent WMH and the subject with punctiform WMH. Finally, on the subcortical and global CBF images from the subject with diffuse confluent WMH, there is relative hypoperfusion of the right thalamus, where a lacunar infarct is present (*arrow*). (Reproduced with permission from Am J Neuroradiol 2008; 29:1296–1301)

4.5 Iron Accumulation

Iron accumulates in the brain during the process of usual ageing. Most of it corresponds to intraneuronal ferritin. The accumulation of neuronal iron may also occur in conjunction with other compounds, such as lipofuscin. An excessive accumulation of iron in the brain may be associated with various neurodegenerative disorders. Iron accumulation may generate local oxidative stress by increasing the concentration of oxygen free radicals and may lead to lipid peroxidation and neurotoxicity.

Iron accumulation in the brain prominently involves areas related to motor functions, such as the basal ganglia, in particular the globus pallidus, as well as the striatum, subthalamic nucleus, substantia nigra, and the dentate nucleus of the cerebellum. However, iron deposition is also found in the cerebral white matter and cortex.

From a neuroimaging point of view, iron accumulation in the brain is seen as a hypointensity on T2-WI, but such hypointensity is also very much dependent on the field-strength and pulse-sequence. Confirmation that this hypointensity is due to iron rather than other minerals, such as calcium, came from MR-pathological correlations and *ex vivo* studies. SWI can differentiate iron deposition from calcification – iron is associated with a paramagnetic phase shift, whereas calcification causes an opposite diagmagnetic phase shift.

At birth, the brain contains very little iron. Increases in brain iron first occur in the globus pallidus, dentate nucleus, substantia nigra, and the red nucleus. Hypointensity on T2-WI of these structures is therefore a normal finding by the end of the third decade of life. To a lesser extent, there is also iron accumulation in the putamen, caudate, thalamus, cerebral cortex, and white matter. Hypointensity of the putamen usually does not appear until the seventh decade of life and it usually starts posteriorly, progressing anteriorly with age (Fig. 4.12).

4.6 Age-Related Changes in 'Normal Appearing' Brain Tissue

Neuronal shrinkage, rather than cellular death, probably accounts for most of the cortical volume loss in normal ageing, whereas demyelination and axonal changes account for hidden abnormalities of the white matter. Advanced MR techniques, such as proton MR spectroscopy (MRS), magnetization transfer (MT) imaging, T1 and T2 relaxation time measurements, and diffusion tensor imaging (DTI) are more sensitive to metabolic or micro-structural age-related changes in the brain tissue. Functional MR imaging (fMRI) and perfusion-weighted techniques, such as arterial spin labelling (ASL), are sensitive to the functional consequences of such changes.

4.6.1 Metabolic and Microstructural Changes

Proton MRS studies have shown an age-related decrease in the N-acetylaspartate (NAA) peak both in grey and white matter, although there is some debate as to whether this remains true after correction for the volume of grey matter. One study assessing the whole

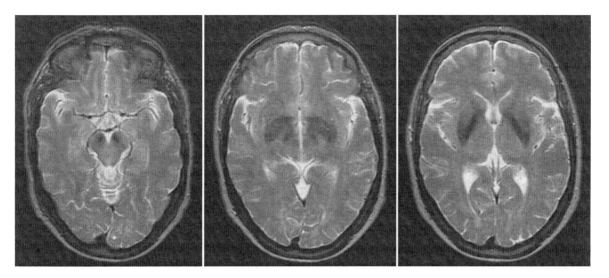

Fig. 4.12 Axial T2-weighted images of a 69-year-old healthy subject showing hypointensity in the globus pallidus, posterior part of the putamina, substantia nigra, and in the red nucleus. Such hypointensity is attributable to iron accumulation

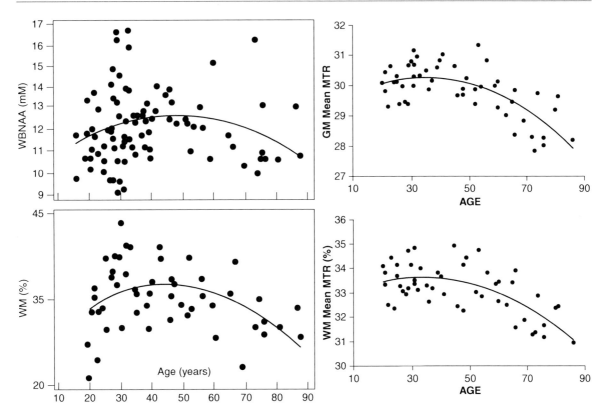

Fig. 4.13 Scatterplots displaying quadratic patterns of change over time in the N-acetylaspartate (NAA) and magnetization transfer ratios (MTR). (Reproduced with permission from Top Magn Reson Imaging 2004;15:355–363)

brain NAA quantification found a quadratic pattern of change over time (Fig. 4.13) similar to the one observed for the white matter volume. Age-related increases of the creatine (Cr) peak have also been reported, which is particularly relevant considering that the concentration of the corresponding metabolites is usually expected to be constant and generally used as the reference for semi-quantitative purposes.

Most proton MRS studies in patients with AD or frontotemporal dementia have shown a decrease of the NAA peak and an increase of the myo-inositol peak. Some studies also refer an increase of the Cr and choline levels in patients with AD. Given that an increase of the Cr levels may occur in normal ageing, the use of this metabolite as an internal reference for quantification should be cautiously considered when trying to differentiate patients with neurodegenerative diseases from control subjects on the basis of MRS.

Age-related magnetisation transfer ratio (MTR) changes do not show a linear pattern of modification over the adulthood. MTR values of both grey and white matter show a slight increase during early adulthood,

approximately up to the age of 40 years, and then start to decline. These findings also suggest continued brain maturation (e.g. myelination) until the middle age of life and are consistent with the previously mentioned quadratic patterns of change over time in both the white matter volume and global brain NAA. However, contrary to the linear pattern of grey matter volume loss, MTR values of grey matter have also a quadratic pattern of modification, which probably reflects changes of myelin content in the grey matter as the age increases (Fig. 4.13).

Age-related MTR decreases at older ages particularly occur in the corpus callosum and in the white matter of the frontal lobe. More pronounced MTR decreases occur in WMH and extend beyond the macroscopic lesions visualised on conventional MR images. MTR values are therefore helpful to evaluate hidden abnormalities.

The age-related quadratic pattern of MTR modification is also consistent with age-related changes of T1 relaxation time measurements. T1 relaxation times decline throughout adolescence and early adulthood, achieving a minimum value in the fourth to sixth decade of life, and then start to increase. This sequence of T1 relaxation time changes over time is different for the various brain structures and generally occurs later in the grey matter than in the white matter. It is believed to reflect both the change of water and lipid content during the process of myelination at younger ages, and possibly degeneration, demyelination, increase of extracellular space, or iron accumulation at older ages. By contrast, T2 relaxation times appear not to be influenced by age.

Diffusion-weighted imaging studies show that during the myelination process, there is an increase of fractional anisotropy (FA) and a decrease of the apparent diffusion coefficient (ADC) values in the white matter, which continues until the third decade of life. Conversely, at older ages, there is an age-related decline of both global brain and white matter FA, as well as an increase of the ADC values. The age-related decline of FA in the white matter particularly occurs after the age of 40 years in the genu of the corpus callosum and centrum semiovale. Diffusion changes at older ages are suggestive of demyelination and axonal loss, with some confirmation on post-mortem examination. Moreover, increases in diffusion both in lesions and in normal-appearing white matter are found in subjects with WMH, again indicating a generalised cerebrovascular disease process, rather than one confined to lesions visualised on conventional MR images.

Microstructural abnormalities on diffusion MRI techniques seem to be correlated with measures of motor and cognitive functioning and may be useful in identifying subjects with dementia (Fig. 4.14).

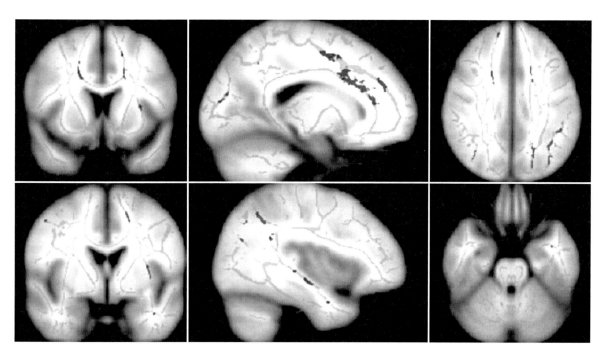

Fig. 4.14 Differences in white matter tract integrity displayed in blue, for older compared with younger healthy subjects, and in red, for patients with AD compared with healthy elderly control subjects. Coronal, sagittal, and axial images represent t-maps ($p < 0.05$, corrected). For visualisation purposes, the differences between patients with AD and healthy elderly control subjects is also displayed in pink, at a more liberal statistical threshold ($p < 0.001$, uncorrected). In green, the mean fractional anisotropy 'skeleton' is overlaid on the MNI152 standard brain. Note that normal ageing is associated with reductions in FA mostly in frontal tracts; the temporal tracts are more strongly affected in AD. (Reproduced with permission from Hum Brain Mapp 2009; 30:1051–1059) (*continued on next page*)

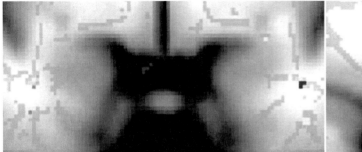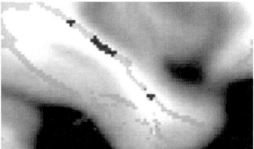

Fig. 4.14 (continued)

4.6.2 Functional Changes in Normal Ageing

Memory task-related fMRI studies have shown conflicting results regarding the transition from healthy ageing to early AD. Some studies showed enhanced activation in subjects at risk of developing AD (e.g. those carrying an APOE-ε4 allele), whereas others have shown that subjects at high risk of AD have decreased responses to encoding tasks. Of particular interest is the so-called default-mode network. This network has been consistently observed to be active during rest, but it is suspended (deactivated) during goal-directed behaviour. Its activity may therefore serve as a measure of integration of intrinsic brain activity and, therefore, of basal functional connectivity.

Reduced resting-state fMRI activity in the default-mode network (anterior frontal, precuneus, and posterior cingulate cortices) in 'usual' ageing is correlated with executive dysfunction and slowing of processing speed. In line with this, ineffective deactivation of the default-mode network during active memory tasks may represent an early marker of AD.

ASL shows a decline in the grey matter to white matter perfusion ratio as the age increases, mainly due to a reduction in the grey matter perfusion: grey matter perfusion decreases by 0.45% per year in healthy adult subjects, predominantly in the frontal cortex. The ageing effects seen in perfusion mirror those observed in terms of metabolism on fluorodeoxyglucose – positron emission tomography (FDG-PET). Whereas successful ageing is accompanied mostly by reductions in frontal metabolism, hypometabolism in the posterior cingulate cortex may herald pre-symptomatic AD, especially in subjects who have the APOE-ε4 genotype. Likewise, amyloid-PET shows abnormal uptake in approximately 30% of the normal elderly subjects (Fig. 4.15), which appears to be a subclinical manifestation of Alzheimer's pathology (long-term follow-up is needed for these cases); this has led to suggestions that preclinical and certainly pre-dementia AD may be diagnosable with some certainty.

4.7 Is It Normal or Not?

Usual ageing phenomena may be difficult to distinguish from abnormalities occurring in pathological ageing. Their clinical correlates can be surprisingly weak. An impressive burden of silent brain lesions and changes may be found as incidental findings in subjects without symptoms. An understanding of the range of age-related changes is, therefore, essential for the interpretation of imaging at the borderlands between usual and pathological ageing. Certain changes carry greater significance than others:

- MTA is more suggestive of pathological ageing than GCA.
- High rates of global/regional volume loss are more suggestive of pathology.
- Diffuse confluent WMH (Fazekas grade 3) are always abnormal.
- Whereas VRS in general can be normal, 'état criblé' is always pathological.

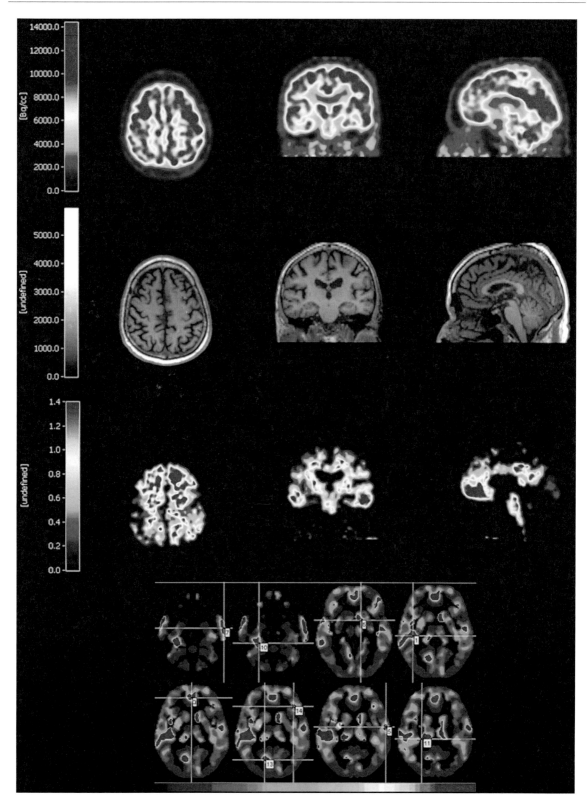

Fig. 4.15 Silent amyloid accumulation. This 74-year-old man complained of mild memory problems. His MMSE was 29/30, and extensive neuropsychological testing revealed no objective abnormalities. MRI was normal for age, and CSF analysis was normal. FDG-PET was normal by visual analysis (*upper row*) and by voxel-wise analysis (very few red areas on the *bottom panel*). However, amyloid-PET (third *row*) revealed extensive abnormal binding in most of the neocortex, perhaps signalling preclinical Alzheimer pathology

Finally, all of these distinctions are also a matter of degree that certainly depends on the awareness of the observer in recognising the typical range of the very variable 'normal ageing' brain. A standardised imaging protocol is essential to appreciate the spectrum of normal age-related changes, since they may appear quite differently on for example FLAIR and heavily T2-weighted MRI.

Suggested Reading

Agartz I, Saaf J, Wahlund LO, Wetterberg L (1991) T1 and T2 relaxation time estimates in the normal human brain. Radiology 181:537–543

Bartzokis G, Beckson M, Lu PH, Nuechterlein KH, Edwards N, Mintz J (2001) Age-related changes in frontal and temporal lobe volumes in men: a magnetic resonance imaging study. Arch Gen Psychiatry 58:461–465

Bastos-Leite AJ, van Waesberghe JH, Oen AL, van der Flier WM, Scheltens P, Barkhof F (2006) Hippocampal sulcus width and cavities: comparison between patients with Alzheimer disease and nondemented elderly subjects. AJNR Am J Neuroradiol 27:2141–2145

Bookheimer SY, Strojwas MH, Cohen MS et al (2000) Patterns of brain activation in people at risk for Alzheimer's disease. N Engl J Med 343:450–456

Brass SD, Chen NK, Mulkern RV, Bakshi R (2006) Magnetic resonance imaging of iron deposition in neurological disorders. Top Magn Reson Imaging 17:31–40

Fazekas F, Chawluk JB, Alavi A, Hurtig HI, Zimmerman RA (1987) MR signal abnormalities at 1.5 T in Alzheimer's dementia and normal ageing. AJR Am J Roentgenol 149:351–356

Fleisher AS, Podraza KM, Bangen KJ et al (2009) Cerebral perfusion and oxygenation differences in Alzheimer's disease risk. Neurobiol Aging 30:1737–1748

Fox NC, Crum WR, Scahill RI, Stevens JM, Janssen JC, Rossor MN (2001) Imaging of onset and progression of Alzheimer's disease with voxel-compression mapping of serial magnetic resonance images. Lancet 358:201–205

Good CD, Johnsrude IS, Ashburner J, Henson RN, Friston KJ, Frackowiak RS (2001) A voxel-based morphometric study of ageing in 465 normal adult human brains. Neuroimage 14:21–36

Inglese M, Ge Y (2004) Quantitative MRI: hidden age-related changes in brain tissue. Top Magn Reson Imaging 15: 355–363

Inzitari D, Simoni M, Pracucci G et al (2007) Risk of rapid global functional decline in elderly patients with severe cerebral age-related white matter changes: the LADIS study. Arch Intern Med 167:81–88

Kapeller P, Schmidt R, Fazekas F (2004) Qualitative MRI: evidence of usual ageing in the brain. Top Magn Reson Imaging 15:343–347

Parkes LM, Rashid W, Chard DT, Tofts PS (2004) Normal cerebral perfusion measurements using arterial spin labelling: reproducibility, stability, and age and gender effects. Magn Reson Med 51:736–743

Savva GM, Wharton SB, Ince PG, Forster G, Matthews FE, Brayne C (2009) Age, neuropathology, and dementia. N Engl J Med 360:2302–2309

Villemagne VL, Pike KE, Darby D et al (2008) Aβ deposits in older non-demented individuals with cognitive decline are indicative of preclinical Alzheimer's disease. Neuropsychologia 46:1688–1697

Primary Grey Matter Loss

5

Contents

5.1 Introduction: Primary Grey Matter Loss

Many dementias are characterised by grey matter (GM) abnormalities; these are usually, but not exclusively, atrophy. AD is the most prevalent and the prototypic cortical GM dementia. GM atrophy reflects a loss of neurons irrespective of the underlying protein defect (amyloid, tau, alpha-synuclein); atrophy may be generalised or focal, and the pattern of atrophy may be diagnostic in itself.

The current chapter focuses on disorders principally affecting GM, encompassing, besides AD, the growing class of frontotemporal lobar degeneration syndromes, including FTD and PPA, CBD and PSP, as well as PD with dementia as in DLB, Huntington's disease and many less common diseases. A special chapter is dedicated to the prion diseases that typically affect the more basal grey matter structures as well.

F. Barkhof et al., *Neuroimaging in Dementia*,
DOI: 10.1007/978-3-642-00818-4_5, © Springer-Verlag Berlin Heidelberg 2011

5.2 Alzheimer's Disease

Synonyms

Senile Dementia of Alzheimer type; Senile Dementia;
Presenile Dementia; Alzheimer's disease (AD);
Degenerative Dementia

5.2.1 History

The disease is named after Aloïs Alzheimer, who presented his observations on a case of a 51-year-old
woman in 1906. In her brain, Alzheimer found the hallmarks of the disease that now bears his name, the senile
plaques and neurofibrillary tangles – findings that had
been described before, but not in the context of a progressive dementia. The disease that was named after
him by Kraepelin in 1914 remained a rare disorder until
large autopsy series carried out in the USA and UK
found that the same changes were present in elderly
individuals that presented with cognitive decline.
Subsequently, Alzheimer's disease became synonymous
for any type of dementia occurring at any age. It was not
until 1984 that a work group under the auspices of the
NINCDS formulated the first clinical criteria.

5.2.2 Histopathology

The diagnosis of AD is ultimately made by pathological examination of the brain (at autopsy or by brain
biopsy in exceptional cases). Macroscopic examination reveals gross atrophy, also reflected by a severely
decreased brain weight (usually below 1,000 g), affecting mostly the temporo-parietal regions (Fig. 5.1),
sometimes however, quite prominent at the frontal
poles. The characteristic lesions at the microscopic
level are extracellular neuritic plaques, consisting of a
core composed of beta amyloid (abeta), and intracellular neurofibrillary tangles (NFT) consisting of hyper-
phosphorylated tau-protein (Figs. 5.2–5.4).

On a semantic level, the term 'senile plaques' means
amyloid plaques for US scientists, but neuritic plaques
for European neuropathologists, and probably also for
Aloïs Alzheimer himself, when dealing with silver
staining of brain tissue sections. Therefore, senile

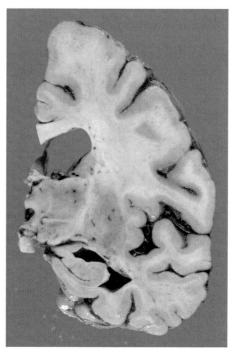

Fig. 5.1 Coronal cut of the brain of a patient with AD. Note
mild ventricular dilatation including the inferior horn, secondary
to hippocampal atrophy

plaques can be considered as 'neuritic plaques', that is
amyloid plaques surrounded by degenerating neurites
filled with tau pathology. This is why neuritic plaques
(amyloid + tau) are the best histological markers of AD.
Amyloid plaques are found in both non-demented and
demented patients, while neuritic (senile) plaques are
only found in demented patients. The staging system
developed by Braak and Braak (see Fig. 5.5) describes
the extent, location and the presumed sequence of accumulating neurofibrillary tangle pathology, which in AD
is thought to start in the transentorhinal and entorhinal
areas, before spreading to the hippocampus, the association cortices, and the rest of the cortex.

Current histological criteria for the diagnosis of AD
are based on the density of neuritic plaques and NFTs
in the neocortex and limbic areas. Various sets of histopathological criteria exist (NIA-Reagan, CERAD)
that may not be in agreement. While histopathological
diagnosis is still considered the gold standard, considerable discrepancies between pathological diagnoses
at post-mortem and the clinical diagnoses during life-
time exist. A prospective, population-based study on
the prevalence of AD in people over 85 years found

Fig. 5.2 Histopathology in cortex of an AD patient. *Left* image stained with tau (*brown*), Congo-red (*red*) and haematoxylin blue. The *long arrow* points at a neuritic plaque with a red nucleus of amyloid with a tau-positive corona; *arrowheads* point at tangles. Magnification on the *right* shows the amyloid core with a dense corona of tau-filled neurites

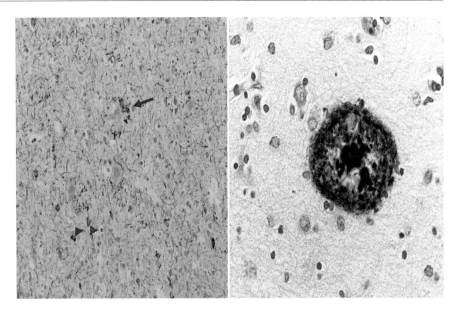

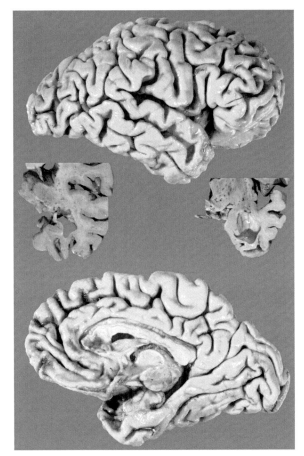

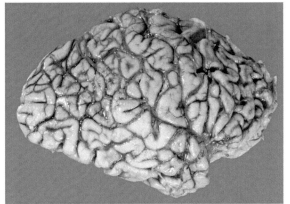

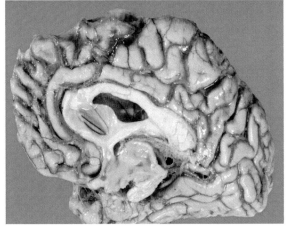

Fig. 5.3 Gross specimen of familial AD patient. Lateral and medial view showing prominent temporal as well as parietal atrophy with widening of the Sylvian fissure. Note the marked mediotemporal atrophy on the inserts

Fig. 5.4 Atypical distribution of atrophy in AD. Macroscopic image of an AD patient showing remarkable frontal lobe atrophy

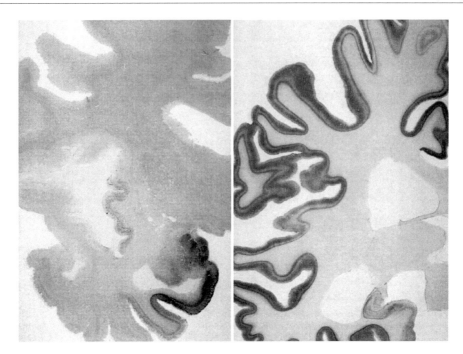

Fig. 5.5 Braak stages of cortical spreading in AD. Neurofibrillary tangles start to accumulate in the entorhinal region and spread from there to the amygdala, hippocampus, and finally involves virtually every subdivision of the cerebral cortex. The anatomical specimen on the *left* shows stage IV, in which destruction of the entorhinal area has spread into the amygdala, hippocampal formation and, in particular, from the transentorhinal region into the adjoining association areas of the temporal neocortex. At this stage the disease is becoming clinically manifest. The anatomical specimen on the *right* illustrates stage V–VI of the progress of Alzheimer's disease, with a high density of neurofibrillary changes in the entire cortex. This stage represents full-blown AD. (Courtesy of H. and E. Braak, Goethe University, Frankfurt/Main, Germany)

that 55% of the individuals who met the neuropathological criteria for AD were either not demented during life-time or classified as vascular dementia. Conversely, 35% of patients with clinical AD did not fulfil the pathological criteria. In spite of this, the prevailing theory regarding the origin of AD is reflected by the Amyloid cascade hypothesis, originally coined by Selkoe and Hardy (Fig. 5.6).

Amyloid pathology is not confined to AD, but is also found in other diseases, like Lewy body dementia and cerebral amyloid angiopathy. The majority of subjects with clinical, late-onset AD show co-existing vascular pathology at post-mortem and it is quite plausible that the two pathologies interact (e.g. vessel wall changes and hypo-perfusion accelerating AD); see also Sect. 6.7.

5.2.3 Genetics

The cause of AD is still not fully understood, except for the familial autosomal dominant inherited cases with early onset, associated with mutations in the amyloid precursor protein (APP) and presenilin genes (PSEN1 and PSEN2). However, the familial form of AD is extremely rare, with prevalence below 1%. The far more common, sporadic form of AD is genetically associated with the apolipoprotein E4 allele (APOEe4), although APOEe4 is neither necessary nor sufficient to cause AD. The APOEe4 allele increases the risk of the disease threefold in heterozygotes and by 10–15 times in homozygotes. Apart from genetic risk factors, the most important risk factor for AD is age. It is estimated that both the incidence and the prevalence double with every five year increase in age. Other risk factors for AD include female sex and vascular risk factors, such as diabetes, hypercholesterolemia and hypertension. Whether these vascular risk factors are causally related to the neuropathological process of AD or induce cerebrovascular damage that coincides with or adds to Alzheimer-type neuropathology remains to be established and are important issues with regard to potential preventive measures.

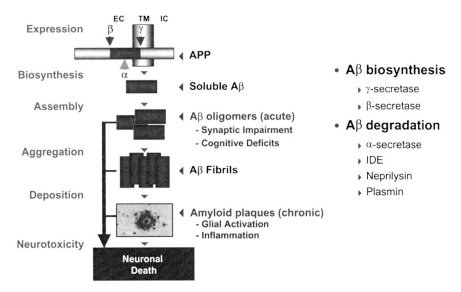

Fig. 5.6 Amyloid cascade. Abnormal cleavage of APP leads to amyloid aggregation and plaque formation. (Reprinted with permission from Biochem Soc Trans 2005;33:553–558)

5.2.4 Clinical Presentation

Typically, AD is characterized by an insidious onset of cognitive decline, starting with deficits in episodic memory. Patients and their family complain, for example, of forgetting recent personal and family events, losing items around the house, and repetitive questioning. As the disease progresses, other deficits such as aphasia, apraxia, agnosia, visuo-spatial difficulties and executive dysfunction, arise gradually. In Fig. 5.7, a typical example of the clock drawing test is given. This test relies on visuospatial, praxis and executive abilities. Short of a full neuropsychological examination,

a simple bed-side test called the "mini-metal state examination (MMSE)" can be used. Normal subjects score 27 or higher on the 30-point MMSE scale, although cut-off values are somewhat dependent on education.

Psychiatric and behavioural problems such as mood disorders, psychosis, agitation and sleep disorders occur more frequently as the disease progresses. The patient becomes increasingly dependent on others. Clinical diagnosis is made using criteria, of which the Mckhann criteria published in 1984 are the most widely validated and used. The criteria discern three categories of certainty: definite AD (established by post-mortem or

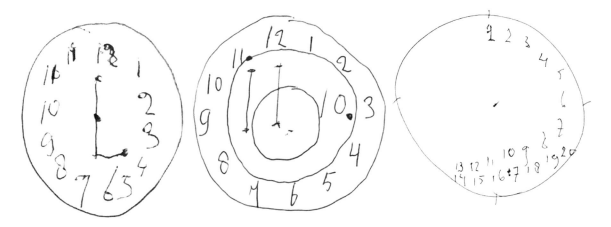

Fig. 5.7 Clock drawings by patients with AD. Patient is asked to draw a clock and set the time for 10 min past 11 o'clock. Shown are three examples of different patients with various degrees of severity of visuospatial and executive functions. Note that none of the clocks has the hands placed correctly or at all

Table 5.1 NINCDS-ADRDA criteria for probable AD

1. Dementia established by clinical examination and confirmed by neuropsychological tests
2. Deficits in two or more areas of cognition, including memory impairment
3. Progressive worsening of memory and other cognitive functions
4. No disturbances of consciousness
5. Onset between ages 40 and 90
6. Absence of systemic disorders or other brain disease that in and of themselves could account for the progressive deficits in memory and cognition

Source: McKahnn et al. Neurology 1984;34:939–944

biopsy), probable AD (Table 5.1) and possible AD (when there are other explanations for the cognitive syndrome that are as likely). The average survival in AD is typically about 8–13 years from the onset of symptoms.

Since the publication of the NINCDS-ADRDA criteria, the elucidation of the biological basis of AD has advanced greatly, allowing a better understanding of the disease process. The clinical phenotype of AD is no longer described in exclusionary terms, but can be characterized more definitively on a phenotypic basis. Distinctive markers of the disease are now recognized including structural brain changes on MRI with early and extensive involvement of the medial temporal lobe, functional neuroimaging changes on PET with hypometabolism or hypoperfusion in temporo-parietal areas, amyloid imaging with specific PET ligands and changes in CSF biomarkers. A driving force behind this emerging identity of AD has been the intense research interest in characterizing the earliest stages of AD that predate the crossing of the dementia threshold, defined by functional disability. From this, a need was felt to identify prodromal AD that must be distinguished within the broad and heterogeneous state of cognitive functioning that falls outside normal aging described by a wide range of terms including Age-Associated Memory Impairment, Age-Related Cognitive Decline, Age-Associated Cognitive Decline, Mild Cognitive Disorder, Mild Neurocognitive Disorder, Cognitively Impaired Not Demented, and Mild Cognitive Impairment (MCI).

5.2.4.1 From MCI to Prodromal AD

This designation of MCI has been the most widely used diagnostic label referring to individuals who have subjective memory and/or cognitive symptoms accompanied by objective evidence of isolated memory and/or other cognitive impairment and whose activities of daily living are considered to be generally normal. Progression to clinically diagnosable dementia occurs at a higher rate from MCI than from normal (typically 10–15% per year – compared to rates of ~1% with normal ageing), but is clearly not the invariable clinical outcome at follow-up. A more refined definition of AD is then required, to reliably identify individuals with the disease at its earliest stages. A large group of EU–US investigators has formulated new criteria for this earliest stage of AD, starting from the presentation with a memory complaint in typical AD (see also Chap. 2) and adding biomarker information from MRI, PET or CSF or genetic confirmation. The proposed criteria are listed in Table 5.2.

Table 5.2 Prodromal Alzheimer's disease criteria. (A plus one or more supportive features B, C, D, or E)

Core diagnostic criteria
A. Presence of an early and significant episodic memory impairment which includes the following features:
(i) Gradual and progressive change in memory function reported by patient or informant over more than 6 months
(ii) Objective evidence of significantly impaired episodic memory on testing: this generally consists of memory performance that does not improve significantly with cueing or recognition testing and after effective encoding of information has been previously controlled.
(iii) The episodic memory impairment can be isolated or associated with other cognitive changes at the onset of AD or as AD advances.
Supportive features (B and/or C and/or D and/or E)
B. Presence of medial temporal lobe atrophy Volume loss of hippocampus, entorhinal cortex, amygdala evidenced on MRI with:
• Qualitative ratings using visual scoring (referenced to well-characterized population with age norms) or quantitative volumetry of regions of interest (referenced to well-characterized population with age norms).
C. Abnormal CSF biomarkers:
• Decreased Aβ 1–42 and/or increased total tau and/or increased phospho-tau.
• Other well-validated markers to be discovered in the future.
D. Specific pattern in functional neuroimaging with PET:
• Reduced glucose metabolism in bilateral temporal parietal regions.
• Other well-validated ligands including those that foreseeably will emerge such as PiB or FDDNP.
E. Proven AD autosomal dominant mutation within the immediate family.

Source: Dubois B et al. Lancet Neurol 2007;6:734–746

Table 5.3 Atypical presentations of AD

- Posterior presentations of AD (includes 'posterior cortical atrophy'; 'biparietal AD' and 'visual variants')
  - Balint's like syndrome (optic ataxia, simultanagnosia, optic apraxia)
  - Visual disorientation and navigation problems
  - Aperceptive visual agnosia
  - Limb apraxia and a corticobasal syndrome
- Frontal presentations
  - Behavioural disturbances resembling FTD; apathy
- Language presentations
  - Fluent or non-fluent aphasia; logopenic variant

Table 5.4 Clinical Red flags and alternative diagnostic considerations

Red flag	Alternative diagnosis
Abrupt onset	VaD
Stepwise deterioration	VaD
Prominent behavioural changes	FTD, VaD
Profound apathy	FTD, VaD
Prominent aphasia	SD, PNFA, VaD
Progressive gait disorder	VaD, NPH
Prominent fluctuations in level of consciousness	Delirium due to infection, medications, or other causes
or cognitive abilities	DLB, Temporal lobe epilepsy, OSAS, metabolic disturbances
Hallucinations or delusions	Delirium due to infection, medications, or other causes, DLB
Frequent falls	PSP, DLB
Extrapyramidal signs or gait	Parkinsonian syndromes, VaD
Eye-movement abnormalities	PSP, Wernicke's encephalopathy

Source: Modified from Kawas CH. N Engl J Med 2003;349: 1056–1063. For abbreviations see list on page XV

5.2.4.2 Atypical Presentations of AD

Besides the typical neuropsychological profile of AD presenting with early memory deficits as mentioned above, there is evidence from clinico-pathological studies that AD patients may present with different neuropsychological profiles (Table 5.3). These atypical variants of AD suggest that the distribution of neuropathological changes (Fig. 5.4) rather than the nature of the disease are reflected in the clinical syndrome and that in clinical practice AD should be considered as diagnosis in a broad range of focal cognitive syndromes. Atypical presentations are more often seen in young onset AD patients (arbitrarily defined as before the age of 65) and in patients not carrying an APOE4 allele.

5.2.5 Differential Diagnosis of AD

When a patient fulfils clinical criteria as mentioned in the previous paragraph, there is still a chance that the patient has in fact a different underlying pathological substrate. Although this can never be ruled out completely in vivo, there are certain clinical features that render the diagnosis of AD less likely. In Table 5.4, these clinical features, or red flags, are listed, with the other diagnostic considerations listed along side.

In addition to the above, the differential diagnosis of early onset dementia is much wider than AD alone. In fact, in the young, AD and FTD have been reported to be equal in prevalence (see also Sect. 2.3). A careful and detailed history and the use of ancillary investigations are needed to disentangle the various underlying possible pathologies.

5.2.6 Neuroimaging Strategy and Findings in AD

All current criteria stipulate that structural imaging needs to be done at the initial evaluation of a patient suspected to have dementia. Although mainly based on the assumption that this is needed to exclude other brain diseases amenable to (surgical) treatment, this has evolved over time to include imaging to identify the earliest and most specific changes that would allow a diagnosis of AD.

5.2.6.1 CT and MRI Findings

The choice between CT and MRI depends on many factors, including clinical suspicion (a priori chance), contraindications, costs and quality of scanners, claustrophobia, and the ability of the patient to keep still for the time needed for the MRI; see Table 5.5.

Table 5.5 Neuroimaging in AD: modalities and common findings

Option	Modality	Result
Rule out structural lesion	CT = MRI	Tumour, hydrocephalus, subdural hematoma
Hippocampal atrophy	MRI > CT	Symmetric atrophy; slight asymmetry sometimes
Cerebral atrophy	MRI = CT	Biparietal atrophy; precuneus atrophy
White matter changes	MRI > CT	None to moderate changes, symmetric, frontal > parietal
Microbleeds	MRI	None to many, usually lobar
Lacunes	MRI > CT	None to a few, most often in basal ganglia
Hypometabolism	FDG-PET	Temporal, parietal, post. cingular usually symmetrical
Hypoperfusion	SPECT = PET	Biparietal-temporal
Amyloid plaques	Amyloid-PET > FDDNP PET	Binding in frontal, temporal and parietal lobes
Dopaminergic transport and receptors	SPECT/PET	normal

The sign = denotes modalities equally effective; > denotes one superior over the other. For abbreviations see list on page XV

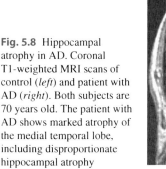

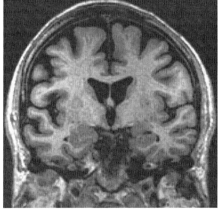

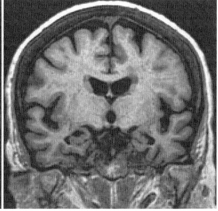

Fig. 5.8 Hippocampal atrophy in AD. Coronal T1-weighted MRI scans of control (*left*) and patient with AD (*right*). Both subjects are 70 years old. The patient with AD shows marked atrophy of the medial temporal lobe, including disproportionate hippocampal atrophy

The microscopic histological changes in the neurodegenerative diseases are inevitably associated with progressive regional and global brain atrophy, which may be assessed in vivo using MRI. In AD, focal atrophy in the medial temporal region, including the hippocampus, has been the focus of extensive study. It reflects the typical pattern of progression of (tangle) neuropathology, spreading from the entorhinal cortex and hippocampus to the association cortices, as described by Braak and Braak. Neuropathological studies have shown that hippocampal volumes, as measured using MRI, correlate well with the neuropathological burden at post-mortem. Many studies initially using CT and later MRI and more recently again using multislice CT (see Chap. 3) have assessed the diagnostic value of hippocampal atrophy for AD. In a meta-analysis of studies using visual and linear measurements of medial temporal lobe atrophy (MTA) on MRI, the overall sensitivity and specificity for detection of AD compared with controls was estimated to be 85% and 88% respectively. In clinical practice, simple visual rating scales estimating hippocampal atrophy are useful (see Table 3.5). A striking example is shown in Fig. 5.8.

As a cautionary note, it should be emphasized that hippocampal atrophy may occur in other diseases as well; so rating MRI scans for MTA should not be done out of clinical context (see Table 5.6 for differential diagnosis)

Besides the existence of medial temporal lobe atrophy, the most important structural imaging feature of AD is *progression* of atrophy. A yearly decline in hippocampal volume approximately 2.5 times greater in patients with AD than in normal aged subjects is reported and a relationship exists between memory loss and hippocampal damage across the spectrum from normal aging to dementia.

However, neuroanatomical changes over time may be too mild, diffuse, or topographically complex to be detected by simple visual inspection or even with manually traced measurements of regions of interest.

Table 5.6 Differential diagnosis of hippocampal atrophy

Side	DD	Other MR findings	Clinical clues	Additional tests
Bilateral	AD	Temporo-parietal atrophy	Episodic memory loss	CSF, FDG, PIB
	Hippocampal Sclerosis	Isolated finding	Episodic memory loss	None
	FTLD	Temporal pole or frontal atrophy	Behavioural, language	FDG – frontal hypoperfusion
	DLB (late)	Diffuse cortical atrophy	Visual hallucinations, extrapyramidal	Dopamine PET/SPECT
Unilateral	FTLD	Anterior more than frontal, temporal pole, frontal lobe atrophy	Behavioural, language	FDG frontal hypoperfusion
	Mesial temporal sclerosis	High signal hippocampus	Epilepsy	EEG

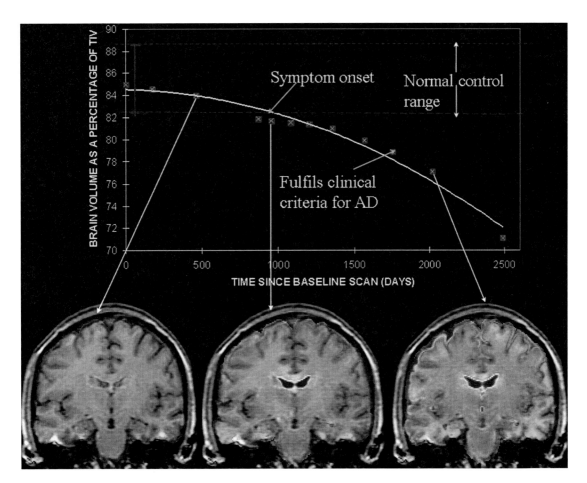

Fig. 5.9 Serial MRI in pre-symptomatic familial AD. Graph shows brain volume changes (derived from serial T1-weighted brain MRI) over a 7 year period spanning symptom onset in an individual with familial AD (PS1 mutation). *Lower* images show scans registered to baseline with fluid *overlay – green*=loss of brain volume; *yellow–red*=expansion of CSF regions. Note the pre-symptomatic hippocampal atrophy and prominent loss over time

New serial volumetric imaging techniques developed in the last few years represent an added value to identify subtle structural brain changes, which have brought extensive neocortical changes to the fore. Studies of MCI subjects and asymptomatic individuals at risk for familial AD show that hippocampal, cingulate and generalised neocortical losses are all present at an early and even presymptomatic stage (see Fig. 5.9).

In clinical practice, the pattern of atrophy across the entire brain should be taken into account, rather than an isolated evaluation of the medial temporal lobe. Usually, AD is characterized by global atrophy with prominent atrophy of the medial temporal lobe. However, prominent posterior atrophy is prevalent among younger AD patients and more often in APOE4 non-carriers (Figs. 5.10 and 5.12). Atrophy of either the parietal lobe or the precuneus (including the posterior cingulate) may be the only finding in young-onset AD, and the finding of a normal hippocampus should not distract from the diagnosis of AD. The differential diagnosis of posterior/parietal atrophy is listed in Table 5.7.

Besides atrophy, cerebrovascular pathology has been associated with AD, especially in the late onset form. As such, overlap with vascular dementia (VaD) may occur and patients may actually fulfil both criteria for AD and VaD. Unfortunately, no operational criteria for so-called mixed dementia exist, so it is left to the judgement of the clinician, what label fits best with the clinical picture of the patient. Further, use of PET or CSF may help to tease out the relevant pathologies.

In AD, most often, signs of small vessel disease are present on MRI in the form of white matter hyperintensities (WMH), lacunar infarcts (lacunes) and microbleeds (Fig. 5.11.) Microbleeds have been associated with amyloid angiopathy, but their clinical relevance is still uncertain in AD (see Box 5.1 and Fig. 5.11).

As with clinical findings there are numerous 'red flags' on structural imaging that should prompt the clinician to reconsider a diagnosis of AD. In Table 5.8 the most prominent ones are summarized.

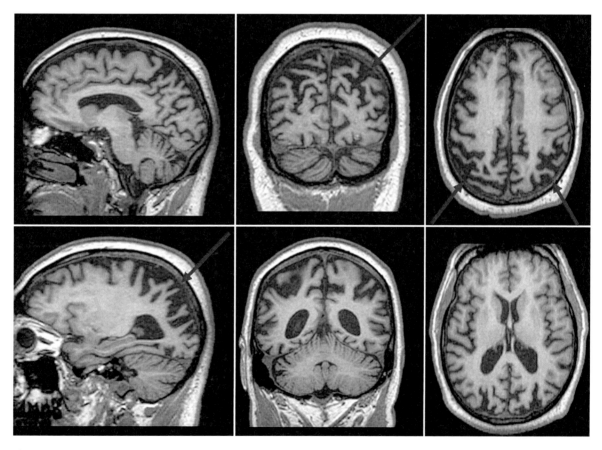

Fig. 5.10 Posterior cortical atrophy in AD. This 58-year-old presented with cognitive decline characterised by marked visuospatial and praxis difficulties with relatively preserved memory – she had gone to the opticians several times to change her glasses (with no benefit) because she was aware of her difficulties in 'seeing' things. Note the prominent posterior cortical atrophy (*red arrows*) with preservation of the hippocampus (*green arrow*)

Table 5.7 Differential diagnosis of posterior/parietal atrophy

DD	Other MRI findings	Clinical clues	Additional tests
AD	Symmetrical (usually) biparietal atrophy (+/−occipital); hippocampal atrophy relatively late feature	Memory not completely normal, apraxia, visuospatial deficits	CSF, PIB
DLB	Generalised atrophy; parietal and occipital	Extrapyramidal signs, hallucinations, fluctuations	Dopamine imaging
CBD	Asymmetric parietal (and frontal) atrophy	Asymmetrical limb praxis, myoclonus	Dopamine imaging, FDG-PET
CJD (Heidenhain variant)	FLAIR and DWI abnormal – cortical ribbon or striatum. May have generalised cerebral and cerebellar atrophy	Rapid decline, myoclonus	EEG (may be normal), CSF (tau &14-3-3)
Cerebrovascular	FLAIR/T2 signal change, watershed distribution	Subcortical clinical features; stroke	Vascular risk factors

For abbreviations, see list on page XV

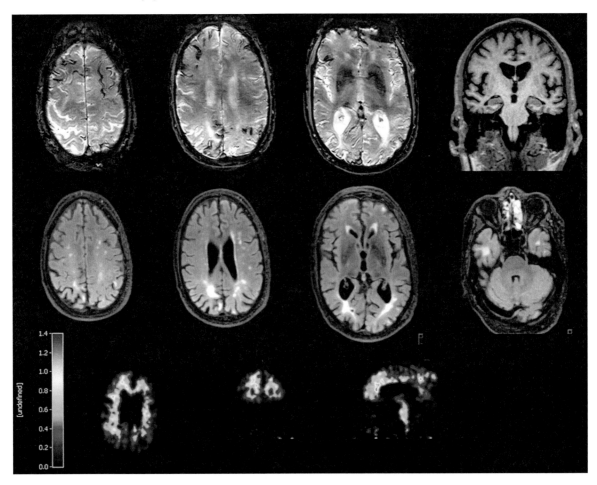

Fig. 5.11 Amyloid angiopathy in AD. This 58-year-old man presented with memory complaints, loss of orientation, inability to use house-hold tools and perform activities of daily living. His MMSE was 24/30. MRI showed minimal hippocampal (*upper right*) and parietal atrophy, expected for age; in addition, moderate white matter changes on FLAIR (*middle row*) and multiple microbleeds and sub-pial hemosiderin (low signal in *upper row*). The clinical diagnosis was early-onset AD and PIB-PET revealed increased cortical binding (*lower row*)

Box 5.1 Facts about microbleeds (MBs)

- Incidental findings on MRI, occur more frequently in AD
- Prevalence related to MR field strength and MR acquisition parameters
- Lobar MBs probably related to (amyloid) angiopathy
  - Accompanying superficial siderosis suggests amyloid angiopathy
- Central MBs related to hypertensive angiopathy
- Related to worse outcome in dementia
- Caution when oral anticoagulants are needed

5.2.6.2 Positron Emission Tomography (PET)

Metabolic changes may precede structural brain changes and can be visualised using the metabolic tracer [18F]fluorodeoxyglucose (FDG). In AD, temporal, parietal and most notably, posterior cingulate hypometabolism is found, discriminating AD patients from controls with good discriminatory power (sensitivity and specificity in the range of 85–90%) (Fig. 5.12). In current clinical guidelines, routine FDG-PET is not advised, since the added value over clinical diagnosis and structural imaging is uncertain. FDG-PET can be useful in the differential diagnosis between AD and FTLD.

An exciting novel application of PET is the in vivo imaging of amyloid. The amyloid β protein is considered essential in the pathogenesis of AD. Several PET tracers have been developed for this purpose. The Pittsburgh compound B ([11C]PIB) (see Sect. 3.5.3.2) is the most widely studied amyloid tracer in AD patients (Figs. 5.12 and 5.13). The future development of ^{18}F-labelled PIB or other compounds such as AV45 or Florbetaban with a slower radioactive decay will greatly facilitate clinical implementation of amyloid imaging, which is currently restricted to centres with a cyclotron for on-site production of the tracer.

In vivo amyloid imaging may considerably add to our understanding of the underlying pathophysiological mechanisms of AD. Furthermore, imaging of amyloid may prove to be a sensitive diagnostic marker and enable prognoses in the earliest stages of formation of neuropathology (see Fig. 5.14). It should be noted, however, that amyloid deposition is not exclusively confined to AD and also occurs in dementia with Lewy bodies and congophilic amyloid angiopathy (CAA).

In familial AD, patients may be asymptomatic for a long time, while carrying a mutation in one of the known dominant genes. Amyloid imaging (with PET-PIB) may help to identify the earliest signs of plaque deposition and may thus help to identify the best time to install future treatments aiming at removing plaques. Some amyloid imaging studies in presenilin 1 mutation carriers suggest the striatum to be the earliest site of amyloid deposition – present over a decade younger than anticipated age at onset of symptoms. When symptomatic, FAD patients additionally have neocortical and thalamic PIB deposition, although this does not appear to reach the density observed in the striatum. Cerebellar PIB deposition

Table 5.8 Red flags on MRI/CT in patients suspected of AD

Findings	Consider alternative diagnoses
No hippocampal atrophy	Normal aging; very mild AD, young onset AD, DLB, PCA, MCI; FTD, PSP, CBD
Unilateral hippocampal atrophy	Semantic dementia (FTLD)
Anterior but not posterior hippocampal atrophy	FTLD
Extreme hippocampal atrophy	Argyrophilic grain disease, FTLD: semantic dementia; tau mutations (anterior > posterior)
Abundant small vessel disease	VaD; CAA; amyloid angiopathy
Unilateral cerebral atrophy	Corticobasal syndrome, FTLD

For abbreviations, see list on page XV

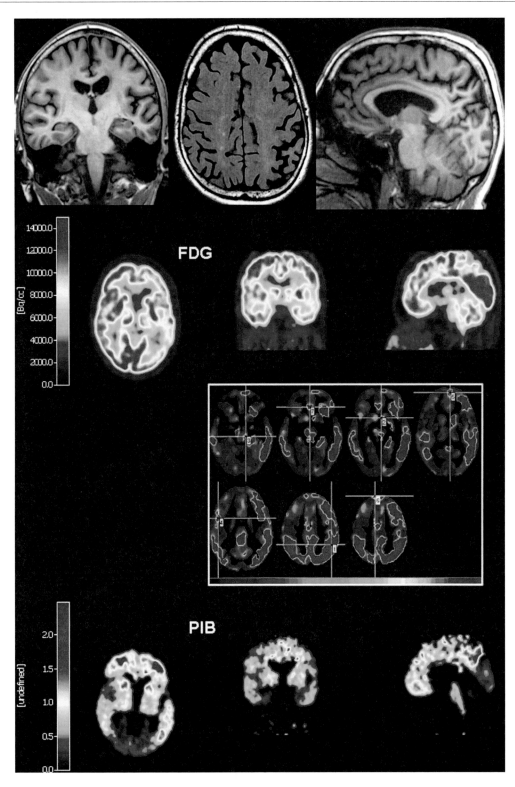

Fig. 5.12 FDG-PET in clinically diagnosed Alzheimer's disease. This patient with presenile onset of dementia had no hippocampal atrophy, but clear parietal atrophy, asymmetric to the left hemisphere (*upper row*). FDG PET revealed reduced metabolism in the posterior cingulate and parietal association cortices, which was confirmed by voxel-wise mapping (insert) – note that also the less atrophic right parietal cortex is now clearly abnormal. The diagnosis was further corroborated by abnormal cortical PIB-binding (*lower panel*)

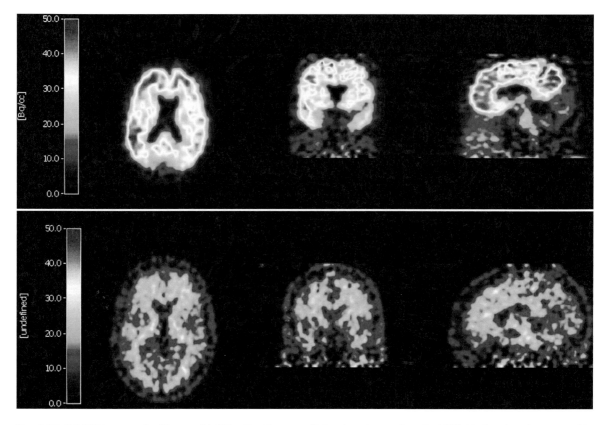

Fig. 5.13 ^{11}C-PIB images of a 63-year-old AD patient (*top panel*) showing abnormal cortical PIB binding in red, compared to images of a 64-year-old healthy individual (*lower panel*), showing only mild non-specific white matter binding

has also been noted in PSEN1 mutation carriers leading to the recommendation that the pons, rather than cerebellum, is used as the reference region in FAD PIB studies (Fig. 5.15).

5.2.6.3 Single Photon Emission Tomography (SPECT) in AD

The most widely used tracer to study regional cerebral blood flow is [99]TcHMPAO (Fig. 5.16). In AD bilateral temporo-parietal hypoperfusion is typically seen (as with FDG-PET). The application of SPECT in clinical routine has been hampered by false positive findings and insufficient added value over MRI. More promising and partly included in the routine clinical setting are *neuroreceptor* studies using [123]IFP-CIT (also referred to as 'DAT-scan') which allows visualization of the degeneration of the nigrostriatal dopaminergic neurons. Scintigraphically, it allows the distinction between patients with essential

tremor and patients with Parkinson's disease or PSP and MSA. In dementia, the distinction between AD and DLB may be relevant, especially when there are no extrapyramidal features. For this, the use of DAT can certainly be helpful, being abnormal in DLB, but normal in AD (see Sect. 5.4). A ^{123}I-IBZM-SPECT shows the integrity of the post-synaptic dopamine receptor. It shows the distinction between idiopathic Parkinson's disease and diseases with parkinsonism like PSP and MSA.

5.2.7 Recommendations for Further Imaging

A clinical diagnosis can be made when clinical criteria are met and MRI fits well, for instance, by showing hippocampal atrophy. The need for further work up may come up when normal findings on MRI are reported and there is clinical suspicion of AD, or in

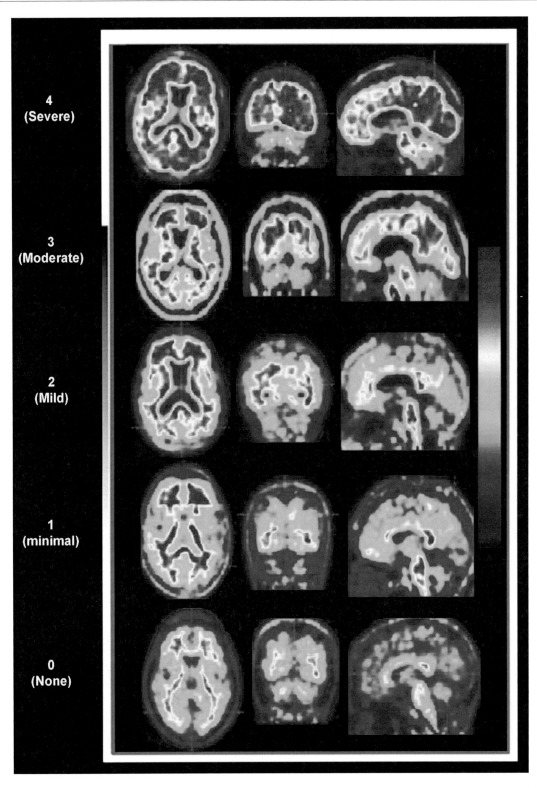

Fig. 5.14 Amyloid imaging score. PET scans obtained with a fluorinated ligand (AV-45). This ligand allows staging of severity of binding illustrated by stages 0–4, ranging from only white matter (with no cortical) to intense neocortical binding. (Courtesy of C. Clark, M.D.)

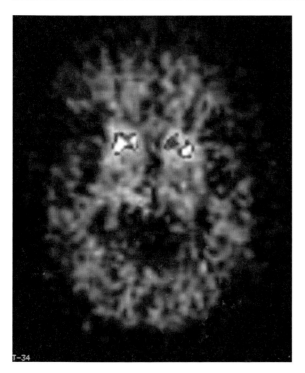

Fig. 5.15 Prominent striatal PIB deposition (in *red*) in an asymptomatic PSEN1 mutation carrier approximately three years prior to anticipated age at onset

the case of early AD, a need arises to treat the patient in a research (randomized clinical trial) setting. MR spectroscopy in AD may show mildly reduced NAA and increased myo-inositol, consistent with axonal damage and glial proliferation. Whilst these findings are robust on a group-level, diagnostic value is not robust enough in individual cases. As outlined above, FDG-PET and especially amyloid imaging may be more sensitive in the early stage. A clear example is given in Fig. 5.12, where no hippocampal atrophy is seen in a patient suspected of early-onset AD. FDG visual and voxel-wise analyses show metabolic deficits and amyloid PET is suggestive of amyloid deposition. Also in the distinction of AD with other dementias, a dopamine SPECT scan may help to identify DLB, and an FDG PET showing most prominent frontal or temporal metabolic deficits may indicate FTD rather than AD.

5.2.7.1   Repeated MRI

When to repeat a structural examination, especially MRI? There are no guidelines yet, but the 2010 EFNS guidelines mention that serial MRI may be helpful to document disease progression in a clinical setting, and may aid in the management of the patient and caregiver. An example is seen in Fig. 5.17. Changes over a

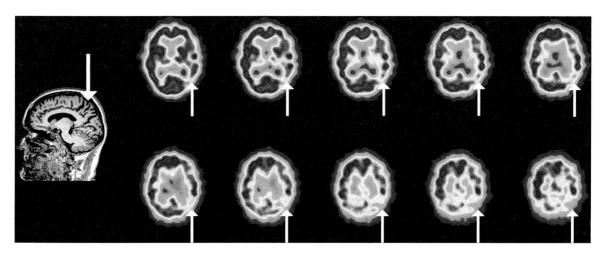

Fig. 5.16 SPECT in AD. A 51-year-old woman that was admitted depressive symptoms, visuospatial deficits, apraxia, and minor memory loss. Parasagittal T1-weighted image (*left*) showing atrophy of the precuneus (*large arrow*) – a pattern of posterior cortical atrophy consistent with the diagnosis of early-onset AD. 99mTc-HMPAO single-photon emission computed tomography images (*right*) revealing marked brain hypoperfusion involving the left parietal lobe (*small arrows*), far beyond the regions of brain volume loss. (Reproduced with permission from *Psychiatry Res* 2010;182:287–288)

Fig. 5.17 Serial MR images of an early-onset AD patient with rapid progression. Over 18 months time (the interval between the left and right images) this is gross loss of cortical grey matter with marked sulcal widening. Note also ventricular widening and only minimal progression of medial temporal atrophy

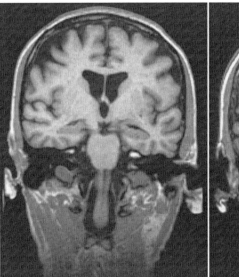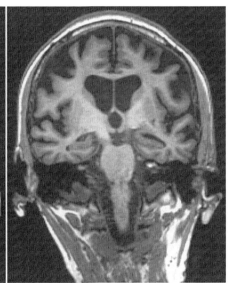

time period of one year that are visible to the naked eye result from volume changes that exceed those of normal ageing, and may thus strengthen the clinical diagnosis. For more subtle changes, advanced image registration techniques are needed (see Fig. 5.9).

However, in most cases, more quantitative procedures are needed to pick up changes, but these have been restricted to research use only and are being used routinely in clinical trials. Results from various groups have shown consistent loss of whole brain volume in the order of around 1.5–2% per year depending on severity, while normal ageing losses are usually below 0.5% annually.

5.3 Frontotemporal Lobar Degeneration

Jonathan D. Rohrer

5.3.1 Introduction and Synonyms

The terms frontotemporal lobar degeneration (FTLD) and frontotemporal dementia (FTD) describe a group of clinical syndromes which may be produced by a number of histopathologically distinct neurodegenerative causes with both sporadic and genetic aetiologies.

Three *overlapping* classification systems for FTLD/FTD can be considered:

1. The clinical syndrome, for which the term FTD is often used, which reflects topography of neuronal loss/pathology and maps to a particular pattern of atrophy and hypometabolism:
   a. Behavioural variant frontotemporal dementia (bvFTD)
   b. Progressive nonfluent aphasia (PNFA)
   c. Semantic dementia (SD)
2. The histopathological syndrome, for which the term FTLD is most frequently used (tau-positive, TDP-43-positive, FUS-positive): the morphology and (especially) the immunological staining of protein inclusions.
3. The molecular genetic syndrome (*MAPT, GRN, VCP, CHMP2B, TARDP, FUS*): the gene and the type of genetic defect.

It is important, therefore, when discussing FTLD/FTD to be clear whether one is referring to a particular clinical syndrome, a specific pathology or to a genetically determined condition. The terminology can be confusing partly because pathological diagnoses, clinical syndromes and genetics are often used interchangeably and also because the nosology of FTLD has changed a number of times over recent years. Furthermore, there is an overlap between FTLD and atypical parkinsonian disorders (progressive supranuclear palsy and

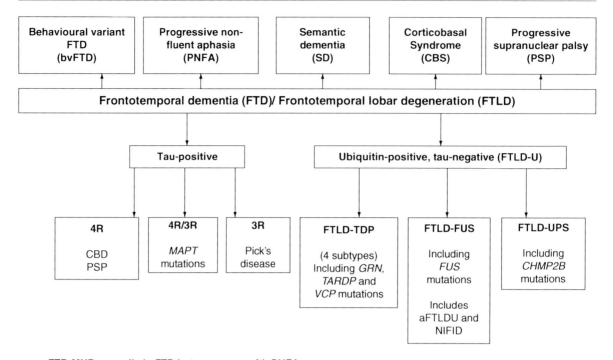

FTD-MND - usually bvFTD but can occur with PNFA

Fig. 5.18 Summary of clinical syndromes (FTD) and genetic/pathological syndromes (FTLD). *4R* 4-repeat-tau, *3R* 3-repeat-tau, *FUS* fused in sarcoma, *UPS* unknown pathological substrate. For further abbreviations, see text

corticobasal degeneration syndromes) and also motor neuron disorders (see also chapters on parkinsonian disorders). The relationship between clinical, histopathological and genetic classifications is illustrated diagrammatically in Fig. 5.18.

Synonyms of FTLD

Pick's disease; Frontotemporal degeneration; Frontotemporal dementia (FTD); Frontotemporal dementia with parkinsonism linked to chromosome 17 (FTDP-17); Frontotemporal lobar degeneration with ubiquitin-positive inclusions (FTLD-U)

Synonyms of the Clinical Subtypes of FTLD

Behavioural variant frontotemporal dementia (bvFTD); Behavioural variant frontotemporal lobar degeneration (bvFTLD); Frontal variant frontotemporal lobar degeneration (fvFTLD); Progressive non-fluent aphasia (PNFA); Primary progressive aphasia (non-fluent variant); Semantic dementia (SD) Primary progressive aphasia (fluent variant); Right-temporal lobe variant of FTLD;

5.3.2 History and Nosology

Arnold Pick first described patients with focal atrophy of the frontal and temporal lobes in 1892 including patients both with personality change and language impairment. Although there were a few other reports of such cases in the early part of the twentieth century, in the Western world, behavioural and progressive language problems were forgotten about for many years. In Japan however, the disorder Gogi (literally 'word meaning') aphasia was described in the 1940s. The first modern accounts of the progressive language disorders in the Western literature were by Warrington (1975) and Mesulam (1982). Warrington described the selective impairment of semantic memory in a group of patients. She did not however use the term semantic dementia which was coined a number of years later. Mesulam independently described a group of patients who presented with progressive language problems including those with impairments of production and those with comprehension problems and collectively calling them primary progressive aphasia (PPA). He distinguished this from aphasia seen in other degenerative diseases such as AD by the isolated nature of the language disintegration, with many patients remaining without other cognitive or behavioural deficits for up to 10 years.

The Lund-Manchester criteria (1994, updated in 1998) later defined two variants of progressive aphasia: progressive non-fluent aphasia (PNFA) and semantic dementia (SD) as well as the behavioural variant of FTLD which they called frontotemporal dementia. Other clinical criteria for FTLD were later described by McKhann et al. 2001 with separate criteria for the progressive aphasias described by Mesulam in 2001. New criteria for both behavioural variant frontotemporal dementia and the language syndromes are in preparation. Arguments over the nosology of the progressive aphasias continues, although most authorities now agree that there are two main language syndromes that occur with FTLD pathology, namely SD and PNFA. A third variant called logopenic aphasia or the logopenic variant of primary progressive aphasia (LPA) has recently been described but appears to be more commonly associated with Alzheimer pathology rather than FTLD pathology. Overlap is seen between FTLD and two atypical parkinsonian disorders, namely corticobasal degeneration (CBD) and progressive supranuclear palsy (PSP). This led Kertesz to suggest that this group of disorders should be named the Pick Complex. There is also overlap with motor neurone disease (MND). Although only about 5–10% of patients with FTLD will have MND (termed FTD-MND) and only 5% of patients with MND will have FTLD, up to 50% of MND patients have at least mild executive dysfunction.

5.3.3 Epidemiology, Aetiology and Genetics

There are few epidemiological studies of FTLD and no clear environmental aetiological factors have been determined. FTLD is the third most common degenerative cause of dementia after AD and Dementia with Lewy Bodies, accounting for about 5–10% of all cases of dementia. However, FTLD commonly occurs in younger patients (45 to 65) and in this age group it is second in frequency only to AD. Despite being commoner in younger people, a wide range of ages of onset has been described from 21 years of age to mid-80s. Studies are limited, but for bvFTD prevalence probably lies somewhere between 3 and 15 per 100,000 in people over the age of 50. The prevalence of SD and PNFA is less clear, but for both syndromes it is probably similar and about half that of bvFTD. The duration of disease is also extremely variable; however, on average is around 6–10 years for FTD and PNFA but longer for SD (average 10–15 years) and much shorter for FTD-MND (average 3 years). There appears to be no difference in prevalence between men and women.

Around 30–50% of patients with FTLD have an autosomal dominant family history. Some of these families are linked to a locus on chromosome 17 and this disease group is termed FTDP-17 because an association exists between FTD and parkinsonism. Mutations in the microtubule-associated protein tau (*MAPT*) gene are causative of FTDP-17 and over 40 mutations have now been described in this gene (www.molgen.ua.ac.be/FTDmutations). Most patients present with a behavioural syndrome and/or parkinsonism (including syndromes similar to corticobasal degeneration or progressive supranuclear gaze palsy). Primary language impairment is rarely seen in *MAPT* mutations and the two prototypical syndromes of PNFA and SD are not described as occurring in this group. Mutations in another gene, progranulin (*GRN*), also on chromosome 17, have more recently been described as causing FTLD. Patients tend to present either with bvFTD or with language output impairment, commonly PNFA. Unlike *MAPT* mutations, parietal lobe deficits have also been described in *GRN* patients including presentation with a CBD syndrome. FTD associated with chromosome 3 is caused by mutations in the *CHMP2B* (charged multivesicular body protein 2B) gene and is a very rare cause of FTLD (seen mainly in a single large Danish family) presenting most commonly with bvFTD. Mutations in the *VCP* (valosin-containing protein) gene on chromosome 9 cause the syndrome of inclusion body myopathy, Paget's disease and frontotemporal dementia (IBMPFD), and similarly are a rare cause of FTLD. Rare case reports of FTLD patients with mutations in the TARDP (TAR-DNA binding protein) and FUS (fused-in-sarcoma) genes have also been described. Other loci on chromosome 9 are associated with FTD-MND but the genes are yet to be identified.

5.3.4 Histopathology

Macroscopically, bilateral frontal and/or temporal lobe atrophy with relative sparing of the parietal and occipital lobes is seen in FTLD (see Fig. 5.19). This may be symmetrical (particularly when just affecting frontal lobes), but is often asymmetrical and can be strikingly so; an antero-posterior gradient is typical.

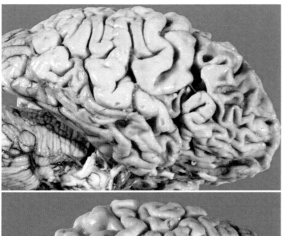

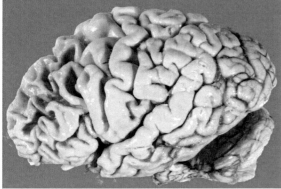

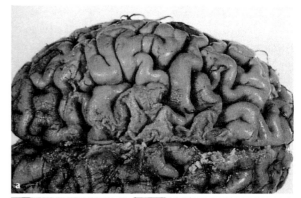

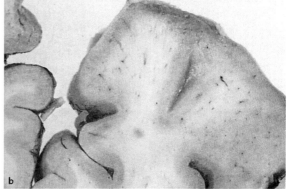

Fig. 5.19 Macroscopy of FTLD. Note the selective severe atrophy of the frontal gyri

Fig. 5.20 Pathology of CBD. Macroscopic findings in a pathological specimen of a patient with corticobasal degeneration (CBD). (**a**) At gross inspection there is circumscribed cortical atrophy in the superior frontal and anterior parietal gyrus. (**b**) The coronal section shows thinning of the cortical grey matter and underlying white matter. (Courtesy of T. Revesz, M.D., and S.E. Daniel, M.D.)

Neuropathological criteria have been described by Cairns et al. (2007) and updated recently by Mackenzie et al. (2010). There are three main subdivisions of histological findings: tau-positive inclusions, TDP-43-positive inclusions and FUS-positive inclusions (Fig. 5.18). Only a small minority of patients with FTLD do not have one of these three pathologies including patients with CHMP2B mutations and those lacking any inclusions. The tau-positive group can be subdivided into three groups on the basis of the relative proportions of four or three repeat tau (Fig. 5.18). The tau-positive includes classical Pick's disease, a diagnosis which historically required the presence of Pick bodies which are ubiquitin and tau-positive inclusions – Pick bodies turn out to have predominantly three-repeat tau. Also, within the tau-positive group are the *MAPT* mutations as well as CBD (Fig. 5.20), PSP and the relatively rare argyrophilic grain disease. The TDP-43-positive group includes four subtypes; patients with *GRN*, *VCP* and TARDP mutations have TDP-43-positive pathology. The recently described FUS-positive pathological group includes cases previously described as atypical FTLD with ubiquitin-positive

inclusions and cases with neuronal intermediate filament inclusion disease (NIFID). Clinico-pathological correlation remains complex, but certain patterns are starting to emerge through large studies of clinically well-defined patients; e.g. TDP-43-positive pathology is associated with the clinical syndromes of SD and FTD-MND, while most clinical PSP cases are associated with tau pathology.

It is also important to recognize that rarely each of the FTD clinical syndromes can be associated with AD pathology i.e. a prominent and primary behavioural or language syndrome indistinguishable from bvFTD, PNFA or SD may occur. Also, the term 'frontal AD' is used by some people to indicate a clinical syndrome in which patients have features of episodic memory impairment characteristic of AD and also early behavioural symptoms characteristic of bvFTD. Some patients that present with the clinical picture of bvFTLD may turn out to have AD (Fig. 5.21).

5.3.5 Clinical Features and Imaging Strategy

While CT can be helpful to exclude other pathology involving the fronto-temporal region (e.g. a meningioma), MRI is the method of choice to demonstrate (early) atrophy (e.g. in the orbitofrontal region and/or temporal pole). Coronal images are particularly relevant as they allow assessment of asymmetry in the (medial) temporal lobe structures (Fig. 5.22).

5.3.5.1 Behavioural Variant Frontotemporal Dementia (bvFTD)

BvFTD presents with insidious change in personality and behavioural symptoms (Box 5.2). Common presenting symptoms include apathy or social withdrawal, disinhibition or inappropriate social behaviour and loss of empathy. Loss of insight is characteristic and other behavioural symptoms are seen: perseverative or stereotyped behaviour, mental inflexibility, obsessiveness, hoarding, poor self-care, distractibility and a change in eating behaviour, in particular development of a sweet tooth. Patients may often initially be thought to have non-organic psychiatric disorders. Executive dysfunction with poor planning and problem-solving is also seen although this can dissociate from behavioural changes.

Language symptoms include a decrease in the amount of speech (although commonly without errors), perseverative use of certain phrases and echolalia; many patients will become mute as the disease progresses. MMSE scores can be entirely normal early in the disease because it does not test executive function and also because behavioural features may predominate without significant cognitive impairment. Although there are no curative treatments, management of the behavioural features is important.

Atrophy of the frontal and temporal lobes is best seen on (coronal) T1-weighted images (Fig. 5.23 and 5.24). Characteristically in bvFTD, there is fronto-temporal atrophy with an antero-posterior gradient i.e. relative sparing of the parietal and occipital lobes, and this atrophy, although commonly bilateral, is often asymmetrical. T2-weighted and FLAIR imaging is useful to exclude concurrent vascular disease as well as assessing for subcortical white matter changes which have occasionally been described in pathologically confirmed cases of FTD. Although most areas of the

Box 5.2 Criteria for behavioural variant frontotemporal dementia (from Neary et al. 1998)

Character change and disordered social conduct are the dominant features initially and throughout the disease course. Instrumental functions of perception, spatial skills, praxis, and memory are intact or relatively well preserved.

I. Core diagnostic features

Insidious onset and gradual progression
Early decline in social interpersonal conduct
Early impairment in regulation of personal conduct
Early emotional blunting
Early loss of insight

II. Supportive diagnostic features
Behavioural disorder

1. Decline in personal hygiene and grooming
2. Mental rigidity and inflexibility
3. Distractibility and impersistence
4. Hyperorality and dietary changes
5. Perseverative and stereotyped behaviour
6. Utilization behaviour

Speech and language

1. Altered speech output
   (a) Aspontaneity and economy of speech
   (b) Press of speech
2. Stereotype of speech
3. Echolalia
4. Perseveration
5. Mutism

Physical signs

1. Primitive reflexes
2. Incontinence
3. Akinesia, rigidity, and tremor
4. Low and labile blood pressure

Investigations

1. Neuropsychology: significant impairment on frontal lobe tests in the absence of severe amnesia, aphasia, or perceptuospatial disorder
2. Electroencephalography: normal on conventional EEG despite clinically evident dementia
3. Brain imaging (structural and/or functional): predominant frontal and/or anterior temporal abnormality

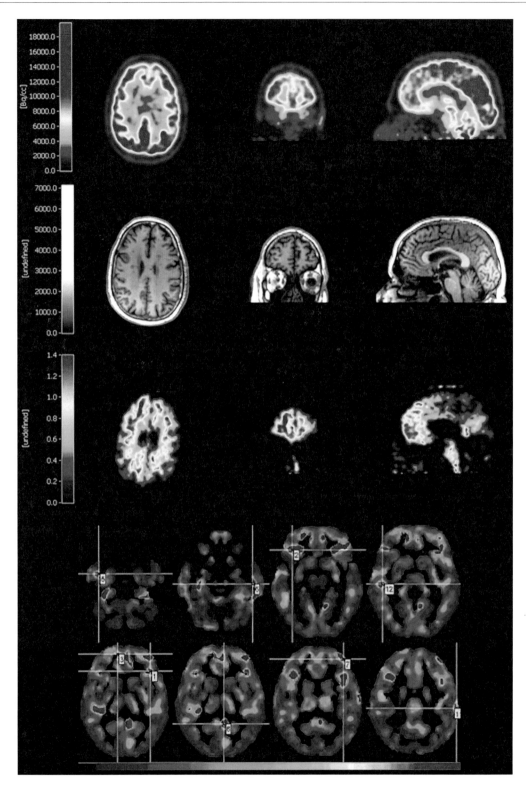

Fig. 5.21 Behavioural variant FTD syndrome caused by Alzheimer pathology. This 63-year-old man presented with slowly progressive change in character, apathy and memory disturbances. Cognitive testing revealed mainly executive dysfunction, and less prominent memory disturbances. His mother developed AD at the age of 63. FDG-PET showed reduced fron- tal metabolism (*upper row*), which was confirmed using voxel- wise analysis (*bottom panel*), suggesting FTLD rather than AD. MRI showed frontal but not hippocampal atrophy (*second row*). Amyloid PET (*third row*), however, showed abnormal cortical PIB binding, consistent with AD, despite normal CSF amyloid- beta levels (slightly increased p-tau)

Fig. 5.22 Comparison of coronal CT and MR in SD. Note that the coronal reformats of the multi-detector CT (*top*) are quite comparable to the coronal MR (*bottom*) images in showing left more than right anterior temporal atrophy in the same patient. (Images courtesy of Mike Wattjes)

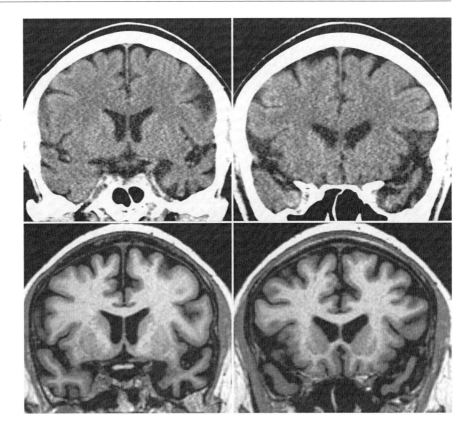

frontal and temporal lobes can be affected during the disease, the earliest area of sulcal widening is probably the orbitofrontal cortex, followed by the mesiofrontal (interhemispheric) cortex. In more advanced cases, the dorso-lateral prefrontal cortex will also become involved. Bilateral hippocampal and amygdalar atrophy are also features of FTD, again often asymmetrical. The key point to note here though is that the medial temporal lobe is more affected anteriorly (e.g. the amygdala being more affected than the hippocampus and posterior hippocampus often appearing normal).

It should be noted that in early disease structural imaging is often normal. Although many patients will progress to show frontotemporal atrophy later in the disease, a small number of cases appear to clinically progress slowly in terms of behavioural symptoms with little or no atrophy on longitudinal MR scanning. There are few studies looking at this 'benign' variant of bvFTD, and its status within the FTLD spectrum remains unclear: some of these cases are likely to be non-neurodegenerative phenocopies and serial imaging may be helpful in distinguishing these.

The features of diffusion-weighted imaging in FTD are unclear. While no signal increase on DWI is expected in the cortex, increased mean diffusivity (MD) in various white matter regions including bilateral superior frontal gyri, right orbitofrontal gyrus, bilateral anterior temporal lobes and left middle temporal lobe have been found using DTI in the adjacent white matter.

5.3.5.2 Progressive Non-Fluent Aphasia (PNFA)

PNFA presents with impairment of speech production, commonly hesitant, effortful speech or stuttering (apraxia of speech) and/or agrammatism (breakdown of syntax). Other characteristic features are phonemic paraphasias (sound-based errors), anomia and impaired repetition of polysyllabic words. There is preserved single-word comprehension early on although impaired sentence-level comprehension. Reading is also nonfluent and effortful. Writing is preserved early in the disease but becomes more affected later on with spelling and grammatical errors. Language disintegration progresses inexorably to mutism. Behavioural symptoms are rare early in the disease and there may

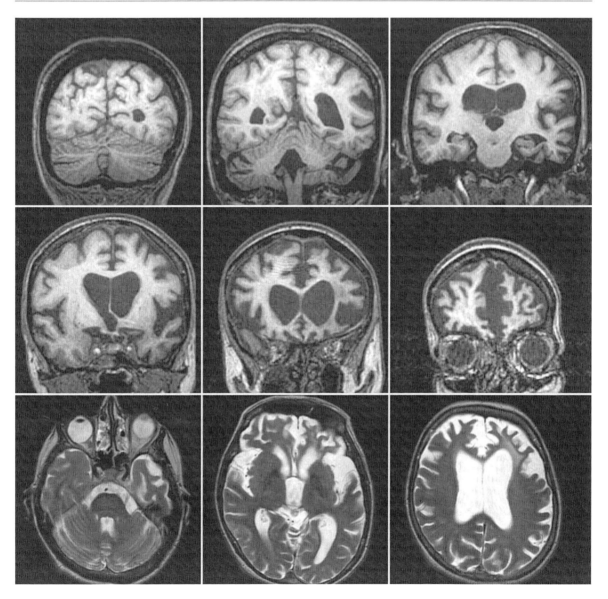

Fig. 5.23 Behavioural variant frontotemporal dementia. Severe asymmetric frontal atrophy in a patient with confirmed TDP-43-positive inclusions at post-mortem; also note less severe involve-ment of the temporal lobes, with the lateral gyri more affected than the medial part

be no other features (either cognitive or behavioural) beyond the speech production disorder many years into the disease. Late behavioural changes include mental rigidity and apathy in particular. The diagnostic criteria are summarized in Box 5.3.

PNFA is heterogeneous and some patients initially present with one of agrammatism, apraxia of speech or anomia in the absence of other features. There is some

evidence that the initial or predominant language pheno-type predicts the underlying pathology e.g. apraxia of speech appears to be commonly associated with tau-positive pathology. A number of rare variants of progres-sive speech production impairment are described although their nosological status is unclear: progressive aphemia (severe speech apraxia), cortical anarthria and progressive dynamic aphasia (impaired spontaneous

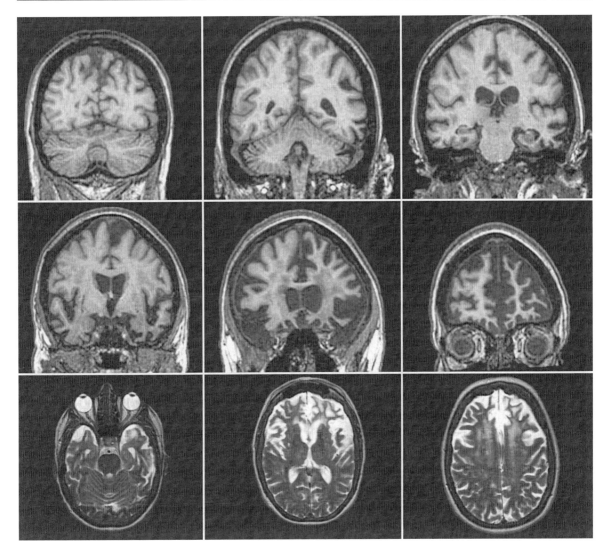

Fig. 5.24 Behavioural variant frontotemporal dementia. End-stage bilateral severe frontal atrophy was seen in this patient with confirmed tau-positive Pick's disease at post-mortem. Note the anterior–posterior gradient of atrophy and the signal change in the very atrophic areas. Note relative sparing of the temporal lobe (especially posteriorly), and importantly relative preservation of the hippocampus

propositional speech). Such disorders may represent different stages of a common disease continuum reflecting a specific pattern of anatomical involvement or alternatively, may constitute distinct disease entities with different pathophysiological and histopathological correlates.

Imaging findings in PNFA are heterogeneous implicating a number of regions in the dominant (left) hemisphere although classically it is the left perisylvian regions, particularly the left inferior frontal and insular cortices, which are affected (Fig. 5.25) – in addition there may be involvement of left superior frontal, superior temporal and inferior parietal lobes. Variation between patients likely reflects heterogeneity of the neurolinguistic deficits in PNFA – although relation with different pathological findings in PNFA has not yet been studied in detail.

Box 5.3 Criteria for progressive non-fluent aphasia (from Neary et al. 1998)

Disorder of expressive language is the dominant feature initially and throughout the disease course. Other aspects of cognition are intact or relatively well preserved.

I. Core diagnostic features

Insidious onset and gradual progression

Non-fluent spontaneous speech with at least one of the following: agrammatism, phonemic paraphasias, anomia

II. Supportive diagnostic features

Speech and language

1. Stuttering or oral apraxia
2. Impaired repetition
3. Alexia, agraphia
4. Early preservation of word meaning
5. Late mutism

Behaviour

1. Early preservation of social skills
2. Late behavioural changes similar to FTD

Physical signs: late contralateral primitive reflexes, akinesia, rigidity, and tremor

Investigations

1. Neuropsychology: non-fluent aphasia in the absence of severe amnesia or perceptuospatial disorder
2. Electroencephalography: normal or minor asymmetric slowing
3. Brain imaging (structural and/or functional): asymmetric abnormality predominantly affecting dominant (usually left) hemisphere

5.3.5.3 Semantic Dementia (SD)

SD is characterized by loss of semantic or conceptual knowledge (see Box 5.4). It commonly presents as a fluent aphasia with empty, circumlocutory speech, loss of word meaning, anomia and impaired single-word comprehension. Patients often complain of 'word-finding difficulty' as an initial symptom. Semantic paraphasias are seen both in spontaneous speech and tasks of confrontational naming. Although language progressively decreases in amount, patients often maintain a small repertoire of stereotyped phrases. Non-verbal semantic loss also occurs and a visual

Box 5.4 Criteria for semantic dementia (from Neary et al. 1998 modified by Adlam et al. 2006)

Semantic disorder (impaired understanding of word meaning and/or object identity) is the dominant feature initially and throughout the disease course. Other aspects of cognition, including autobiographic memory, are intact or relatively well preserved.

I. Core diagnostic features

Insidious onset and gradual progression

Language disorder characterized by

1. Progressive, fluent, empty spontaneous speech
2. Loss of word meaning, manifest by impaired naming and comprehension
3. Semantic paraphasias and/or

Impairment on tests of non-verbal associative knowledge

Preserved single-word repetition

Preserved ability to read aloud and write to dictation orthographically regular words

II. Supportive diagnostic features

Speech and language

1. Press of speech
2. Idiosyncratic word usage
3. Absence of phonemic paraphasias
4. Surface dyslexia and dysgraphia
5. Preserved calculation

Behaviour

1. Loss of sympathy and empathy
2. Narrowed preoccupations
3. Parsimony

Physical signs

1. Absent or late primitive reflexes
2. Akinesia, rigidity, and tremor

Investigations

Neuropsychology

1. Profound semantic loss, manifest in failure of word comprehension and naming and/or face and object recognition
2. Preserved phonology and syntax, and elementary perceptual processing, spatial skills, and day-to-day memorizing

Electroencephalography: normal

Brain imaging (structural and/or functional): predominant anterior temporal abnormality

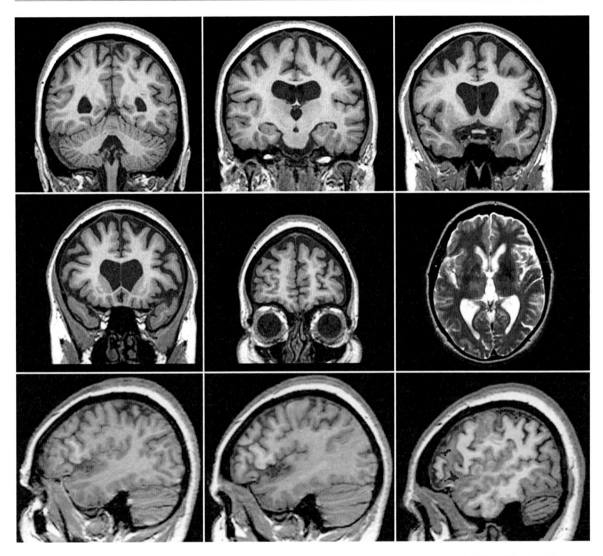

Fig. 5.25 Progressive non-fluent aphasia. This 55-year-old woman with PNFA had serial MR imaging as part of a longitudinal study of individuals at risk of FTLD. At the time of the imaging shown in the *top two rows* (51 months from her first, and asymptomatic, scan), she had become symptomatic with difficulties with expressive language. She had speech production difficulties and made phonemic errors in spontaneous speech and in reading. She also had the beginning of orofacial apraxia at this time. The coronal T1-weighted images show frontal atrophy with asymmetrical (left > right) perisylvian fissure atrophy and more posteriorly also seen on the single T2-weighted axial image. *Bottom row* of sagittal images show deformation maps from serial MRI – green represents regions of contraction. *Left* image (0–14 months, presymptomatic) with focal antero-lateral left frontal lobe loss, particularly centred around the pars opercularis (Broca's area) presaging speech production problem; *middle* image (14–26 months), region of increased atrophy rate now spreading; *right* image (45–56 months), much more generalised atrophy. (Adapted with permission from J Neurol Neurosurg Psychiatry 2005;76:162–168)

agnosia may occur with the inability to recognize the nature of an object in the presence of normal perception. Behavioural features include disinhibition and abnormal eating behaviour in particular.

Structural imaging in SD generally reveals left greater than right temporal lobe atrophy. In particular, the temporal lobe atrophy is mainly anterior, with an antero-posterior gradient and knife-edge or 'razorback' atrophy of the anterior temporal lobe can be seen (Fig. 5.26). Temporal lobe atrophy is also mainly inferior (often severe involvement of the fusiform gyrus) with relative sparing of the superior temporal gyrus. As the disease progresses the right temporal lobe becomes more involved. Amygdalar and hippocampal atrophy occurs but is asymmetrical with normally left greater than right atrophy and again

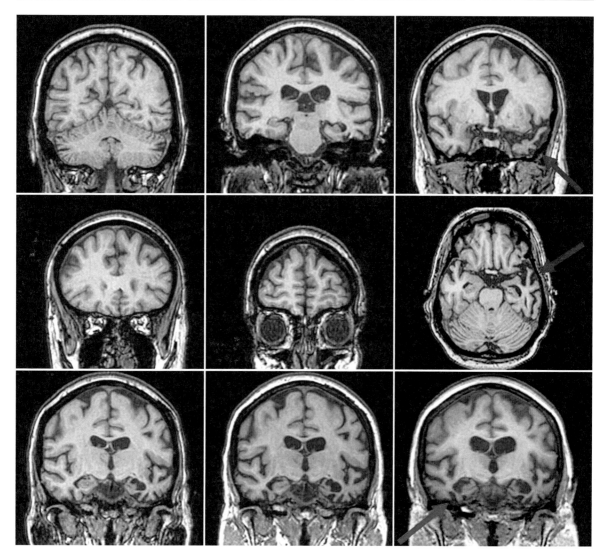

Fig. 5.26 Semantic dementia. Patient with confirmed TDP-43-positive inclusions (type 1) at post-mortem. *Top rows* show isolated left-sided atrophy of the temporal pole (*red arrows*) at presentation.

Registered serial images (*bottom*) from baseline to 2 years (*middle*) and 4 years (*right*) show spread to the right side (*blue arrow*); note the fusiform and inferior anterior temporal lobe atrophy

a marked anterior greater than posterior atrophy gradient – so the amygdala is more affected than the hippocampal head which is in turn more affected than the body of the hippocampus. These features are very useful in distinguishing SD from AD. The clinical syndrome of SD can sometimes occur in patients with right greater than left temporal lobe atrophy.

5.3.5.4 Logopenic Aphasia (LPA)

Patients with LPA are most commonly found to have Alzheimer pathology at post-mortem. They present with long word-finding pauses in their speech and may have anomia and impaired phonological working memory. Parietal lobe features (e.g. limb apraxia and dyscalculia) are also seen. Imaging shows an asymmetrical pattern of atrophy with predominant left posterior temporal cortex and inferior parietal lobule involvement with progression to involve the posterior cingulate and more anterior temporal lobe areas including the hippocampus.

5.3.5.5 Corticobasal Degeneration (CBD)

CBD is covered in more detail under Sect. 5.4 and the inclusion of CBD in more than one place in this book is symbolic of the position that CBD occupies, with overlap (and clinical uncertainty) involving FTLD and AD as well as Parkinson's disease and other extrapyramidal disorders. Although cognitive and behavioural symptoms were originally thought to be rare in patients

with a CBD syndrome (also called corticobasal syndrome or CBS to distinguish the clinical syndrome from the pathological syndrome of CBD), it is now clear that they form a major part of the syndrome and the overlap with both bvFTD and PNFA is well-described.

Imaging often reveals asymmetrical fronto-parietal atrophy with relative sparing of the temporal and occipital lobes with either hemisphere being predominantly affected. However, imaging findings are heterogeneous which is likely to represent at least in part the variable pathological causes of a CBS: although most commonly a tauopathy (CBD or rarely PSP pathology), CBS can also be caused by Alzheimer pathology and FTLD-TDP pathology. Detailed imaging studies of pathologically confirmed tau-positive CBD are limited (Fig. 5.27).

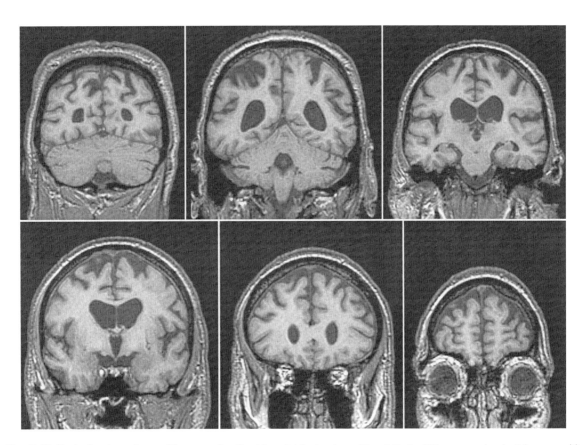

Fig. 5.27 Corticobasal syndrome. Note superior frontal–parietal atrophy, with relatively little asymmetry in this case with confirmed tau-positive CBD pathology at post-mortem

5.3.5.6 FTD with Motor-Neuron Disease (MND)

The association of MND with FTLD associates most commonly with bvFTD but also with PNFA. Patients may present initially with just fasciculations but often proceed to amyotrophy with involvement of upper and lower motor neurones later in the illness. Imaging studies are limited but suggest that FTD-MND is associated with relatively symmetrical prefrontal and medial temporal lobe atrophy.

5.3.5.7 'Right Temporal Lobe Variant' of FTLD

Rarely, patients may present with a progressive neurodegenerative disorder in which symptoms are attributed to right temporal lobe atrophy, in particular prosopagnosia. It remains unclear, however, whether this is a distinct clinical syndrome. Although some of these patients subsequently develop the clinical syndrome of semantic dementia (with concomitant left temporal lobe atrophy) and similarly have FTLD-TDP pathology, others do not. Further clinical and particularly pathological characterization is required to further the nosological status of the right temporal lobe variant (Fig. 5.28a and b).

5.3.5.8 Genetic Subgroups

GRN mutations have been described to be associated with strikingly asymmetrical atrophy of either cerebral hemisphere. Furthermore, unlike most other cases of

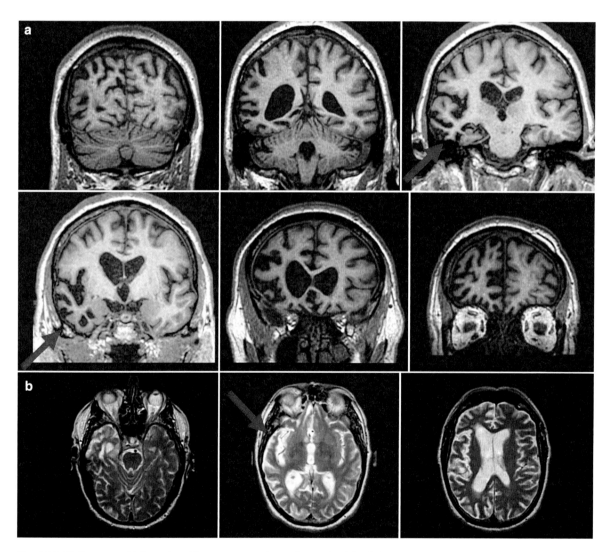

Fig. 5.28 Right temporal lobe variant. Note severe atrophy (*red arrows*) of the right temporal lobe (lateral more than medial), with associated asymmetric frontal lobe atrophy

FTLD, early parietal lobe involvement is described. *MAPT* mutations are associated with frontotemporal atrophy with an antero-posterior gradient. Atrophy is commonly bilateral and is usually relatively symmetrical or only mildly asymmetrical (Fig. 5.29a and b).

5.3.6 Differential Diagnosis, Imaging Follow-up (Including PET/SPECT), Other Tests (CSF/EEG)

5.3.6.1 Other Imaging

Functional imaging may be useful in FTLD (exemplified by the fact that reimbursement for PET scanning in permitted in the USA if FTLD is being considered). FDG-PET and (HMPAO-)SPECT characteristically show frontal and temporal hypoperfusion/hypometabolism with relative sparing of parietal and occipital blood flow in comparison with AD where posterior changes are more prominent. A pattern of bilateral frontal hypoperfusion in the absence of bilateral parietal hypoperfusion has been reported to have a sensitivity and specificity of 0.8 for a diagnosis of FTD. It can be useful particularly in separating patients who have a behavioural syndrome due to either FTLD or AD pathology. Little is known about the use of PET/SPECT in PNFA or SD – absence of hypoperfusion of temporo-parietal association cortex favouring FTLD over AD, but unilateral hypoperfusion may be associated with either type of pathology. SPECT findings in SD have not been studied in detail. As an alternative strategy, perfusion MRI may demonstrate hypoperfusion in temporal or frontal regions (Fig. 5.30).

[11]C-PIB PET imaging is designed to specifically image amyloid plaques. Although it is yet to enter clinical practice, most studies have shown a high specificity for AD, with mostly negative scans in FTLD (Fig. 5.31 and 5.32). Amyloid imaging may become important in patients presenting with progressive aphasia when it is important to distinguish patients with likely Alzheimer pathology for management decisions – most commonly positive amyloid imaging is associated with LPA and only very rarely with PNFA or SD.

The use of magnetic resonance spectroscopy in FTLD is of limited value. N-acetylaspartate (NAA)/creatine (Cr) levels, a marker of neuronal loss, are lower in FTLD compared to controls but similar to Alzheimer's disease, mostly regional, occurring in the frontal and temporal but not the parietal lobes (Fig. 5.33). Myoinositol (mI)/Cr levels, a marker of gliosis, are higher than controls but again similar to Alzheimer's disease.

5.3.6.2 CSF

Traditional CSF constituents are generally normal or negative i.e. white cell count, protein, glucose and oligoclonal bands. Current CSF biomarkers are not very useful in making a positive diagnosis of FTLD. Total tau may be increased relative to controls but to a lower level than in AD, but may also be normal, reflecting the heterogeneous nature of FTLD. Aβ42 levels in FTLD may be low but not as low as in AD – with considerable overlap. Large autopsy-proven studies are needed to assess the utility of CSF in FTLD.

5.3.6.3 EEG

EEG in FTLD has traditionally been thought to be normal compared to the loss of alpha rhythm, generalized slowing and excess theta rhythm seen in AD. However, many patients in fact may have abnormal EEGs. Cases with a 'temporal' presentation (SD and right temporal lobe atrophy) are probably more likely to have an abnormal EEG compared to 'frontal' cases.

5.4 Dementia with Parkinsonism

5.4.1 Introduction: Neurodegeneration, Parkinsonism and Dementia

A number of neurodegenerative diseases present with a combination of cognitive impairment and parkinsonism (bradykinesia, rigidity, tremor and loss of postural reflexes) as prominent aspects of their phenotype. The degree and distribution of cognitive deficits varies according to the underlying pathological diagnosis, but a sub-cortical pattern of cognitive impairment is a common feature. The relative prominence and order of appearance of the parkinsonian features and the cognitive decline varies considerably between diseases. The balance between motor and cognitive deficits also varies within a specific disease and over time. Patients may be referred to movement disorders or to dementia

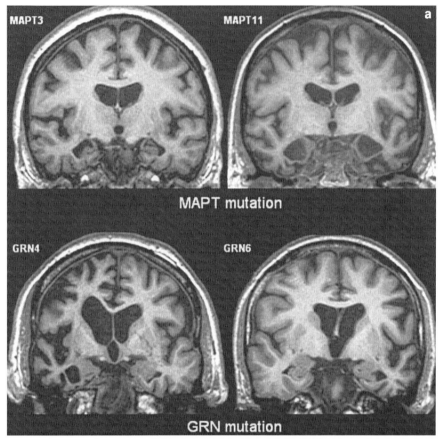

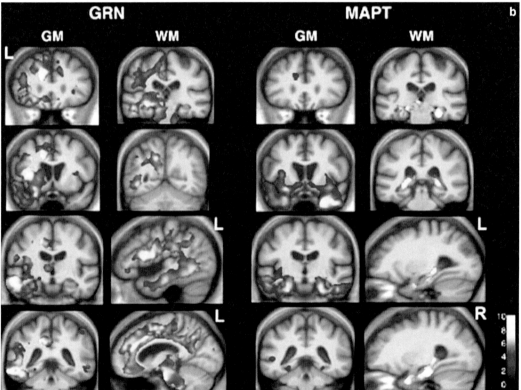

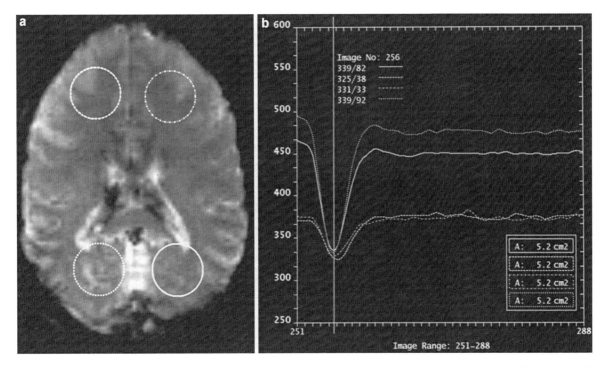

Fig. 5.30 Perfusion MRI in FTD. MR perfusion study in a patient with probable frontotemporal dementia. (**a**) On the EPI image of the perfusion series the regions of interest are indicated: two in the frontal lobes and two in the occipital lobes.

(**b**) The graphic analysis of the perfusion in these areas. It is clear that the perfusion in the frontal lobes is reduced to about one-third of the occipital perfusion. This finding supports the diagnosis of frontotemporal dementia

clinics – and fall into the 'parkinson's-plus' or the 'dementia-plus' diagnostic sieve depending on the specialism of the investigating physician. In many of these conditions, additional phenotypic features such as dysautonomia, gaze palsies, cerebellar and pyramidal signs may suggest one pathological process over another, but none are pathogonomic. Histopathologically as well as phenotypically there is overlap with FTLD: e.g. progressive supranuclear palsy (PSP) and corticobasal degeneration (CBD) are both tauopathies; see also Sect. 5.3.

Lewy body disorders are the most common causes of progressive cognitive decline with parkinsonism. Once considered separate entities, dementia with Lewy bodies (DLB) and Parkinson disease dementia (PDD) are now best thought of as ends of a continuous spectrum – the former with an initial presentation with dementia, the latter presenting with parkinsonism followed by dementia after more than a year but ultimately both have widely distributed Lewy bodies in the cerebral cortex. These disorders are therefore discussed together in the next section.

Fig. 5.29 (**a**) Impact of genetic mutation on atrophy pattern. Two patients with MAPT mutations both presenting with behavioural variant frontotemporal dementia (*top*); two patients with progranulin mutations presenting with a progressive aphasia (predominant left hemisphere atrophy) and with behavioural variant frontotemporal dementia (predominant right hemisphere atrophy) – in both these progranulin cases the atrophy is very asymmetric. Note the wide dispersion of imaging phenotype by genotype. (**b**) Voxel-based morphometry studies in FTLD. VBM analysis on grey matter (GM) and white matter (WM) regions in GRN- and MAPT-

associated FTLD relative to healthy controls. Statistical parametric maps (SPMs) have been thresholded at $p < 0.001$ after false discovery rate correction over the whole-brain volume and rendered on a study-specific average group T1-weighted MRI template image in DARTEL space. The colour bar (*lower right*) indicates the t score. *Left* (*L*) and *right* (*R*) markers are shown for ease of reference. GM panels show the same series of coronal MR sections in the GRN and MAPT cases; WM panels show coronal (*above*) and sagittal (*below*) sections based on zones of maximal white matter loss in each disease group

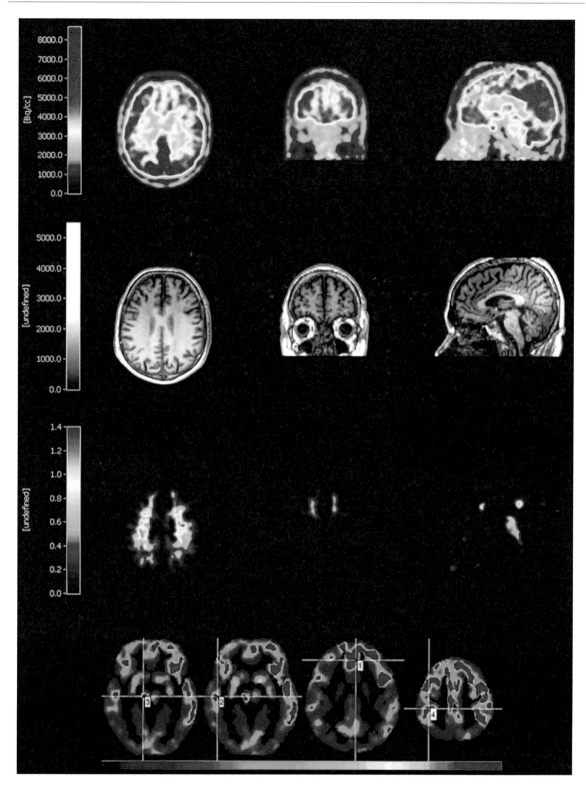

Fig. 5.31 Behavioural variant FTD. This 68-year-old man with a psychiatric history presented with recent onset of altered behaviour and lack of initiative. There were no speech or language abnormalities. Co-registered FDG-PET (*upper row*) shows frontal hypometabolism and concomitant mild frontal atrophy on MRI (*second row*). Normal CSF findings and normal amyloid-PET (PIB, third row with only white matter, but no cortical uptake) ruled out AD. Note that the voxel-wise map of the FDG scan (*bottom row*) shows abnormal metabolism mostly in frontal areas, but not in the posterior cingulate, as in AD

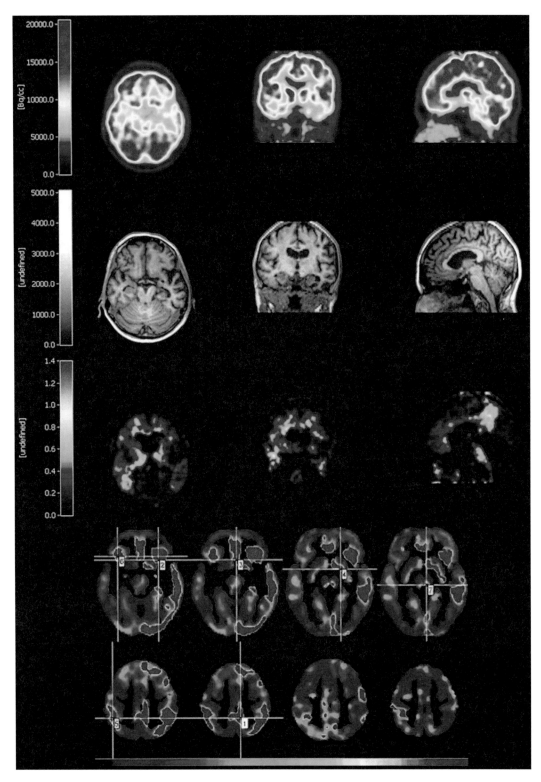

Fig. 5.32 PET in semantic dementia. This 62-year-old woman presented with slowly progressive impaired word-finding which developed over the course of 1 year. Examination revealed anomia, fluent language and mild memory disturbances (MMSE 24/30). MRI showed marked left temporal pole atrophy. FDG-PET (*upper row*) also showed reduced temporal metabolism, and additional frontal hypometabolism (red areas on the voxel-wise map in *bottom panel*). Amyloid-PET (*third row*) showed hardly any cortical PIB-binding, ruling out atypical AD

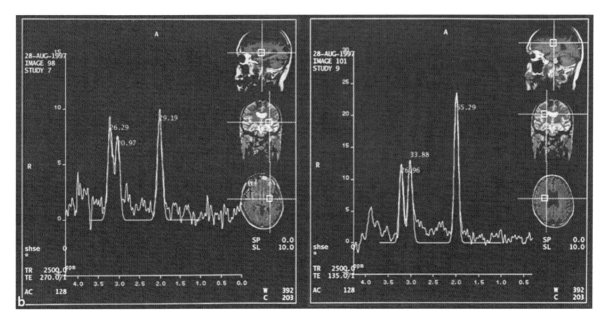

Fig. 5.33 MR spectroscopy in FLTD. The spectrum of the left insular region (*left panel*), despite some broadening of the spectral lines, shows decreased N-acetylaspartate and some increase in cho- line. Lactate is not present. The spectrum of the right paraventricu- lar region (*right panel*) is normal. Note that the two spectra were not corrected for signal intensities as displayed along the x-axis

5.4.2 Dementia with Lewy Bodies and Parkinson's Disease Dementia

John O'Brien and Michael Firbank

Synonyms

Dementia with Lewy bodies (DLB); Diffuse Lewy body disease; Dementia associated with cortical Lewy bodies; The Lewy-body variant of Alzheimer's disease; Lewy body dementia; Senile dementia of Lewy-body type; Parkinson's disease dementia (PDD);

5.4.2.1 History and Nosology

Lewy bodies (LB) are spherical, neuronal inclusions that were first described by the German neuropatholo- gist Friedrich Lewy in 1912, examining cases of 'paral- ysis agitans'. In 1961, a case report was published about two elderly men who presented with dementia and died shortly thereafter with severe extrapyramidal rigidity. Autopsy showed LB in their cerebral cortex. Over the next 20 years, Japanese researchers reported a further 34 similar cases. Lewy body disease thus came to be considered as a rare cause of dementia until a series of

studies in Europe and North America in the late 1980s identified LB in the brains of 15–20% of elderly demented cases reaching autopsy, establishing it as a common form of degenerative dementia in old age. Dementia with Lewy bodies (DLB) is now the preferred term for such cases. Its recent recognition is largely due to the widespread use of improved neuropathological techniques that enhance visualization of cortical LB, in particular anti-ubiquitin immunocytochemistry. DLB is responsible for around 15% of cases of late-onset dementia and is therefore the second most common degenerative dementia in the elderly after AD.

Parkinson's disease (PD) is a major risk factor for dementia. Parkinson's disease dementia (PDD) has a prevalence of 20–30% among PD subjects, but detailed longitudinal studies have shown that the majority of PD subjects will develop PDD if they live long enough and that age is a major risk factor for the development of dementia in PD. PDD and DLB share cognitive, neuropsychiatric, neurochemical, patho- logical and imaging similarities, suggesting they should be considered the same disorder or different ends of the same spectrum. The distinction between DLB and PDD is based on the sequence of symptom appearance. If parkinsonism precedes dementia by <1 year, or follows the dementia, the diagnosis is

DLB. If dementia develops >1 year after parkinsonism, then PDD is used. The clinical and cognitive features of PDD are similar to DLB; parkinsonism is (by definition) invariably present in PDD, while both fluctuation and visual hallucinations are slightly less frequent than in DLB. While arbitrary, this distinction allows a common definition to facilitate research into the boundaries between the two.

5.4.2.2 Clinical Findings and Treatment

Clinically, the characteristic features that distinguish DLB from AD include fluctuating cognitive impairment, prominent (vivid) and recurrent visual hallucinations and spontaneous extrapyramidal symptoms. Other important features are neuroleptic (antipsychotics) hypersensitivity and REM sleep-behaviour disorder. According to the current International Consensus Criteria, the diagnosis of DLB requires at least two of these three features (Table 5.9).

Men appear to be more susceptible to DLB than women. The extrapyramidal features are typically bilateral and include rigidity, bradykinesia, mask-like facies, stooped posture, and a slow, shuffling gait. In many patients the only parkinsonian sign is an abnormal gait. Resting tremor is less common. The extrapyramidal features typically show modest response to anti-parkinson treatments. The visual hallucinations are typically recurrent and detailed and may involve people and animals. The visual hallucinations most reliably distinguish the disorder clinically from AD or

Table 5.9 Consensus criteria for the clinical diagnosis of *probable* and *possible* dementia with Lewy bodies (DLB)

1. The *central feature* required for a diagnosis of DLB is progressive cognitive decline of sufficient magnitude to interfere with normal social or occupational function. Prominent or persistent memory impairment may not necessarily occur in the early stages but is usually evident with progression. Deficits on tests of attention and of executive function, and visuospatial ability may be especially prominent.

2. Two of the following *core features* are sufficient for a diagnosis of probable DLB, one is sufficient for possible DLB
   (a) Fluctuating cognition with pronounced variations in attention and alertness
   (b) Recurrent visual hallucinations which are typically well formed and detailed
   (c) Spontaneous features of parkinsonism

3. Features *suggestive* of the diagnosis. If one or more of these is present in the presence of one or more core features, a diagnosis of probable DLB can be made. In the absence of any core features, one or more suggestive features is sufficient for possible DLB. Probable DLB should not be diagnosed on the basis of suggestive features alone.
   (a) REM sleep behaviour disorder
   (b) Severe neuroleptic sensitivity
   (c) Low dopamine transporter uptake in basal ganglia demonstrated by SPECT or PET imaging

4. Features *supportive* of the diagnosis (which are commonly present, but not proved to have diagnostic specificity)
   (a) Repeated falls and syncope
   (b) Transient, unexplained loss of consciousness
   (c) Severe autonomic dysfunction e.g. orthostatic hypotension, urinary incontinence
   (d) Systematized delusions
   (e) Depression
   (f) Hallucinations in other modalities
   (g) Relative preservation of medial temporal lobe structures on CT/MRI scan
   (h) Generalized low uptake on SPECT/PET perfusion scan with reduced occipital activity
   (i) Abnormal (low uptake) MIBG myocardial scintigraphy
   (j) Prominent slow wave activity on EEG with temporal lobe transient sharp waves

5. A diagnosis of DLB is *less likely*
   (a) In the presence of cerebrovascular disease evident as focal neurologic signs or on brain imaging
   (b) In the presence of any other physical illness or brain disorder sufficient to account in part or in total for the clinical picture
   (c) If parkinsonism only appears for the first time at a stage of severe dementia

6. *Temporal sequence*
   DLB should be diagnosed when dementia occurs before or concurrently with parkinsonism (if it is present). The term Parkinson's disease dementia (PDD) should be used to describe dementia that occurs in the context of well-established Parkinson's disease. In a practice setting the term that is most appropriate to the clinical situation should be used and generic terms such as LB disease are often helpful.

Source: McKeith et al. (2005) Neurology 65;1863–1872

vascular dementia (VaD). Although hallucinations and delusions occur in AD, visual hallucinations occur in 80% of patients with DLB; 45% have auditory hallucinations, 9% have delusions. Depression is more common than in AD. Recognition of the disorder is particularly important because of the potentially dangerous adverse reaction to neuroleptics.

Treatment of DLB is symptomatic only. Overall, patients with PDD and DLB respond to cholinesterase inhibitors (AChEI) at least as well as patients with AD. Hallucinations in particular can often be very responsive to AChEI treatment. The benefit of memantine in DLB is less clear with some reports of improvement in attention but also worsening of delusions.

5.4.2.3 Neuropathology, Aetiology and Genetics

Macroscopically, the brain of a patient with DLB has less cortical and less hippocampal atrophy than the brain of a patient with AD with a similar level of dementia severity. In addition to the presence of Lewy bodies (which may also be present in AD), the pathological feature that distinguishes DLB from AD is the presence of ubiquitin-positive neurites in the CA2–CA3 region of the hippocampus. Prominent neurofibrillary tangles in the cortex are absent. Psychiatric symptoms are reflected in the prominent involvement of the limbic system: amygdala, limbic cerebral cortex, including the cingulate, entorhinal and temporal cortex. Extra-pyramidal features are due to the involvement of the substantia nigra. Dysphagia is caused by the involvement of the dorsal motor nucleus; autonomic dysfunction is the result of the involvement of the intermedio-lateral cell column and the sympathetic ganglia.

Neurochemically, DLB is characterized by alpha-synuclein, a presynaptic protein with unknown function. In DLB, a cholinergic deficit exists greater than in AD, possibly explaining the beneficial response to cholinesterase inhibitors. Similar findings are reported in PDD, though PDD has been associated with greater nigrostriatal loss than DLB, and DLB with greater cortical pathological burden of both Alzheimer and synuclein pathology.

Familial forms of DLB have been reported, but typically the disease occurs in a sporadic fashion and no known genetic mutations exist. In line with this, the aetiology of Lewy body pathology remains elusive.

5.4.2.4 Neuroimaging Strategy

Neuroimaging findings are similar in DLB and PDD; so both are considered here together. Structural imaging can be used to rule out vascular disease as the cause of Parkinsonism, and also to render a diagnosis of AD less likely (absence of medial temporal lobe atrophy). Molecular imaging (SPECT/PET of dopamine transporter uptake) has greater specificity for DLB/PDD and is most helpful in the differential diagnosis with AD. The imaging findings are summarized in Table 5.10.

MRI Findings

Anatomical T1-weighted MR in DLB/PDD most consistently found that in the medial temporal lobe, there is less atrophy in DLB/PDD than AD on visual inspection – best assessed on thin slice coronal imaging (see

Table 5.10 Neuroimaging findings in DLB/PDD

Atrophy pattern
- Mild generalized atrophy, with no specific predilection
  - Atrophy rates lower than in AD
    - Severe occipital atrophy more likely in Balint variant of AD
- Hippocampal atrophy initially mild/absent
  - Prominent in late-onset AD, but may be absent in presenile AD
- Mild midbrain atrophy only

Vascular white matter lesions
- Slightly more than in normal ageing
  - Extensive lesions may suggest vascular dementia

Cerebral blood flow and metabolism
- General reduction, most marked in occipital region
  - Quite similar to AD
- Focal deficits favour vascular dementia rather than DLB

Dopaminergic imaging – most important marker
- Abnormal striatal uptake
  - Putaminal before caudate abnormality
    - May be asymmetric initially

Amyloid imaging
- Cortical uptake may be similar to AD

Differential diagnosis and red flags
- AD – severe hippocampal/precuneus atrophy early in disease
- PSP – humming bird sign (mesencephalon)
- MSA – 'cross sign' or pontine atrophy
- Vascular dementia – severe white matter lesions, lacunes, infarcts
  - may have brain stem involvement

Fig. 5.34 Medial temporal lobe appearance in DLB versus AD. Coronal T1-weighted MRI in an Alzheimer patient (*left*) shows marked medial temporal atrophy (MTA: *red arrows*). Note normal hippocampi (*green arrows*) in DLB patient on the *right*

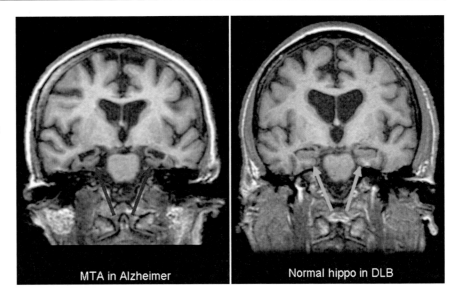

MTA in Alzheimer | Normal hippo in DLB

Fig. 5.34) or on volumetric analysis. However, this is a matter of degree which needs to take into account the severity of the patient – medial temporal atrophy does not rule out a diagnosis of DLB since it will become prominent in more severe disease and older patients. In fact, atrophy of the hippocampus is related to Alzheimer-type pathology in both disorders.

Voxel-based methods show increased atrophy in DLB/PDD relative to controls (see Fig. 5.35) in insular cortex, frontal, inferior parietal & temporal lobes, with relative sparing of the sensory-motor cortex; some studies found little cortical atrophy in DLB, with only changes in the midbrain region. In the midbrain, substantia innominata atrophy has also been reported in DLB, though midbrain changes can be found in AD as well. Putaminal volume reduction (of about 15%) has also been observed in DLB and advanced PD, which may be a result of increased striatal synuclein pathology.

Longitudinal rates of whole brain atrophy in PDD/DLB are probably intermediate between healthy ageing and AD. The atrophy rates in DLB and PDD are probably similar, and vary with disease duration. In summary, atrophy is widespread but modest in DLB/PDD, with *initially* preserved medial temporal lobes in comparison to (incipient) AD.

White matter hyperintensities on T2-weighted images are perhaps more severe in DLB/PDD than healthy subjects, but no more so than in AD. They probably reflect concomitant vascular problems. However, evidence of extensive cerebrovascular pathology (stroke,

severe white matter lesions) makes the diagnosis of DLB less likely (Tables 5.9 and 5.10).

The few studies using MR spectroscopy have found less change than in AD, with relatively normal NAA and myo-inositol levels. Using fMRI, one study in DLB found reduced activity in response to moving stimuli. Studies of Parkinson's disease with hallucinations have found reduced cortical responsivity to visual stimuli.

PET/SPECT Findings

Perfusion/Metabolism

FDG-PET and SPECT using a variety of tracers (^{99m}Tc-HMPAO, ^{123}I-IMP and ^{99m}Tc-ECD) show a similar pattern of hypometabolism/perfusion in DLB and PDD particularly involving occipital cortex, midline and inferior parietal regions, lateral temporal cortex, inferior and medial frontal lobe (Fig. 5.36). In comparison to AD there is lower occipital metabolism in DLB/PDD (Fig. 5.37). Focal hypometabolism/perfusion is indicative of stroke/cerebrovascular disease, and hence makes the diagnosis of DLB/PDD less likely.

Dopaminergic System

The ^{123}I labelled dopaminergic presynaptic ligand FP-CIT has been extensively studied in PDD/DLB. In

Fig. 5.35 Grey matter atrophy in DLB. Voxel-wise mapping of grey matter loss on MRI in DLB relative to healthy control subjects of similar age

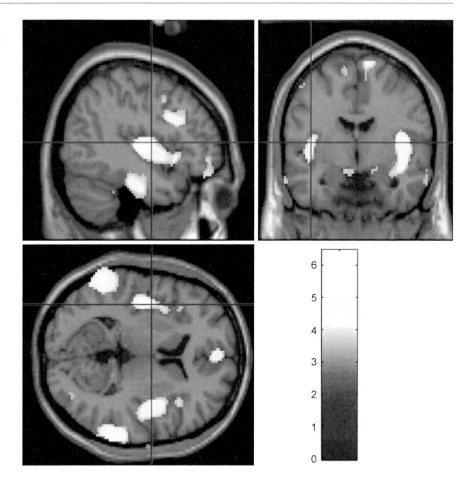

normal subjects and those with AD, the ligand is taken up in the caudate and putamen, whereas in PDD and DLB uptake is almost absent in the putamen, and reduced in the caudate (Fig. 5.38). Abnormal FP-CIT imaging has good diagnostic accuracy for DLB/PDD with sensitivity ~75% and specificity of ~90% for distinguishing PDD/DLB from non-DLB dementia (e.g. AD). PDD is associated with more pronounced overall striatal dopaminergic striatal loss than DLB, consistent with autopsy findings of greater nigrostriatal dopaminergic cell loss.

Amyloid Imaging

PET imaging with the ^{11}C-labelled Pittsburgh Compound B (PIB) is positive (suggesting amyloid deposition) and appears similar to AD in 40–80% of subjects with DLB (less so in PDD).

Cardiac Imaging

Cardiac scintigraphy using iodine-123 meta-iodo-benzylguanidine (^{123}I-MIBG), an analogue of noradrenaline, has been used for estimating local myocardial sympathetic nerve damage. The ratio of uptake in heart to mediastinum uptake is calculated using a region of interest. This ratio has been found to be reduced in DLB and PDD in comparison to AD and healthy subjects with a diagnostic accuracy for DLB/PDD vs. AD of approximately 90% though currently, studies with larger group sizes are needed to confirm the diagnostic accuracy.

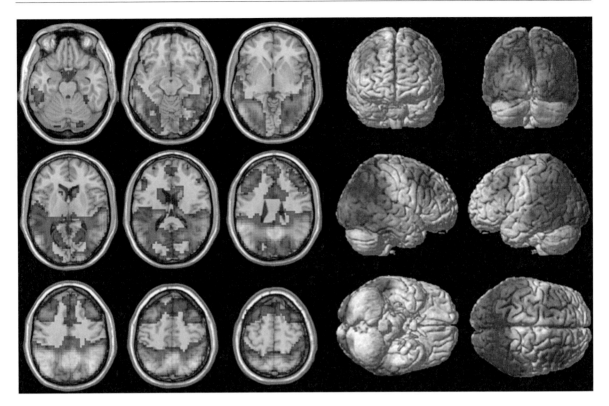

Fig. 5.36 Hypoperfusion in DLB/PDD. Voxel-wise analysis (from HMPAO-SPECT) of DLB/PDD group versus healthy controls. Regions of reduced perfusion are overlaid in *red/yellow* onto a standard MRI anatomical scan on the *left*; surface rendering on the *right*

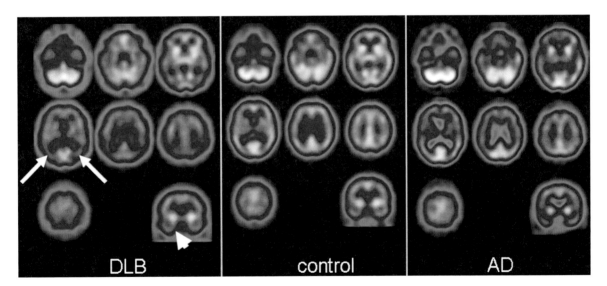

Fig. 5.37 Hypoperfusion in DLB versus AD. Axial slices HMPAO SPECT scan of DLB subject (*left*), healthy control (*middle*) and AD patient (*right*). Bottom right image in each panel is a coronal section through the medial temporal lobe. Note low signal in occipital lobe (*arrows*) and preserved medial temporal lobe (*arrow head*) in DLB

Fig. 5.38 Dopaminergic imaging in DLB. Axial images from FP-CIT scan of control and DLB subject showing reduced dopamine transporter density in the caudate (*arrow*) and putamen (*asterisk*)

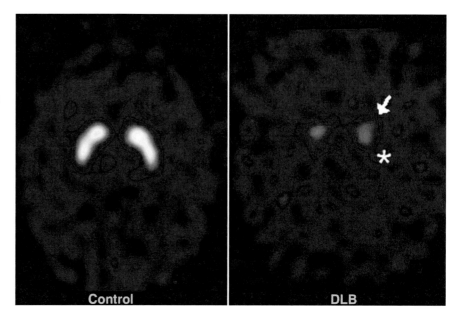

5.4.2.5 Differential Diagnosis and Imaging Follow-up

No sensitive or specific blood, cerebrospinal fluid (CSF), or urine tests are currently available for DLB. In a typical case, there may be limited need for anything more than structural (MR) imaging. However, the differential diagnosis in DLB/PDD is wide and atypical cases are common. The differential may be divided into two broad categories: First, where cognitive decline is the prominent feature and with some parkinsonism. In this situation the differential relates to DLB versus AD +/− VaD. The second category is distinguishing DLB/PDD from atypical parkinsonian syndromes which includes PSP, CBD, MSA and FTLD. The key imaging modalities are a combination of MRI to assess hippocampal atrophy (AD versus DLB); focal fronto-temporal atrophy (FTLD, CBD), midbrain atrophy (PSP), pontine/cerebellar atrophy (MSA) and vascular load; and dopamine transporter imaging to assess the dopaminergic deficit.

5.4.3 Progressive Supranuclear Palsy (PSP)

Dominic Paviour

Synonyms

Steele-Richardson-Olszewskisyndrome; Richardson's syndrome; PSP-P;

5.4.3.1 History

In June 1963, Dr Clifford Richardson presented a clinical report of eight cases collected over approximately ten years, of 'heterogenous system degeneration' with supranuclear ophthalmoplegia, pseudobulbar palsy, nuchal dystonia and dementia. He described a previously unrecognized disorder occurring in the seventh and eighth decades of life which progressed to death occurring within 9 years. The following year, together with Dr John Steele, they published a seminal paper entitled 'Progressive supranuclear palsy'.

5.4.3.2 Clinical Presentation

When the typical features of PSP are present, it provides a striking clinical picture. Patients often present with falls, typically backwards, as well as mild symmetrical Parkinsonism. The classic supranuclear ophthalmoplegia results initially in slowed vertical saccades and progresses through impaired vertical and horizontal pursuit movements (correcting with vestibulo-ocular reflexes) to a complete ophthalmoparesis. Axial rigidity is a striking feature of the typical clinical syndrome resulting in an upright posture and a 'growling' dysarthria is frequently described.

Atypical clinical presentations of PSP include so-called PSP-Parkinsonism cases present with asymmetric Parkinsonism and tremor with the ophthalmoplegia occurring later. This is distinct from the more typical Richardson's syndrome.

Cases presenting with pure akinesia without rigidity, gait freezing, rest tremor, isolated dementia, parkinsonism without dementia, limb apraxia and asymmetric parkinsonism as well as a number of cases dying without recorded evidence of the distinctive supranuclear ophthalmoplegia have been reported.

Despite these observations, the current operational criteria for the diagnosis of PSP include very few clinical features that were not recognized in the original description. The high specificity of these criteria is important for clinical research studies, but their sensitivity is suboptimal for clinical care and descriptive epidemiological studies. Patients are generally considered as 'clinically possible' PSP if they suffer from a gradually progressive disorder of more than 12 months duration, with age at onset after 40 years, and with a tendency to fall within the first year of disease onset, in the absence of defined exclusion criteria Table 5.11. There should be no clinical features suggestive of Creutzfeldt–Jakob disease or any other identifiable cause for their postural instability.

In terms of prognosis, the absence of supranuclear gaze palsy, early falls and early bulbar dysfunction, with a positive response to levodopa, convey a better prognosis. There are no definite biological markers for the ante-mortem diagnosis of PSP and the 'definite' diagnostic category has traditionally been reserved for cases that are pathologically confirmed. The cases that remain undiagnosed in life make up at least 20% of the pathologically diagnosed cases of PSP and most have unusual or atypical clinical pictures. There is no specific treatment for PSP.

5.4.3.3 Neuropathology and Genetics

Like FTLD, PSP is a tauopathy (see Sect. 5.3). Tau is a microtubule-associated protein, which in the normal human brain is distributed mainly in axons. By supporting cytoskeletal structure and sustaining axonal transport, tau plays a fundamental role in neuronal survival. In normal neurones, tau is soluble, binds to microtubules reversibly and has a rapid turnover. In neurodegenerative diseases such as PSP, tau loses its affinity for microtubules and collects as insoluble aggregates in the form of proteolytic resistant filaments.

PSP is characterised by the presence of neurofibrilliary tangles (NFTs) and neuropil threads (NTs) resulting from aggregation of hyperphosphorylated tau protein filaments. The NFTs are basophilic and globose, their structure resembling a ball of string. However, flame-s haped tangles (characteristic of AD) can also occur in PSP.

The pathological criteria for the diagnosis of PSP are well established and reflect many of the findings described by Olszewski, specifically neuronal loss with gliosis and NFTs in the subcortical and brainstem nuclei and the cerebellar dentate nucleus. New immuno-histochemical methods have allowed further characterization of the pathological changes and some

Table 5.11 NINDS-SPSP Diagnostic criteria for PSP (after Litvan et al. 1996)

Diagnostic categories	Inclusion criteria	Exclusion criteria	Supportive criteria
For possible and probable:	*For possible and probable:* Gradually progressive disorder with age at onset at 40 or later;	*For possible and probable:* Encephalitis; alien limb; cortical sensory deficits; hallucinations or delusions unrelated to dopaminergic therapy; prominent, early cerebellar symptoms or unexplained dysautonomia	Symmetric akinesia/rigidity, retrocollis; poor/absent response to levodopa; early dysphagia & dysarthria; early onset of cognitive impairment including > 2 of: apathy, impaired abstract thought, decreased verbal fluency, utilization or imitation behaviour, or frontal release signs
Possible	Either vertical supranuclear palsy or both slowing of vertical saccades & postural instability with falls <1 year disease onset		
Probable	Vertical supranuclear palsy and prominent postural instability with falls within first year of disease onset		
Definite	All criteria for possible or probable PSP are met and histopathologic confirmation at autopsy		

regional pathological variation has been reported. The characteristics of the pathological accumulation of the microtubule protein tau into filamentous deposits of abnormally phosphorylated protein have also been examined in PSP. Alternative splicing of the tau gene transcript from Chromosome 17 yields six different tau isoforms. The selective tau deposition in clinically and pathologically distinct diseases differs by the relative amounts of tau containing three (3R-tau) or four (4R-tau) microtubule-binding domains. Normal brain has 3R-tau isoforms; in PSP, hyper-phosphorylated 4R-tau is abundant (Fig. 5.18).

Braak has postulated that in idiopathic Parkinson's disease, the pathology begins in the lower brainstem structures and spreads caudally to involve progressively upper brainstem structures and eventually the neo-cortex. This theory does not explain the diversity in disease presentation whereby as an example patients may present with predominant tremor or with much more prominent bradykinesia or even cognitive disturbances. Whether a similar topographic evolution of the pathology occurs in PSP and other tauopathies is not clear; however, in typical PSP, or Richardson's syndrome, the brainstem is likely to be involved earlier than cortical sites.

5.4.3.4 Neuroimaging Findings

MRI and CT may show generalized supratentorial atrophy in PSP, sometimes with a frontal predominance.

Most importantly, however, MRI allows better visualisation of the posterior fossa and brainstem structures on sagittal images and is helpful to discriminate 'atypical Parkinsonism' (including PSP and MSA) from PD.

The characteristic qualitative imaging features of PSP include midbrain atrophy, divergence of the red nuclei, dilatation of the third ventricle, atrophy of the superior cerebellar peduncle and frontal cortical atrophy. The term 'hummingbird' is used to describe the midsagittal imaging appearance (Fig. 5.39), while the axial appearance has been referred to as the 'Mickey Mouse' sign; more detailed information is provided in Table 5.12.

Definitive and diagnostically supportive abnormalities may only be seen in about 50% of patients with a clinical diagnosis of PSP and clear MRI abnormalities are more likely to be present in individuals with typical clinical presentations and as it is the atypical cases that need diagnostic clarification, the MRI appearances may only help to reinforce the existing clinical diagnosis rather than dispelling diagnostic uncertainty.

In addition to the brainstem nuclei being severely affected by the pathological process in PSP resulting in midbrain atrophy, the dentate nucleus of the cerebellum is involved. Its main efferent pathway is the brachium conjunctivum or superior cerebellar peduncle (SCP). Its fibres pass into the brainstem and decussate at the level of the inferior colliculi before synapsing in the red nucleus and the ventrolateral nucleus of the thalamus. Signal change in this structure may be seen in PSP but is not particularly sensitive. Qualitative

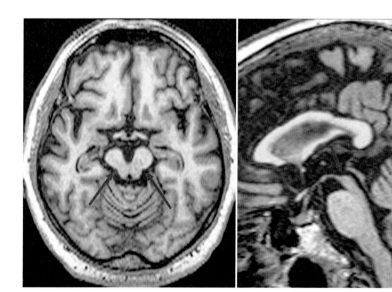

Fig. 5.39 Mesencephalic atrophy. Axial (*left*) and sagittal (*right*) T1 MR demonstrating striking midbrain atrophy in pathologically proven PSP

Table 5.12 MRI features in atypical Parkinsonian syndromes

	MRI technique	Imaging feature	Discriminates
Progressive supranuclear palsy (PSP)	T1 (sag)	Midbrain atrophy (A-P diameter <14 mm)	PSP from MSA
	T1 (sag)	Concave upper midbrain border	PSP from IPD
	T1 (cor/axial)	Superior Cerebellar Peduncle atrophy	PSP from MSA, IPD
	T1 (sag)	Ratio of Midbrain:Pons area – 'hummingbird'	PSP from MSA
Multiple system atrophy (MSA)	T2/PD (axial)	'Hot cross bun' or Pontine cross sign	MSA (cerebellar subtype) from PSP, IPD
	T2/PD (axial)	Putamen, low intensity	MSA from IPD
	T2/PD (axial)	Putaminal slit sign	MSA (parkinsonian subtype) from PSP, IPD
Cortico-basal degeneration (CBD)	T1	Global atrophy	Seen in advanced PSP as well
	T1	Asymmetric cortical atrophy	More suggestive of CBD than PSP or MSA

IPD idiopathic Parkinson disease

judgements of SCP volume on axial and coronal T1 MRI may be more useful suggesting sensitivities and specificities of 75% and 95% and these findings are supported by separate pathological studies (Fig. 5.40).

Biological variation in normal anatomy is bound to contribute to overlap in quantitative data and requires adjustments to be made. Quantitative brainstem measurements harbour potential as diagnostic aids in degenerative Parkinsonian disorders. Especially the midbrain to pons ratio is helpful in differentiating PSP from MSA (Fig. 5.41).

In terms of linear measurements, PSP can be differentiated from MSA-P using a midbrain diameter <14 mm on the sagittal scan (A-P diameter), but between 20% and 40% of patients cannot be correctly classified either because no characteristic changes were seen, or because changes occur in both conditions.

A ratio of linear measurements accounts in part for this variability and this can be improved by employing volumetric MRI to asses the true volume of the brainstem structures. While more labour intensive, resulting measurements more reliably discriminate the degenerative causes of Parkinsonism (Fig. 5.42).

Regional ADC measurements from the middle cerebellar peduncle are capable of discriminating PSP and MSA with a sensitivity and specificity of 91% and 84% on the basis that the ADC in the tissue concerned is abnormal in MSA. Clear abnormalities in PSP have not been established. Diffusion tensor imaging and tractography demonstrates promise in this regard, but at present this remains a research tool (Fig. 5.43).

There is no specific molecular probe for SPECT/PET in PSP. FDG-PET may show hypometabolism in the frontal lobes as in FTLD, and also in the mid-brain. The specificity of these findings is uncertain, and no clear indication exists.

5.4.3.5 Differential Diagnosis

PSP remains a clinical diagnosis and most investigations are conducted to exclude alternative diagnoses (Box 5.5). CSF examination and quantification of Tau protein remains a research tool and may not discriminate different degenerative diseases well, but is still offered in some specialist centres. In case of doubt, follow-up imaging can be used to demonstrate progressive atrophy (Fig. 5.44).

5.4.4 Multiple System Atrophy (MSA)

Dominic Paviour

Synonyms

Shy Drager syndrome; Striatonigral degeneration (SND); Olivopontocerebellar atrophy (OPCA);

Fig. 5.40 Atrophy of the
superior cerebellar peduncle
in PSP. (**a** and **b**) Axial and
coronal T1 section in PSP
showing a thin Superior
Cerebellar Peduncle (*arrow*).
(**c** and **d**) Axial and coronal
T1 section in healthy control
for comparison

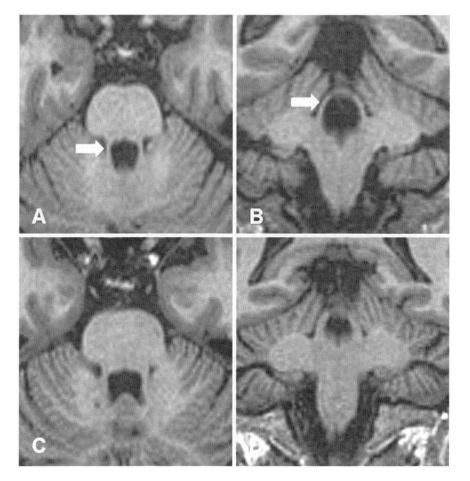

Fig. 5.41 Sagittal
T1-weighted MR illustrating
area measurements of
midbrain and pons in PSP
(*left*) and MSA (*right*). Note
also the flattening of the
superior profile of the
midbrain in PSP

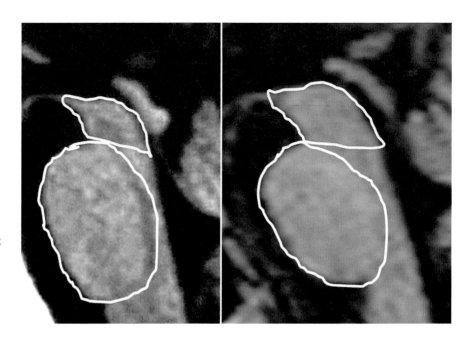

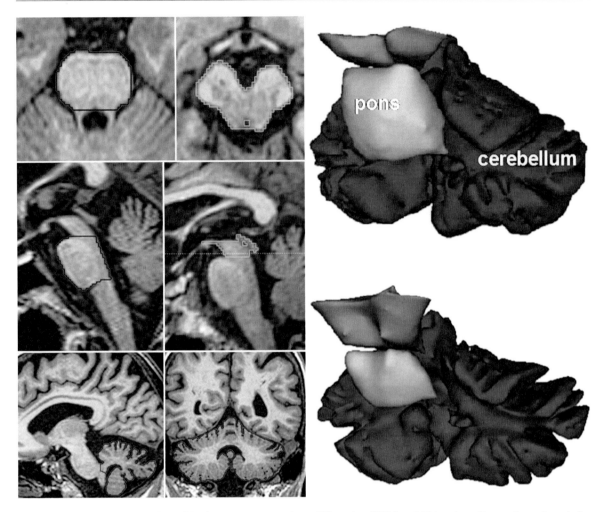

Fig. 5.42 Regional segmentation of brainstem structures in MSA (*blue*) and PSP (*green*) with computer generated 3D volume reconstructions for illustration. Smaller midbrain volumes differentiate PSP from MSA and smaller pontine and cerebellar volumes differentiate MSA from PSP

5.4.4.1 History

Olivopontocerebellar atrophy (OPCA) and striatonigral degeneration (SND), which are the predominant pathological correlates of what is now known as multiple system atrophy (MSA) were described independently in 1900 and 1961. The Shy-Drager syndrome has historically been the term used to describe the less common phenotypic presentation of MSA with predominant autonomic failure. Graham and Oppenheimer first proposed the term MSA based on the finding that sporadic OPCA, SND and Shy–Drager syndrome co-exist both clinically and pathologically.

5.4.4.2 Clinical Presentation

Multiple system atrophy (MSA) is a sporadic adult-onset neurodegenerative disease, which usually presents clinically as a combination of parkinsonism, cerebellar ataxia and autonomic failure. Which of these phenotypes is foremost depends on whether the patient is deemed to have MSA-P, the predominantly parkinsonian phenotype which is the commonest, MSA-C (cerebellar) or MSA-A, (autonomic failure), the last of these being least common. It is a disease of later adult life, with a mean age at onset of 58 years (with a range from mid 30s to 80s). Dementia was historically not thought

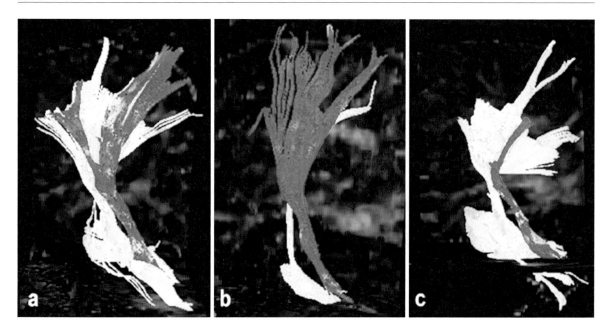

Fig. 5.43 DTI analysis in PSP versus MSA. Tractography of the middle cerebellar peduncle (MCP in *yellow*) and superior cerebellar peduncle (SCP in *purple*) in idiopathic Parkinson disease (**a**), MSA (**b**) and PSP (**c**). Note selective atrophy of the MCP in the patient with MSA and of the SCP in the patient with PSP. (Reprinted with permission from Nilsson Neuroradiology 2007;49:111–9)

Box 5.5 Differential diagnosis of PSP

Degenerative
CBD (alien limb syndrome, cortical sensory deficits, limb apraxia, dystonia, asymmetric bradykinesia)
Parkinson's disease (tremor-dominant disease, levodopa response, less axial rigidity)
Dementia with Lewy bodies (hallucinations, cortical dementia, fluctuations, parkinsonism)
MSA (prominent cerebellar symptoms, autonomic dysfunction, parkinsonism)
CJD (duration <1 year with dementia, myoclonus, abnormal findings on EEG)
HD (family history, findings on genetic test)
Spinocerebellar ataxia (SCA) (family history, cerebellar signs, findings on genetic test)
FTLD-MND/FTLD-U (lower motor neuron signs, abnormal electromyograph [EMG] findings)

Other
Cerebrovascular disease (focal features, imaging findings)
Whipple's disease (ocular-masticatory myorhythmia, polymerase chain reaction [PCR] confirmation)
Neurosyphilis
Normal pressure hydrocephalus (dementia, urinary dysfunction, gait abnormality, imaging findings)
Wilson's disease (Kayser–Fleischer rings, earlier onset, copper metabolic abnormalities)
Iatrogenic secondary to Neuroleptic medication

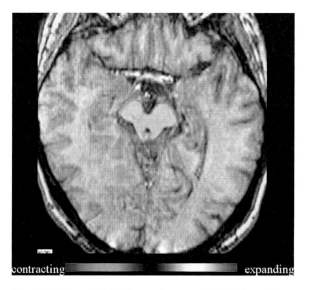

Fig. 5.44 Sequential MRI scans in a case of PSP (1 year interval) registered using Fluid registration. Note high rates of atrophy in the midbrain (*green*) and some inferior frontal degeneration and ventricular expansion (*yellow*)

Table 5.13 Clinical signs and symptoms suggestive of MSA rather than idiopathic Parkinson's disease

Jerky postural component to the tremor
Myoclonic jerks
Rapid disease progression with early instability and falls are common to MSA (and PSP)
Cold, dusky, violaceous extremities
Rapid eye movement (REM) sleep behaviour disorder
Sleep apnoea
Quivery croaky strained element to speech
Facial dystonia
Head drop/antecollis
Early autonomic failure
Cerebellar ataxia
Early uro-neurological dysfunction
Early erectile impotence in men
A lack of sustained levodopa responsiveness.

to be part of MSA – however cognitive decline (usually frontal) can occur and may be severe. The parkinsonian phenotype may be very difficult to distinguish from idiopathic Parkinson's disease, presenting with a typical asymmetric tremor. Red flags suggesting MSA rather than PD are listed in Table 5.13.

In MSA-P particularly, there may be antecollis (marked forward flexion of the neck due to dystonia of the anterior neck muscles). Laryngeal dystonia may result in marked inspiratory stridor and compromise, necessitating tracheostomy. The disease is progressive, with a mean duration to death of 7 years (range 2–16). There is no (curative) treatment for MSA beyond anti-parkinsonian medication.

5.4.4.3 Histopathology and Genetics

The pathological basis of MSA is oligodendroglial cytoplasmic inclusions. These stain heavily for α-synuclein, a protein which when dysfunctional has a pivotal role in the development of MSA. Macroscopically, there is predominant brainstem and cerebellar atrophy; however, marked cerebral atrophy has been reported.

In MSA-C, which is characterized clinically by predominant cerebellar ataxia, the pathological findings are primarily those of neuronal cell loss and gliosis in the inferior olivary nucleus, pontine nuclei, cerebellar hemispheres and vermis, whereas in MSA-P, neuronal

cell loss and gliosis is most severe in the substantia nigra, putamen, caudate nucleus and globus pallidus. Neurodegeneration in the autonomic nervous system (the Shy–Drager syndrome) presents clinically with prominent autonomic failure. It was not until the subsequent discovery of the oligodendroglial cytoplasmic inclusions that the notion of MSA as a single clinico-pathological entity could be confirmed.

A spectrum of pathological phenotypes of MSA exists and is the basis for the variation in the clinical phenotype. Less severe pathological involvement of the nigro-striatal system is required for parkinsonism than involvement of the cerebellum for ataxia. This may explain why the MSA-P phenotype is predominant.

Given the molecular pathology of the cellular inclusions, the α-synuclein gene is the most probable gene responsible for MSA pathogenesis. However, genetic analysis of pathologically confirmed MSA cases has failed to find any pathogenic mutations in the α-synuclein gene.

5.4.4.4 Neuroimaging Findings

As with PSP, the clinical application of MRI in MSA is in discriminating it from the other degenerative parkinsonian disorders (see Table 5.12). In MSA, and particularly the cerebellar subtype, abnormalities on T1-weighted MRI are confined to the posterior fossa with predominant pontine and cerebellar atrophy.

The so-called *hot cross bun* sign on T2-weighted MRI has been proposed as a useful clinical marker of MSA-C (Fig. 5.45). This MR finding is best appreciated on proton-density weighted or FLAIR images. While initially felt to be pathognomonic of MSA-C, it may in fact occur in the advanced stage of other diseases including CNS vasculitis, paraneoplastic syndromes and the spinocerebellar ataxias (SCA).

Signal change in the basal ganglia on T2 MRI is considered a more characteristic finding with low signal evident in the putamen and a thin rim of hyperintensity noted at the lateral posterior putaminal rim, particularly in MSA-P.

As well as these qualitative findings, quantitative MRI studies have confirmed the presence of regional volume loss in MSA as well as increased rates of regional atrophy conforming to the severity of regional pathology, although these rates of atrophy do not allow discrimination of MSA from other degenerative forms

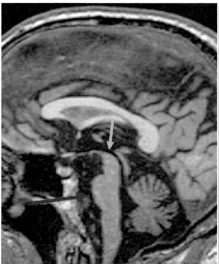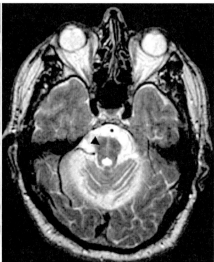

Fig. 5.45 Imaging features of MSA. Mid-sagittal images (*left*) shows flattening of the pons and loss of volume (*red arrow*), transverse T2-weighted image (*right*) a 'hot-cross bun' sign (*arrowhead*). Note the normal appearance of the mesencephalon (*green arrow*), in contrast to PSP

of parkinsonism with 100% sensitivity or specificity. Brainstem volumetry demonstrates that the pons and cerebellum are more severely affected in MSA than in PSP.

Diffusion-weighted imaging allows MSA to be differentiated from PD on the basis of higher regional apparent diffusion coefficients (ADC) measured in the putamen, with a high sensitivity and specificity. However, as already stated, in order to apply this clinically, normal control values need to be measured for a specific scanner.

5.4.5 Cortico-Basal Degeneration (CBD) and Other FTLD Tauopathies

Dominic Paviour

Synonyms

Cortico-basal ganglionic degeneration

5.4.5.1 History

The syndrome of CBD was first recognised formally by Rebeiz in 1967, and at the time was named 'Cortico-dentato-nigral degeneration with neuronal achromasia' on the basis of the pathological findings associated with the clinical features of the disease. It was officially recognised as a specific neuropathological disease entity in

1996; however, the clinical 'Cortico-basal syndrome' can arise as a result of numerous differing underlying pathological processes.

CBD is a tauopathy with hyper-phosphorylation of microtubule associated tau protein (MAPT), like PSP, FTLD. These conditions may have significant overlap in their clinical phenotypes but are regarded as separate clinico-pathological entities albeit with a similar pathological basis.

5.4.5.2 Clinical Presentation

CBD typically presents in the sixth to eighth decades (mean 63 years), and is relatively uncommon although its true incidence and prevalence are unknown. Most of the published literature on CBD comes from movement disorder clinics and thus the documented clinical features at presentation are biased towards the motor manifestations of the disease. However, CBD is possibly as likely to present as a cognitive syndrome, but as these patients present to dementia clinics, this aspect of the clinical syndrome is likely to be under-reported.

The typical clinical features of the cortico-basal syndrome are presentation with a rigid, dystonic, akinetic or dyspraxic arm. Occasionally, patients may describe a striking phenomenon where a limb 'has a will of its own' and it may be observed that the arm may 'wander' and rarely may grasp hold of other body parts, people close by or bedclothes. This is the 'alien limb' phenomenon and it is typically associated with

Differential diagnosis of MSA
Degenerative
Parkinson's disease (tremor-dominant disease, levodopa response)
Dementia with Lewy bodies(hallucinations, cortical dementia with aphasia, parkinsonism)
PSP (earlier falls, no cerebellar signs, sub-cortical dementia with apathy)
Spinocerebellar ataxia (SCA) (family history, cerebellar signs, findings on genetic test)
Other AD/AR cerebellar ataxias
Idiopathic Orthostatic Hypotension and other Autonomic Failure Syndromes
Other
Multiple Sclerosis (typical imaging findings, spasticity, ataxia, minimal extra-pyramidal features)
Cerebrovascular disease (focal features, imaging findings)
Whipple's disease (ocular-masticatory myorhythmia, polymerase chain reaction [PCR] confirmation)
Neurosyphilis
Wilson's disease (Kayser–Fleischer rings, earlier onset, copper metabolic abnormalities)
Iatrogenic secondary to Neuroleptic medication

cortical sensory loss. Patients may also initially complain of a gait disorder. Other presentations with behavioural disturbance, speech and language dysfunction and sensory symptoms are less common. Ultimately, most develop asymmetric parkinsonism (particularly bradykinesia and rigidity) and marked dyspraxia of affected limbs as well as gait impairment.

Many patients have a tremor, but this is different from that seen in Parkinson's disease, in that it is faster, more irregular and jerky and particularly in the later stages of the disease has a stimulus-sensitive myoclonic component. Apraxia may be bilateral but may be more easily elicited in the less parkinsonian upper limb. Gait apraxia may result in an abnormal gait, with the foot sticking to the floor or being dragged along.

Cognitive dysfunction in CBD is common and this is reflected in the revised diagnostic criteria. Aphasia may be seen in as many as 44% of patients. Frontal executive impairment is common and consistent with the known pathological involvement of the parietal lobe calculation and visuo-spatial skills are also commonly impaired. Cognitive impairment may remain the dominant feature of the disease, making clinical diagnosis difficult.

Clinical diagnosis of CBD is less reliable than the other parkinsonian disorders (a sensitivity of around 50% in the late stages of the disease, compared to 80–90% for a clinical diagnosis of PSP when the diagnostic criteria are fulfilled). The clinical diagnosis in life is difficult and when the disease presents as a movement disorder, PSP is the disorder it is most commonly confused with. The commonest cognitive presentations may be difficult to distinguish clinically from the progressive non-fluent aphasia and the FTD subtypes of FTLD. AD (dyspraxic presentations) is a common cause of misdiagnosis of CBD. There is no treatment for CBD.

5.4.5.3 Histopathology and Genetics

A definitive diagnosis of CBD as the basis for the cortico-basal syndrome can only be made at post-mortem. Macroscopically, atrophy is centred on the posterior frontal and parietal cortex with relative sparing of the temporal and occipital regions (Fig. 5.20). Cortical atrophy is classically asymmetric; however, whether this is the case in the more cognitive and less 'asymmetric motor' presentations of the disease is not established.

Normal cortical architecture is destroyed and there is cell loss and gliosis. Similar pathology occurs in the substantia nigra with loss of pigmented cells. The 'neuronal achromasia' originally described is characteristic of the pathology. Affected cortical neurons stain positively for tau and ubiquitin. The distribution and number of 'ballooned neurons' is crucial for the differential diagnosis. The molecular basis of CBD, as in PSP is the accumulation of hyper-phosphorylated tau forming abnormal filamentous inclusions in neurons and glia. As is the case in PSP, CBD is a four repeat tauopathy (see also Sect. 5.3).

No confirmed genetic basis for CBD has been established; however, rare reports of pathologically confirmed familial CBD have been published.

5.4.5.4 Neuroimaging Findings

There have been very few MRI studies of CBD in which pathological confirmation of the diagnosis has been available. This means that in a clinical syndrome with a potentially diverse pathological basis, the documented MRI findings are not likely to be reliable indicators of the underlying neuropathology. Nevertheless, it is likely that in the early stages of the disease, the imaging findings will be normal. As the disease progresses, the macroscopic asymmetric cortical atrophy that is evident at post-mortem may be apparent on volumetric T1-weighted MRI. Asymmetric parasagittal atrophy may be a feature of CBD (Fig. 5.46). Midbrain atrophy and ventricular dilatation are supportive of the clinical diagnosis of PSP and asymmetric cortical atrophy supports the clinical diagnosis of CBD, but no MRI finding is truly specific for CBD. In fact, asymmetric AD is a much more common pathology than

CBD and should always be considered in the differential diagnosis.

Functional imaging studies (perfusion/metabolism on SPECT/PET) may show the asymmetric loss of function which might be predicted in any patient with the cortico-basal syndrome – its diagnostic reliability is unclear.

5.5 Dementia in 'Other' Movement Disorders

5.5.1 Introduction

The previous chapter discussed disorders where the clinical phenotype is predominantly dementia accompanied by parkinsonism. This chapter focuses on

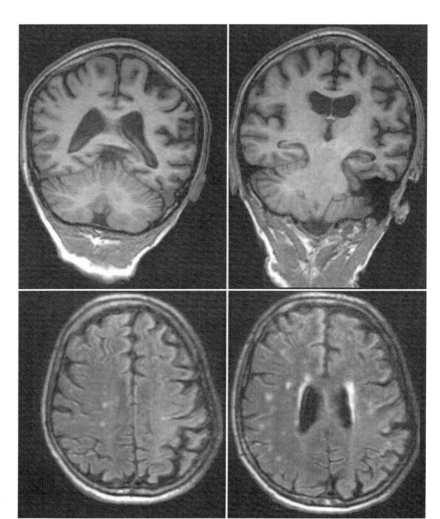

Fig. 5.46 Asymmetric cortical atrophy in CBD. This 67-year-old woman presented with speech difficulties, memory complaints and apraxia. She had difficulties operating household tools and lost control over her right hand. Coronal T1-weighted (*top*) and axial FLAIR (*bottom*) MRI showed markedly asymmetric left hemispheric atrophy, mostly affecting the parietal and temporal lobe

cognitive decline accompanied by non-parkinsonian movement disorders – with or without additional parkinsonism; these groupings do not have clear boundaries and there is inevitably overlap. Nonetheless, the broad clinical and imaging phenotype is key to differential diagnosis and to the imaging strategy and has determined the organisation of this book. Our approach may mean that conditions with common aetiologies but different phenotypes may be dealt with in separate chapters. For example, the trinucleotide repeat disorders of Huntington's disease (HD) and dentatorubral-pallidoluysian atrophy (DRPLA) are covered in this chapter, but another repeat disorder – fragile-X pre-mutation – is covered in Chap. 7, since the key MR feature is white matter hyperintensity; furthermore, the spino-cerbellar ataxias (SCAs) will not be considered at all because cognitive decline is a less prominent or only a late feature. Normal pressure hydrocephalus produces cognitive decline with a movement disorder – characteristically a subcortical cognitive profile and an abnormal gait – and this will be discussed in Chap. 8. As with many chapters, the diverse nature of cerebrovascular disease means it needs to be considered in the differential for a range of presentations – in particular, vascular dementia may be associated by a wide range of neurological features including an abnormal gait, motor signs or a movement disorder depending on the sites of lesions.

Table 5.14 presents a summary (but not comprehensive) list of disorders where dementia and a movement disorder may be part of the presentation or appear sequentially – and shows how multi-system presentation spans a range of aetiologies and phenotypes (and hence chapters).

5.5.2 Huntington's Disease

Ed Wild, Nicola Hobbs, and Susie Henley

Synonyms

Hereditary chorea; Huntington's chorea; Chorea major

5.5.2.1 History

The first description of Huntington's disease (HD) was by George Huntington in 1872. He noted that 'as the disease progresses, the mind becomes more or less impaired, in many amounting to insanity, while in others mind and body both gradually fail until death relieves them of their sufferings'. Huntington's disease usually causes a frontal/subcortical dementia.

Table 5.14 Dementia and movement disorders: an overview

Disorder	Type of movement abnormality	Chapter
Lewy body disease (DLB) and Parkinson's disease dementia	Parkinsonian	5.4
Multiple system atrophy (MSA)	Cerebellar and extrapyramidal	5.4
Progressive supranuclear palsy (PSP)	Falls, parkinsonism, axial rigidity	5.4
Amyotrophic lateral sclerosis (ALS) & FTD-MND	Upper motor neuron & lower motor neuron deficits	5.3
Corticobasal degeneration (CBD)	Dyspraxia, dystonia, myoclonus and parkinsonism	5.4 & 5.3
Huntington's disease (HD) and DRPLA	Chorea, bradykinesia, ataxia	5.5
Neurodegeneration with brain iron accumulation (NBIA)	Parkinsonism, dystonia	5.5
Creutzfeldt–Jakob disease (CJD)	Myoclonus, ataxia	5.6
Alcohol/Drugs	Ataxia, variety of pyramidal/extrapyramidal	5.7
Vascular dementia (VaD) – large vessel	Focal upper motor neuron signs	6.3
Small vessel VaD	Gait abnormalities, lower-body parkinsonism	6.4
Leukodystrophies	Pyramidal and extrapyramidal, ataxia	7
Multiple sclerosis (MS)	Wide range – including ataxia, focal UMN signs	7
Normal-pressure hydrocephalus (NPH)	Slow and unsteady apraxic gait	8

5.5.2.2 Clinical Presentation

The clinical presentation of HD is usually in early adulthood or middle age and characterized by a progressive triad of chorea, psychiatric (behavioural) disturbances and cognitive deterioration. Chorea is the most conspicuous manifestation, but decline of cognitive function becomes evident in nearly all patients and often precedes the onset of motor manifestations. The cognitive dysfunction consists of slowing of mental processes, impaired attention and executive functions, loss of concentration, loss of problem-solving, and ultimately dementia. Psychiatric features of HD include apathy, irritability, depression, anxiety and sometimes compulsive behaviour or aggression. A family history is usually clear but may be obscured: non-paternity or loss of contact with parents or because of censuring and/or anticipation – age at onset becoming earlier in subsequent generations.

There is no cure for Huntington's disease, but relief of symptoms is often possible. Chorea is seldom distressing to the patient but is frequently amenable to treatment with neuroleptics or tetrabenazine. Psychiatric manifestations often respond well to conventional drug therapy. Disease-modifying therapies are being actively sought but none are yet proven.

5.5.2.3 Genetics, Pathogenesis and Histopathology

Huntington's disease is inherited in an autosomal dominant manner, with nearly 100% penetrance and a low spontaneous mutation rate. The causative gene (*HTT*, also known as *IT15* – 'Interesting Transcript') is located on chromosome 4 and contains an excessive number of trinucleotide CAG repeats, encoding a polyglutamine stretch within the protein huntingtin. The protein is widely expressed in both neuronal and non-neuronal tissues. It has also been found in neurofibrillary tangles and neuritic plaques in AD and in Pick bodies, suggesting a neuropathological connection between these neurodegenerative disorders. The normal number of CAG repeats is 8–39, usually less than 35. In patients with Huntington's disease, the number of repeats varies from 40 to over 100. Subjects with over 60 repeats are likely to develop juvenile HD. As with other trinucleotide

syndromes, the number of repeats can increase over generations leading to an earlier age at onset and increased severity – a phenomenon called anticipation.

In terms of pathogenesis, mutant huntingtin is thought to lead to a toxic gain of function through which multiple cellular changes produce the HD pathology. Neuronal damage first appears in the striatum and loss in the striatum occurs in a systematic fashion. The progression is from medial to lateral in the caudate nucleus and in a dorso-ventral direction in the putamen. The projection fibres from the striatum are the most severely affected in Huntington's disease. The medium-sized spiny GABA-ergic projection neurons are particularly vulnerable. Microglial activation is seen before the onset of clinical signs and may contribute to the pathogenic process. The loss of central connection pathways, in particular in the fronto-thalamic area, may well explain the progressive subcortical dementia and the psychiatric disorders.

On macroscopic inspection, the brain may show cortical atrophy, ranging from mild to severe. Atrophy is seen throughout the cortex, but most conspicuous in the caudate nucleus, the putamen and the globus pallidus. The atrophy of the caudate nucleus is most striking on coronal slices (Fig. 5.47). Histologically, the atrophy is associated with loss of neurons and astrocytosis. The severity of atrophy of the caudate nucleus corresponds with both the number of CAG repeats and the clinical severity and progression of disease. Neuronal loss is also found in the hippocampus, hypothalamus and pars reticularis of the substantia nigra.

5.5.2.4 Differential Diagnosis

Clinically, HD closely resembles a number of rare syndromes referred to as Huntington's disease phenocopies or Huntington's disease-like (HDL) syndromes, but the diagnosis of HD in symptomatic patients can be made with certainty on the basis of genetic testing. In neuroacanthocytosis, there is comparable atrophy of the caudate nucleus and putamen as well as abnormal red cell morphology, neuropathy, myopathy, epilepsy, elevated creatine phosphokinase and orofacial dystonia. HDL1 (a familial prion disease), HDL2 (caused by *JPH3* mutations), HDL4 (spinocerebellar ataxia

Fig. 5.47 Gross macroscopy of Huntington's disease. Note ventricular widening due to atrophy of the caudate nucleus, which appear flattened (*red arrow*). As a consequence, the frontal horn of the lateral ventricle has enlarged, while the temporal horn is normal, since there is no hippocampal atrophy (*green arrow*)

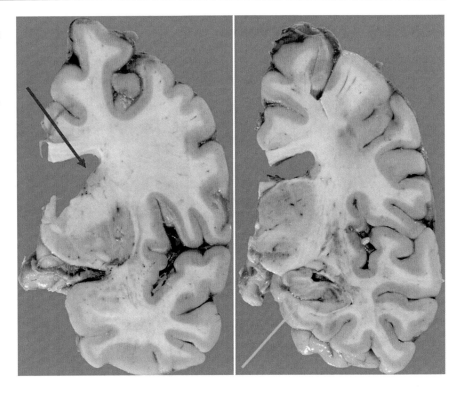

type 17) and neuroferritinopathy all resemble HD and can be diagnosed genetically. The brain iron accumulation disorders NBIA1 and NBIA2, which may mimic HD clinically, invariably produce reduced signal in the basal ganglia on T2*-weighted MRI (Sect. 5.5.3). There are many other diseases that may present with chorea, such as Sydenham's chorea, and any conditions – be they infectious, inflammatory, toxic or vascular – that involve the striatum.

5.5.2.5 Neuroimaging Findings

MRI reveals a variable degree of cortical thinning and atrophy of the striatum, most conspicuous on visual inspection in the caudate nucleus which is best viewed coronally. Striatal atrophy is present prior to clinical manifestation of motor signs and caudate volume has been shown to be a good predictor of motor onset. Striatal atrophy rates may be increased up to 10 years prior to motor onset. Cortical atrophy is present in the pre-motor onset population but becomes more notable and can be fairly widespread in later stages by which time the caudate can be reduced to a

thin rim around the ventricle (Fig. 5.48). In addition, grey matter reductions can be seen in the hypothalamus and opercular cortex.

There is a correlation between the degree of cortical and caudate atrophy and dementia and chorea, respectively. The reduction in volume of the putamen and degree of reduction of dopamine ligand binding are correlated with the degree of cognitive deficits. White matter volume is also reduced and is associated with decreased cortical and striatal glucose uptake and frontal-executive task performance (Fig. 5.49).

Selective atrophy of the caudate nucleus and putamen can also be seen in post-hypoxic ischemic encephalopathy, after bilateral infarctions of the caudate nucleus, in Sydenham's chorea (often unilateral) and in infectious-inflammatory disorders, neurometabolic disorders and toxic encephalopathies. In some of these cases the encephalopathy is static, or the history and other clinical findings will lead to the proper diagnosis. Many other neurodegenerative disorders can be accompanied by atrophy of the caudate. Caudate atrophy has also been described in AD, CJD, HIV-dementia, DRPLA, SLE, MS and selectively in neuroacanthocytosis.

Fig. 5.48 Coronal MRI in HD. Compare the T1-weighted images of a 43-year-old patient with HD on the *left* (duration 8 years) with the 43-year-old gender-matched control on the *right*. Note enlarged ventricles due to volume loss of the putamen and caudate. The latter appear shrunken and the caudate-ventricle border is much less convex in the HD patient. Cortical thinning and sulcal widening is also apparent

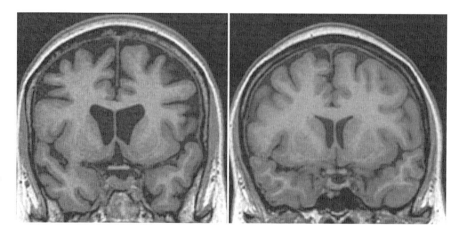

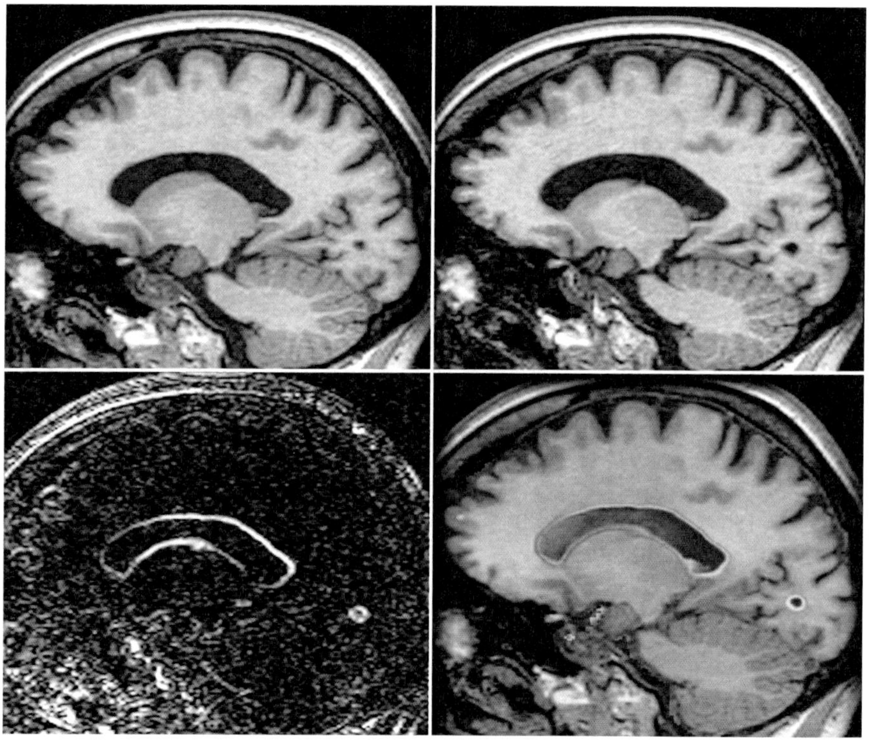

Fig. 5.49 Longitudinal MRI in HD. T1-weighted sagittal images of the patient from Fig. 5.48 showing *top left* baseline, *top right* follow-up 27 months later, *bottom left* difference image and *bottom right* fluid-registered image. Caudate loss, ventricular expansion and sulcal expansion is evident, and confirmed by the difference image; the fluid image illustrates that by this stage in the disease ongoing loss (in *green*) is extensive beyond the caudate – note the frontal white and grey matter atrophy

PET shows promise in the functional analysis of the basal ganglia and has demonstrated striking changes in metabolic rate (FDG), dopamine receptor density and microglial activation in premanifest and early HD (Fig. 5.50). It has also been used to assess striatal graft treatment. fMRI has shown reduced BOLD responses in premanifest gene carriers relative to controls, sometimes in the absence of measurable clinical or cognitive deficits. MR spectroscopy has been inconclusive and is unlikely to be as sensitive as other imaging modalities in HD.

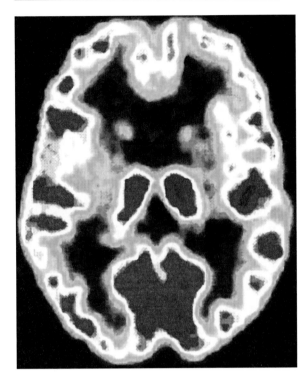

Fig. 5.50 [18F]Fluoro-deoxyglucose PET image in HD. Complete lack of uptake in the caudate nuclei which appear blue rather than red. Notice the preserved metabolism in the thalamus. (Courtesy of Gary Small, M.D., UCLA, USA)

5.5.3 Metal Metabolism Disorders

Synonym

Familial hepato-lenticular degeneration

5.5.3.1 Introduction

Several disorders in the metabolism of metals lead to neuropsychiatric, cognitive and other neurological problems due to metal accumulation in specific brain structures. Those primary disorders of metal metabolisms will be discussed in the current chapter. Other metal metabolism disorders, such as Menkes' disease will not be considered, since dementia is not a lead symptom. Secondary accumulation of metals, especially iron, may occur without obvious clinical correlate in normal ageing (see Chap. 4) and also in many neurodegenerative diseases, for example in AD and Parkinson's disease and is discussed in the relevant chapters.

5.5.3.2 Wilson's Disease

This autosomal-recessive genetic disorder is caused by abnormal accumulation of copper in various organs, including the brain. In Wilson's disease, mutations in the ATP7B gene lead to defects in a cation-transporting P-type ATPase which results in an inability of the liver to excrete copper to the bile and in cerulopasmin. Patients may present in childhood through liver failure or in adulthood with neuropsychiatric symptoms. Excess copper accumulates in the cornea (Kayser–Fleischer rings) and in the basal ganglia of the brain. Neuropsychiatric symptoms include parkinsonism, ataxia, tremor, polyneuropathy, emotional lability, depression and personality changes, and dementia. Treatment is with chelating agents or liver transplantation (for liver failure).

Neuroimaging Findings

MRI tends to be abnormal in all symptomatic individuals with Wilson's disease although the specific involvement can be diverse. High signal lesions are seen on T2-weighted imaging and these are often widespread and particularly involving basal ganglia and white matter; these changes are accompanied by diffuse cerebral, midbrain and cerebellar atrophy (Fig. 5.51). The striatum shows the greatest involvement with putamen caudate and thalamus most frequently involved (and usually bilaterally) – these changes and atrophy increase with neurological severity (rather than disease duration). The classical appearance of the so-called face of the giant panda on T2-weighted axial images at the level of the substantia nigra is due to areas of hypointensity due to iron accumulation (esp. the red nucleus) and areas of high signal in the surrounding tegmentum due to gliosis and demyelination. White matter lesions occur frequently in the frontal lobes, pons or cerebellar peduncles and white matter. In some patients, high signal can be found on T1-weighted images, as in other causes of hepatocerebral

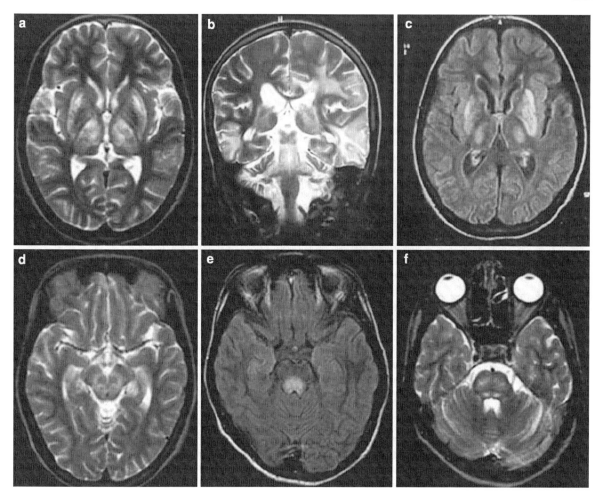

Fig. 5.51 Brain MRI findings in patients with Wilson's disease. (**a**) T2 hyperintensity in bilateral putamen and thalamus; (**b**) Coronal T2 showing signal changes involving white matter, basal ganglia, and brain stem structures; (**c**) FLAIR image revealing signal changes in bilateral caudate, putamen, and thalamus; (**d**) Axial T2 at the midbrain level showing the 'Face of Giant Panda' sign; (**e**) FLAIR through midbrain revealing hyperintense signal changes in the tectum; F: Axial T2 through pons showing high signal similar to central pontine myelinolysis. (Reprinted with permission from Mov Disord. 2010;25:672–8)

syndromes, perhaps caused by manganese deposition. SPECT/PET may show non-specific decrease of perfusion/metabolism in the basal ganglia.

Differential Diagnosis and Work-up

Beyond genetic testing, Wilson's disease can be diagnosed based on a combination of extrapyramidal symptoms, Kayser–Fleischer rings, low ceruloplasmin level, and high urinary copper excretion (with low plasma levels).

5.5.3.3 Neurodegeneration with Brain Iron Accumulation (NBIA)

NBIA refers to a group of disorders that have in common abnormal iron deposition of the brain. This descriptive term has replaced the former eponymous 'Hallervorden-Spatz syndrome'. NBIA encompasses infantile neuro-axonal degeneration which we will not consider, but also diseases that can present in adulthood with cognitive decline and dementia, notably atypical PKAN, idiopathic NBIA and aceruloplasminemia – the latter will be discussed in a bit more detail in this chapter.

NBIA

Early onset, rapid progression	Late onset, sloer progression
• Classic PKAN *(PANK2)*	• Atypical PKAN *(PANK2)*
• Classic INAD *(PLA2G6)*	• Neuroferritinopaty *(FTL)*
• Atypical NAD *(PLA2G6)*	• Aceruloplasminemia *(CP)*
• Idiopathic NBIA (unknown)	• Idiopathic NBIA (unknown)

Disease names are in black; associated gene names are in red
Idiopathic NBIA refers to other froms whose cause are notyet known

Source: http://www.nbiadisorders.org/

Pantothenate kinase-associated neurodegeneration (PKAN) is caused by a mutation in the *PANK2* gene, a regulatory enzyme in the biosynthesis of coenzyme A. The mode of inheritance is autosomal recessive, but many cases present de novo mutations, either in the PANK2 gene or other unknown ones – in which case patients are referred to as idiopathic NBIA. Clinical presentation of the late-onset of the disease, 'atypical PKAN', is in the second or third decade with extrapyramidal symptoms (dystonia and rigidity), speech abnormalities, and psychiatric and cognitive problems which later progress to dementia. Pigmented retinal degeneration is commonly found. Treatment with pantothenate (vitamin B5) may possibly be beneficial.

Aceruloplasminemia is a rare autosomal recessive disorder occurring mostly in Japan, leading to an absence of ceruloplasmin presenting in adulthood. Abnormal deposition of iron occurs in the pancreas (leading to diabetes), liver, retina and the basal ganglia. The clinical triad of findings consists of retinal degeneration, diabetes mellitus, and neurologic symptoms. Among the latter, movement disorders and ataxia are most common, and cognitive problems leading to dementia occur in ~25%. Treatment is with chelating agents like desferrioxamine.

Neuroimaging Findings

In PKAN and idiopathic NBIA, iron deposition leads to bilateral low signal intensity in the globus pallidus, to which in later phases gliosis and necrosis are added, leading to central high signal on T2-weighted images. The combination of low signal surrounding an area of high signal in the medial globus pallidus is referred to as the 'eye-of-the-tiger' phenomenon (Fig. 5.52),

which is closely associated with mutations in the *PANK2* gene – with apparently all PANK2-mutation–positive patients showing the eye-of-the-tiger sign.

In aceruloplasminemia, areas with low signal on T2*-weighted MRI of the brain due to iron deposition can be found in the basal ganglia (including globus pallidus), midbrain nuclei and dentate. Low signal on all sequences, but especially T2-weighted images, is evident on abdominal MRI scans in visceral organs like the liver, spleen and pancreas (as in secondary haemochromatosis)

Differential Diagnosis and Work-up

In patients suspected of NBIA, other disorders to consider are MSA and PSP, and Wilson's disease. Beyond genetic testing, diagnosis is based on clinical findings (retinal degeneration, diabetes and neurological findings), laboratory findings (serum copper, ceruloplasmin, and iron), and neuroimaging findings. Especially in those cases where a genetic mutation cannot be found (~50% of NBIA), the combination of clinical and radiological findings is important.

5.5.4 Dentatorubral-Pallidoluysian Atrophy

Synonyms

Haw River syndrome; Naito-Oyanagi disease

Dentatorubral-pallidoluysian atrophy (DRPLA, OMIM #125370) is an autosomal dominant trinucleotide repeat disorder bearing some similarity to Huntington's disease (HD). It is a rare disorder with most cases having been identified in Japan (prevalence there being ~0.5/100,000). While presentation in younger patients is dominated by myoclonus and epilepsy, adult-onset cases present with ataxia, choreoathetosis, and dementia. Diagnosis is based on clinical findings, family history and genetic testing. Differential diagnosis includes HD and the spinocerebellar ataxias.

DRPLA is characterized neuropathologically by severe generalized brain atrophy. Accumulation of (probably toxic) mutant atrophin-1 leads to intranuclear neuronal inclusions. Atrophy is found in the globus pallidus and subthalamic nucleus

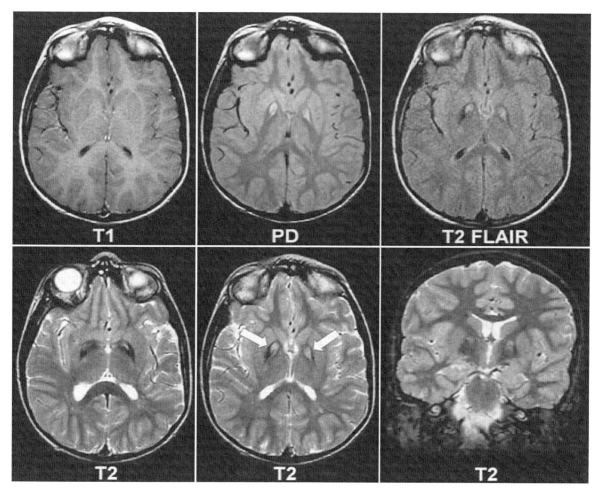

Fig. 5.52 Eye-of-the-tiger sign. This 10-year-old boy with the PANK2 mutation shows MR features characteristic of PKAN. The T1-weighted images appear normal, but high signal intensity can be seen in the globus pallidus on proton density (PD), FLAIR, and T2-weighted images in both the axial and coronal planes. With increasingly heavier T2 weighting (PD/FLAIR/ T2), there is increasingly conspicuous T2 hypointensity at the periphery because of the magnetic susceptibility effects of excess iron, producing the 'eye-of-the-tiger' appearance (*arrows*). (Reprinted with permission from Susan Hayflick (2006) AJNR Am J Neuroradiol. Jun–Jul; 27(6):1230–1233)

(also called the Body of Luys) with associated gliosis. Atrophy also occurs in the dentate and red nucleus.

5.5.4.1 Neuroimaging Findings

Atrophic changes on MRI can be found in cerebellum, brainstem and thalamus. The tegmentum of the pons is particularly affected in DRPLA. The age at MRI and the number of CAG repeats correlate with the degree of atrophy – as in other CAG repeat syndromes increase in expansion length occurs in offspring, a phenomenon referred to as anticipation. Diffuse white matter, thalamic and brainstem hyperintensities can be seen on T2-weighted MRI in patients with adult-onset DRPLA of long duration (Fig. 5.53). Intrafamilial heterogeneity of DRPLA is evident on MRI, as is true for the clinical phenotype. Specific changes in the globus pallidus and subthalamic nucleus can be detected only by means of quantitative MRI (ADC mapping) or MR spectroscopy.

Fig. 5.53 FLAIR findings in DRPLA. This 60-year-old woman had a family history of DRPLA: the number of CAG repeats in the DRPLA gene on chromosome 12 was expanded to 59 (normal allele 10). Note prominent confluent hyperintensities in the periventricular white matter (**a**), and also the bilateral thalami (**b**, *arrow*), areas that are not commonly involved in DRPLA. In the midbrain, the fasciculi surrounding the red nucleus are remarkably hyperintense (**c**). The sagittal image shows hyperintensities in the corpus callosum, cerebral peduncles, decussation of the superior cerebellar peduncles, and the ventrolateral portions of the mesencephalopontine tegmentum, as well as a central region of the pontine base (**d**). (Reprinted with permission from J Neurol Neurosurg Psychiatry. 1998; 65(3):396–399)

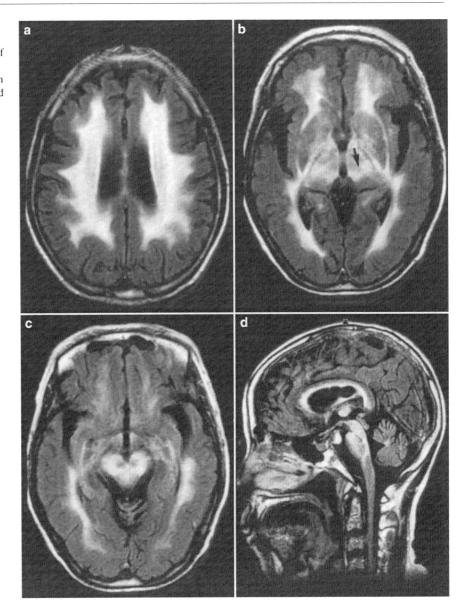

5.6 Prion-Linked Dementias

5.6.1 Introduction – History, Nosology and Transmission

The first transmissible spongiform encephalopathy (TSE) to be recognized was scrapie. This disease of sheep and goats has been known for more than two centuries (the name is derived from the fact that sheep with this disease have a severe itch and scrape off their fleece by rubbing). Icelandic farmers deduced that the disease was infectious and in 1936 scrapie was transmitted by injecting infected sheep brain into uninfected sheep.

The first clinical-pathological description of a TSE in humans was delivered by Jakob (1921), who believed, probably erroneously, that Creutzfeldt had previously described a similar case (1920). The name Creutzfeldt–Jakob disease (CJD) was coined by Spielmayer in 1922. In 1959, Klatzo et al. noted the similarities between the neuropathology seen in CJD and a disease called kuru, a non-inflammatory spongiform encephalopathy found among the Fore people of the eastern highlands of Papua, New Guinea. In 1966,

Gajdusek described transmission of kuru to subhuman primates by intracerebral inoculation of brain tissue, and was awarded the Nobel prize for recognizing this new type of disease in 1976. The proteinaceous infectious particle (PrP) is, in fact, a modified protein (PrP^{sc}). The Nobel prize for medicine in 1998 was awarded to Prusiner, who coined the name prion.

TSE constitutes a group of diseases in humans and animals presenting with similar clinical and neuropathological (spongiform encephalopathy) features. Having in common a long incubation period, they were formerly referred to as slow-virus diseases. In humans, TSE include:

• Creutzfeld–Jakob disease (CJD)	– Sporadic, familial, and iatrogenic
• Variant CJD (vCJD)	– Sporadic (and possibly iatrogenic)
• Gerstmann–Sträussler–Scheinker (GSS) disease	– Only familial
• Fatal Familial Insomnia (FFI)	– Familial and sporadic
• Kuru	– Only sporadic

In animals, TSE encompasses scrapie, transmissible mink encephalopathy, ungulate spongiform encephalopathy ('Chronic wasting disease' in deer and elk), feline spongiform encephalopathy and bovine spongiform encephalopathy (BSE) or 'mad cow disease'. Some of the animal forms are transmissible to humans, e.g. transmissible mink encephalopathy (USA, Russia, Canada) and BSE. With the advent of vCJD, BSE has become known for its transmission to humans on a relatively large scale, now also beyond the United Kingdom. More recently there has been concern about transmission via blood transfusion from donors who may be incubating vCJD. No treatment exists for any prion disease, but diagnosis is important to prevent transmission.

5.6.2 Biochemistry, Genetics and Histopathology

A remarkable feature of human TSE is that both infectious and inherited forms exist. Of all cases of CJD, 15% are inherited in an autosomal dominant way, whereas for GSS syndrome, 100% of cases are inherited. The sporadic, genetic and infectious forms are all due to abnormalities of a 33–35 kDa sialoglycoprotein, the prion protein (PrP). Normal PrP is a cell surface glycoprotein expressed at particularly high levels in nerve cells with unknown function. The infectious particle is composed largely, if not exclusively, of the abnormal isoform PrP^{sc}. Normal cellular PrP molecules convert into pathogenic isoforms (PrP^{sc}) through configurational changes. There is a post-translational modification of the abnormal PrP^{sc}, by formation of an alpha-sheet configuration for a greater part of the molecule with diminished alpha-helical configuration giving rise to four different PrP^{sc} molecular strain types upon Western blot analysis based on type of glycosilation – those subtypes are partly linked with the codon 129 type of homozygosity (M or V).

The prion protein gene maps to the short arm of chromosome 20 and is designated PRNP gene. The methionine to valine (M-V) polymorphism at PRNP codon 129 influences disease expression not only in inherited CJD cases but also in sporadic and iatrogenic forms. In Caucasian patients with sporadic forms of CJD, 95% or more are homozygous for M or V at codon 129. In the normal population the proportions are 12% V/V, 37% M/M and 51% M/V. The vast majority of patients with variant CJD have been M129 homozygous. However, there has been a report of a case of clinically probable vCJD who was heterozygous (MV) at codon 129 – raising the possibility of a second wave of vCJD cases.

Some mutations in the PRNP gene show a close association with particular clinical forms of prion-related disease, but there are also dissociative effects: for example a mutation at codon 178 (replacing asparagine by aspartic acid) in combination with M129 homozygosity is linked to both FFI and a typical CJD syndrome. For unknown reasons, the neuropathology in FFI is largely confined to the mediodorsal and anterior ventral nuclei of the thalamus, whereas in familial CJD the neuropathology is widespread over the cerebral cortex and subcortical nuclei.

GSS syndrome, like CJD, is not a single disease. There are at least six different disorders, each with a separate clinical presentation, neuropathological picture and mutation of the PRNP gene. The codon 102 mutation (P102L) has been linked to the family first described by Gerstmann, Sträussler and Scheinker. The forms linked to the other mutations were named after this syndrome for the presence of the typical GSS-type multicentric prion protein amyloid plaques.

The main histological abnormality is the vacuolated neuron in scrapie and vacuolation of the neuropil between nerve cells in CJD – hence the name spongiform encephalopathy. The gross appearance of the brain in CJD is variable and not diagnostic. There may be cortical, striatal and cerebellar atrophy, with brain weights usually below normal. Histologically, there is a spectrum of intensity and distribution of lesions, and various subtypes have been distinguished histologically: cortical, cortico-striatal with or without visual loss, cortico-striato-cerebellar, corticospinal and corticonigral. It is assumed that the differences are the result of the different isoforms of PrP involved, although this has not been proven.

The prominent pathological features of CJD are spongiform degeneration of neurons, neuronal loss, intense reactive astrogliosis and amyloid plaque formation; the final proof is abnormal PrP staining. The vacuoles are located in the neuropil between nerve cell bodies. This spongiform degeneration can be found in the cerebral neocortex, the subiculum of the hippocampus, putamen, caudate nucleus, thalamus and the molecular layer of the cerebellar cortex. It is minimal to absent in the globus pallidus, Ammon's horn, and the dentate gyrus of the hippocampus, brainstem and spinal cord. Spongiform degeneration of the cerebral cortex occurs in nearly all cases, regardless of the clinical presentation. The amount of vacuolation can vary widely from region to region. The degree of reactive astrocytosis corresponds well with the loss of nerve cells. In CJD and in the other prion diseases, changes in the white matter are mostly secondary to grey matter changes. Particularly in Japan, however, cases have been described with white matter vacuolation. The vacuoles are within the myelin sheath and occasionally axonal. In a minority of CJD cases (5–10%) amyloid plaques are found. These plaques are immunoreactive with PrP antibodies, but are negative with antibodies to beta-amyloid. In CJD, they consist of spherical masses with radiating amyloid spicules at their periphery and are referred to as 'florid' (kuru) plaques. In the differential diagnosis, presenile (frontotemporal) dementia with motor neuron disease is most important, because in that disease vacuolation of the cerebral cortex is

found, whereas in many cases of CJD the lower motor neuron is also involved.

In variant CJD, the findings differ from sporadic CJD. There are numerous florid (kuru-like) amyloid plaques throughout the cerebral and cerebellar cortex, with prominent involvement of the basal ganglia, and posterior nuclei of the thalamus. The amyloid plaques are surrounded by spongiform degeneration. Contrary to sporadic CJD, the lymphoid tissues throughout the body are infected in variant CJD, e.g. the tonsils. In FFI, spongiform lesions are most prominent in the thalamus. In GSS syndrome, bizarre multicentric amyloid plaques are found in the molecular layer of the cerebellar cortex. Plaques can also be found in the basal ganglia and cerebral cortex. The plaques are multicentric with a central larger mass of amyloid surrounded by smaller satellite amyloid deposits. The degree of spongiform changes may be limited, while GSS also features neurofibrillary tangle formation similar to Alzheimer's disease.

5.6.3 Creutzfeldt–Jakob Disease

Sporadic CJD accounts for 85% of human prion-related disease. There are some distinctive variants of CJD: the Heidenhain variant with early cortical blindness and prominent occipital atrophy, the Brownell-Oppenheimer variant with severe cerebellar signs and atrophy but little cognitive impairment, the Stern-Garcia variant with extrapyramidal signs due to prominent involvement of the thalamus and basal ganglia presenting; the panencephalitic (Japanese) variant with disproportionate involvement of the white matter; and finally a poliodystrophic form.

5.6.3.1 Clinical Presentation, Epidemiology and Treatment

Sporadic CJD usually presents in the fifth to seventh decade, but cases with younger and later onset have been described. Worldwide, CJD has an incidence of one per million persons. It is sporadic in 85–90% of cases, inherited in 10–15% of cases and transmissible in less than 5% of cases. The mode of presentation may differ depending on the molecular subtype. The most common (classical) presentation is a triad of subacute dementia, myoclonus and motor disturbances (extrapyramidal or cerebellar) with a characteristic electroencephalographic (EEG) pattern consisting of triphasic waves, and CSF abnormalities (increased CSF tau and/or 14-3-3 protein, the former being more sensitive). Not all these features need to be present, and CJD should be considered in every patient presenting with a rapidly progressive dementia.

In the early stages of sporadic CJD, the symptoms are often ambiguous, with fatigue, insomnia, restlessness and mild anorexia. In several weeks to months, the second stage follows with cognitive deterioration, cerebellar, behavioural and visual signs. The final stage is akinetic mutism, with an almost silent EEG pattern with 'burst suppression'. Median survival is only ~4 months and most patients die within 2 years of diagnosis (survival is influenced by strain type and PRNP 129 genotype). A definite diagnosis can only be obtained by biopsy, but a probable diagnosis according to WHO criteria can be made on the combination of at least two clinical features and typical EEG or CSF findings.

Familial prion disease (*inherited CJD*) accounts for 10–15% of prion disease; point mutations and insertional (repeat) mutations are common and the age at onset varies widely; disease duration is usually longer than in sporadic CJD. Particular phenotypes include fatal familial insomnia (FFI) and Gerstmann–Straussler–Scheinker disease (GSS) described below.

Patients with familial prion disease may also resemble familial AD or sporadic CJD.

Iatrogenic CJD may occur through tissue transfer (e.g. dura mater grafts or cadaveric human growth hormone) or via contaminated instruments used in cerebral operations (e.g. intracerebral electrodes). The incidence of iatrogenic CJD has fallen since the 1990s; remaining cases are due to long incubation periods (up to 40 years). The recent discovery of transfusion-associated variant CJD infections has provoked new concerns about the possibility of further secondary transmissions from operative procedures, blood transfusions and tissue donations.

Variant CJD is a new form of CJD that is associated with BSE, with a very different presentation and different histopathology. Clinical presentation includes prominent psychiatric features at onset, soon followed by dementia but lacking cerebellar signs and occurring at a young age (median age 29 years) and having a longer survival. The salient differences between sporadic and vCJD are summarized in Table 5.15.

5.6.3.2 Imaging Strategy and Findings

The first line of investigation in patients suspected of CJD is MRI, including diffusion-weighted imaging (DWI); images after contrast-material administration

Table 5.15 Clinical and MRI findings distinguishing Classic CJD from Variant CJD

Characteristic	Classic CJD	Variant CJD
Median age at death	68 years	28 years
Median duration of illness	4–5 months	13–14 months
Clinical signs and symptoms	Dementia; early neurologic signs	Prominent psychiatric/behavioural symptoms; painful dysaesthesiasis; delayed neurologic signs
Periodic sharp waves on EEG	Often present	Often absent
Presence of agent in lymphoid tissue	Not readily detected	Readily detected (tonsillar biopsy is a useful diagnostic procedure)
'Pulvinar sign' on MRI on FLAIR/DWI	Rarely	Present in >75% of cases
Caudate/putamen signal on FLAIR/DWI greater than thalamus		Characteristic – unilateral or bilateral
Diffusion changes in neocortex		Often – multifocal

Source: Adapted from Belay E, Schonberger L (2002). Variant Creutzfeldt-Jakob Disease and Bovine Spongiform Encephalopathy. Clin Lab Med 22:849–862

Fig. 5.54 CJD with striatal involvement. This young man developed iatrogenic CJD more than 10 years after receiving human growth hormone. The coronal FLAIR (*left*) and axial diffusion imaging (B1000 – *right*) shows marked striatal involvement characteristic of sCJD and iatrogenic CJD

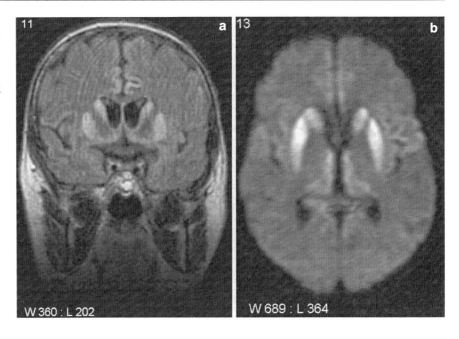

are only useful to rule out alternative disorders. Conventional T2-weighted and especially FLAIR sequences may show increased signal in the striatum, especially the putamen (Fig. 5.54) or the neocortex. However, in up to 20% of cases no abnormal signal is found and a normal conventional MR scan thus does not rule out CJD and may demand the use of other sequences, such as DWI to show striatal or cortical involvement. Abnormal DWI may be especially prominent in early phases of the disease when vacuoles are small, leading to restricted diffusion (which may disappear in late stages). Contrast injection should be performed to rule out other disorders, but will usually not give new information in CJD itself. If the initial imaging is negative, repeat MRI after several weeks is indicated, with emphasis on DWI (which may be progressively abnormal in prion disease).

In spongiform encephalopathies, the patterns of selective involvement of cerebral structures include involvement of either basal ganglia (caudate and putamen – Fig. 5.54), thalamus (medial and dorsal parts) or widespread, irregular involvement of the cerebral

cortex (Box 5.6). Sometimes, there is a combination of neocortical and subcortical involvement. Usually, the pattern of involvement is bilateral, but it may be unilateral as well (initially). In sporadic (and iatrogenic) CJD, involvement of either the striatum (Fig. 5.54) or neocortex (Fig. 5.55) or both is found, with possible determination of the imaging pattern based on the PrP subtype. In vCJD, there is selective involvement of the medial and dorsal (pulvinar) thalamic nuclei, leading to the so-called hockey-stick sign (Fig. 5.56). This pulvinar sign has high sensitivity and specificity for the diagnosis of vCJD. Some atypical cases of probable sporadic CJD with a pulvinar sign have been reported (Fig. 5.57), and the combination of abnormalities may become more evident when DWI is obtained using higher B-values than normal (Fig. 5.58). MR spectroscopy is non-specific, and only in the later phases shows the pattern of brain damage, with lowering of the metabolites, especially N-acetylaspartate, with sometimes high myo-inositol. SPECT may reveal non-specific areas of lower tracer uptake in neocortical areas, with limited specificity.

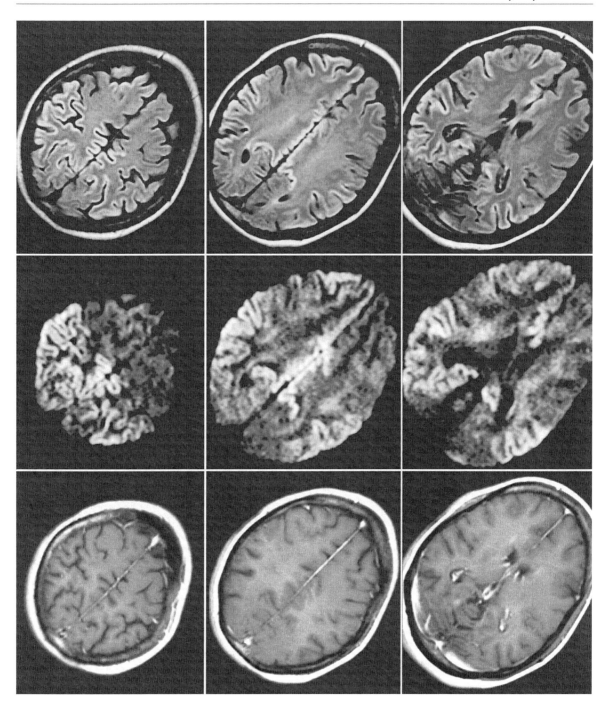

Fig. 5.55 This 39-year-old woman presented with rapid deterioration over half a year, with agitation, hyperactivity, confusion and was initially diagnosed as a psychiatric disorder. EEG findings, CSF analysis and abnormal MRI findings led to the diag-nosis of sCJD. Note the increased signal of multiple cortical areas on FLAIR (*top row*) which are more conspicuous on DWI (*middle row*). As expected there was no enhancement with gadolinium (*bottom row*)

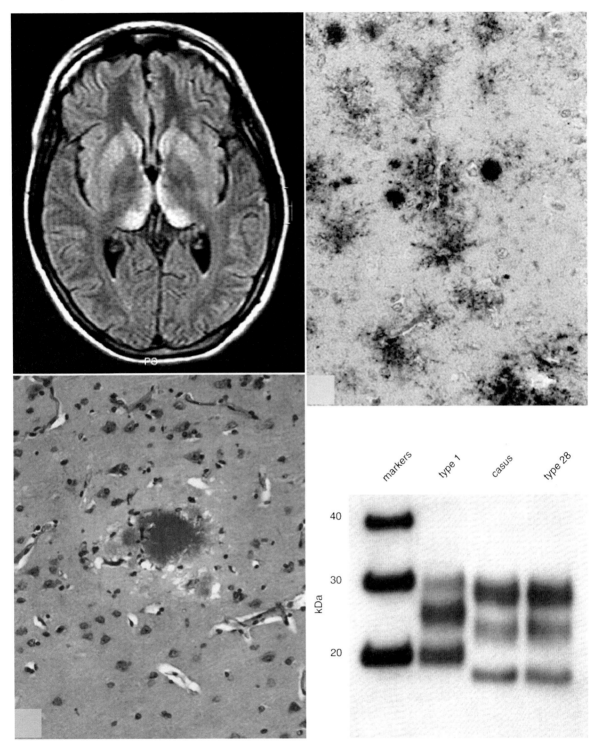

Fig. 5.56 Pulvinar sign in vCJD. A Dutch woman died at the age of 26 years, after a disease duration of 18 months, due to variant CJD. She had never travelled to the UK and there was no history of potential iatrogenic exposure. However, she had worked in the catering and food production industry for the previous 6 years and had frequently consumed raw meat. The disease course showed the classical clinical picture of vCJD, with high signal on FLAIR in the pulvinar on MRI (*upper left*). Post-mortem examination of the brain showed florid plaques with surrounding spongiform changes (*lower left*) and marked immunoreactivity for prion protein (*upper right*). Western blot analysis revealed a protease-resistant prion protein with a glycosylation pattern similar to variant CJD (type 2B) and dissimilar to sporadic CJD (type 1) (Reprinted with permission from Ned Tijdschr Geneeskd 2005;149:2949–2954)

Fig. 5.57 Pulvinar sign in sporadic CJD. Fluid-attenuated inversion recovery (**a**) and diffusion-weighted imaging (**b**) showing a typical pulvinar sign (*arrows*) in a 64-year-old man with a final clinical diagnosis of sporadic CJD. He was initially thought to have possible variant CJD but had a negative tonsillar biopsy. The same imaging 7 months later (**c** and **d**) shows increased signal intensity in the striata bilaterally and disappearance of the pulvinar sign. (Reprinted with permission from Lancet 2010; 375:889–890)

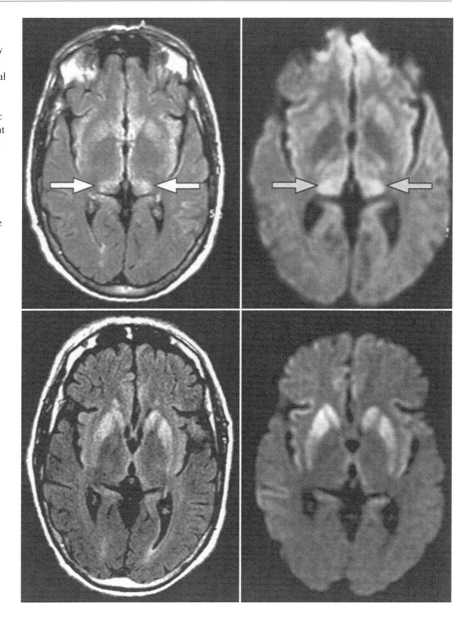

Box 5.6 Key imaging findings in prion disease

Sporadic CJD	Abnormal signal on FLAIR/DWI
	(a) Caudate and putamen signal increase
	(b) Widespread (patchy, ribbon-like) neocortical signal increase
	(c) A combination of the above
Variant CJD	Thalamic hockey-stick or pulvinar sign on FLAIR/DWI
GSS	Abnormal MRI signal in striatum, thalamus or cortex
FFI	Normal MRI and reduced thalamic FDG uptake on PET

... No contrast enhancement in any ...

Fig. 5.58 Comparison of imaging features with different pulse-sequences in a case of sporadic CJD. Note that the mixed pattern of neocortical, thalamic and striatal involvement becomes more prominent on DWI with higher than usual B-values (3,000 rather than 1,000 s/mm^2). (Images kindly provided by Harpreet Hyare, UCL)

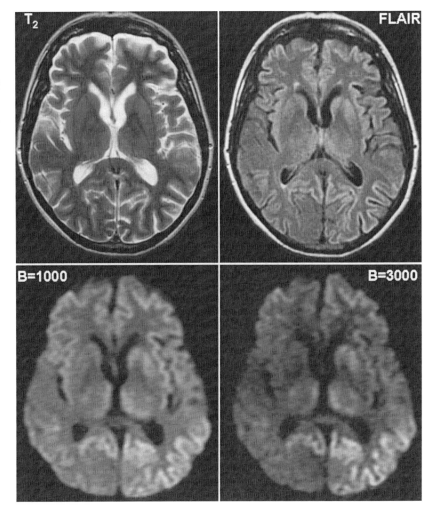

5.6.3.3 Differential Diagnosis and Ancillary Investigations

In patients with a rapidly progressive dementia, several conditions need to be considered, including vasculitis, limbic encephalitis, infections and malignancy. Other degenerative conditions include FTLD with motor-neuron disease and also atypical AD and DLB. Laboratory or other findings that can be helpful include:

- Periodic sharp waves on EEG
  - Also found in herpes encephalitis, SSPE
  - Can be introduced by drugs
  - Lithium, valproate and tricyclic antidepressants
  - Occur in metabolic and post-hypoxic disorders

- 14-3-3 protein is present in the CSF in >90% of sporadic CJD cases
  - Only in ~50% of variant cases
  - Also found in other rapidly progressive neurodegenerative disorders, including AD
- CSF tau is often markedly (>2,000 pmol/l) raised in CJD
  - Much more than in other disorders, e.g. AD
- Tonsilar biopsy in variant CJD shows prion staining
- Brain biopsy may be needed to exclude cerebral vasculitis

MRI abnormalities, especially abnormal DWI probably antedate the onset of typical EEG findings. A pattern of multifocal neocortical DWI abnormalities can be found in mitochondrial disease (e.g. MELAS), herpes infection, venous congestion due to arteriovenous fistula, and perhaps rapidly progressive AD as well.

5.6.4 Gerstmann–Sträussler–Scheinker (GSS) Syndrome

GSS syndrome is a rare familial prion disease. It was first described in a Viennese family in 1936. All cases are inherited via an autosomal dominant mode of inheritance. Worldwide more than 30 families have now been recognized with the disease. Clinically, GSS syndrome presents in the fourth to sixth decades and shows a slow progression over about 6 years. The major symptoms are difficulty in walking and unsteadiness, with variable leg pain and paraesthesias in the early stages, later followed by mental and behavioural deterioration, ataxia and dysarthria.

MRI may only show non-specific atrophy and abnormal signal on DWI in the cortex and striatum, as in sporadic CJD, but also the thalamus. Serial MRI may show progressive cerebellar atrophy as well as cortical loss (Fig. 5.59). Reduced NAA has been reported in spectra from the frontal lobe and cerebellum. SPECT may show reduced tracer uptake, though not in the expected cerebellar location. EEG shows non-specific changes, and 14-3-3 protein is usually absent in CSF.

5.6.5 Fatal Familial Insomnia (FFI)

FFI is an autosomal dominant inherited form of (thalamic) dementia, with some pathological features of CJD. Insomnia, rather than dementia, is the prominent clinical symptom, although this feature may be difficult to apprehend since patients may show excessive daytime sleepiness. A wide variety of neurological features, both pyramidal and extrapyramidal, may occur and the course is fatal in a short period of time. Sporadic cases of FFI occur as well.

MRI may show non-specific atrophy, but no abnormal signal, even in the thalamus. Similarly, EEG shows non-specific alterations and the CSF is negative for 14-3-3 protein. PET is especially important in the diagnosis of FFI, with absent thalamic uptake despite normal appearance on MRI – a finding that is virtually pathognomonic (Fig. 5.60).

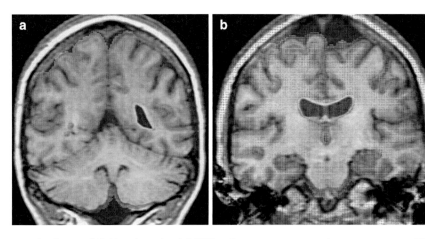

Fig. 5.59 Colour overlay image of fluid-registered serial MRI (2 years apart) of familial prion (GSS) case. This 42-year-old woman with a family history of dementia developed balance problems, a broad-based ataxic gait, dysarthria and minor subtle cognitive and behavioural problems. Hereditary prion dementia (GSS) was confirmed by finding a point mutation at codon 102 of the prion protein gene (PrP Leu 102). The fluid overlays are from two scans spanning symptom onset (the first was acquired when the patient was asymptomatic as part of a research study in at-risk individuals) The figure shows diffuse atrophy involving (**a**) cerebellum, but (**b**) clear preservation of hippocampi (areas of loss are shown in blue/green and areas of expansion are shown in red/yellow e.g. ventricular expansion)

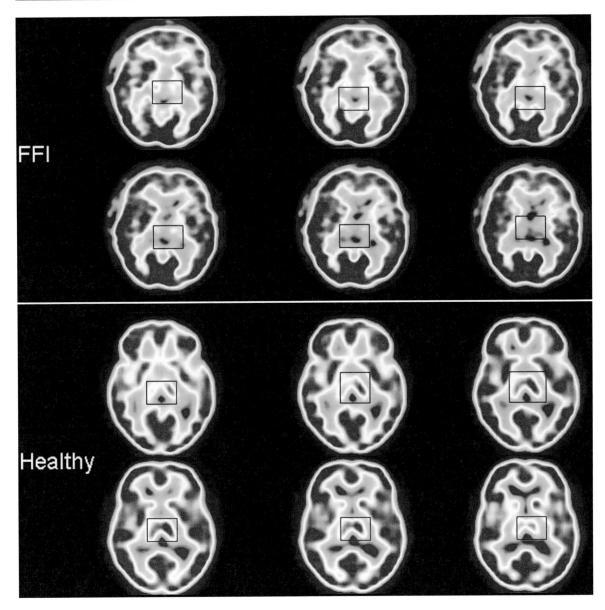

Fig. 5.60 Thalamic hypometabolism in sporadic FFI. FDG-PET images of a 40-year-old male patient who presented with excessive daytime sleepiness, because of nocturnal insomnia that was discovered only later, followed by progressive neurological deficits, such as pyramidal and extrapyramidal features, severe aphasia and ultimately mutism. Tonsillar biopsy was negative. FDG-PET revealed strongly reduced thalamic tracer uptake consistent with FFI (*upper rows*). For comparison, note normal metabolism in the thalamus in a healthy control (*lower rows*)

5.7 Recreational Drugs and Alcohol

5.7.1 Introduction

There are many recreational drugs that have an impact on the central nervous system beyond their presumed recreational effects. In this chapter, we focus on the grey matter damage caused by two more commonly used drugs, ecstasy and alcohol. Some effects of recreational drugs manifest as white matter disease in the brain: Wernicke-Korsakoff syndrome due to B1-deficiency, Machiafavi–Bignami disease and heroin-induced leukencephalopathy. Those disorders will be discussed in Chap. 7. The main findings in recreational substance abuse include:

Alcohol
- Global brain atrophy
- (Hypo)thalamic changes in Wernicke–Korsakoff syndrome
- Cerebellar atrophy
- Corpus callosum changes in Machiafavi–Bignami disease

Cannabis, XTC, etc.
- Hippocampal and amygdala atrophy

Smoking
- Non-specific white matter lesions
- Reduced grey matter density

5.7.2 Alcohol-Induced Dementia

The effect of alcohol on the brain is partly controversial. Population-based studies suggest a possible beneficial effect of moderate alcohol intake on the brain through positive effects on the cardiovascular system – it is even suggested that onset of AD may be delayed. Excessive intake of alcohol over longer periods of time clearly leads to neurological damage and eventually dementia. Memory loss is a central feature of both the Wernicke-Korsakoff syndrome and alcohol-induced dementia. The differential diagnosis between these two entities is possible using neuropsychological testing, with Wernicke-Korsakoff patients having isolated disturbances in encoding, while alcohol-dementia patients display a more global cognitive decline. Diagnosis of alcohol-induced dementia has been operationalised in the DSM-IV (Box 5.7).

5.7.2.1 Imaging Findings in Alcohol Abuse

Intake of alcohol has several effects on the brain, which are partly reversible. Following abstinence, an increase in brain volume has been found (1–2%), with an increase in NAA in spectroscopy, signifying that alcohol-induced changes are partly reversible (Fig. 5.61a). In the long run, irreversible atrophy occurs, with a particular vulnerability of the brainstem, cerebellum, hippocampus, thalamus and frontal lobes (Fig. 5.61b). The findings related to Wernicke-Korsakoff syndrome due to thiamine deficiency are discussed in detail in Chap. 7.5.2,

---

Box 5.7 DSM-IV criteria for alcohol-induced persistent dementia

1. The development of multiple cognitive deficits manifested by both:
   (a) Memory impairment (impaired ability to learn new information or to recall previously learned information)
   (b) One (or more) of the following cognitive disturbances
      - Aphasia (language disturbance)
      - Apraxia (impaired ability to carry out motor activities despite intact motor function)
      - Agnosia (failure to recognize or identify objects despite intact sensory function)
      - Disturbance in executive functioning (i.e. planning, organizing, sequencing, abstracting)
2. The cognitive deficits in criteria A1 and A2 each cause significant impairment in social or occupational functioning and represent a significant decline from a previous level of functioning
3. The deficits do not occur exclusively during the course of a delirium and persist beyond the usual duration of substance intoxication or withdrawal
4. There is evidence from the history, physical examination or laboratory findings that deficits are etiologically related to the persisting effects of substance use (e.g. a drug of abuse, a medication)

---

and include high signal lesion in medial thalamus, mamillary bodies and midbrain.

5.7.3 Marchiafava–Bignami Disease

Marchiafava–Bignami disease is a rare complication of chronic alcohol consumption and is characterised by primary demyelination and necrosis of the central part of the corpus callosum. Clinical features include

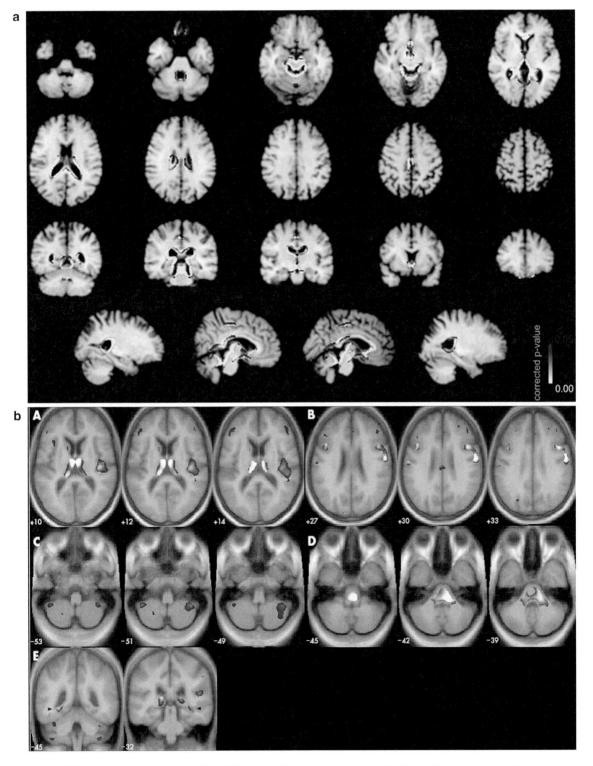

Fig. 5.61 (**a**) Increase in brain volume (*yellow*) following withdrawal in chronic alcoholics. (Reproduced with permission from Bartsch et al. 2007). (**b**) VBM study of brain volume loss in chronic alcoholism. Changes are most prominent in thalamus and brainstem. (Reproduced with permission from Mechtcheriakov et al. 2007)

Fig. 5.62 Marchiafava–Bignami disease. Classic finding of layered necrosis, degeneration and cystic cavitations of the corpus callosum (*blue arrows*). In addition, extensive involvement of the dorsal part of the external capsule (*red arrows*) was seen in this patient. (With permission from Geibprasert et al. 2009 Eur Radiology)

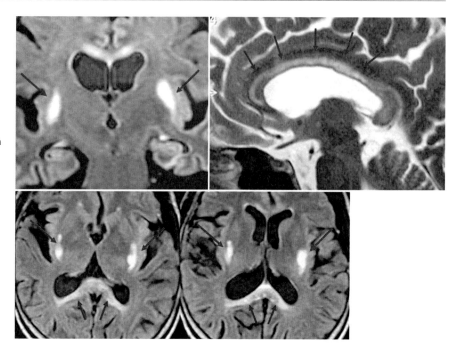

a variety of cognitive and systemic neurological features. Although the true aetiology of this rare condition is still unknown, toxic agents in low quality red wine and/or vitamin B complex deficiencies have been put forward as potential causes. The characteristic MR imaging findings are high T2 signal without significant mass effect within the corpus callosum, which may extend to the genu and adjacent white matter (Fig. 5.62).

Ecstasy (active ingredient 3, 4-methylenedioxymethamphetamine or MDMA) is also widely used for recreational purposes. In animal studies, MDMA causes damage to axons of serotonergic cells. Neuroimaging studies on this relatively novel drug are scarce and confounded by polydrug usage. In frequent users, reduced serotonergic activity has been found using SPECT, while more widespread damage in grey and white matter have been found using diffusion tensor imaging.

5.7.4 Cannabis, Ecstasy and Cocaine

Cannabis (active ingredient Δ9-tetrahydrocannabinol or THC) is often considered to be a harmless stimulant and widely used for recreational purposes. In animal studies, however, cannabis has a neurotoxic effect, especially to the hippocampus. Neuroimaging studies on the effect of cannabis in the brain have yielded conflicting results, but do suggest that hippocampal and amygdalar volumes in heavy cannabis users (who are not polydrug addicts) are reduced. Such a reduction may not only affect memory, but also lower the threshold for psychosis. In fact, in patients with schizophrenia, the rate of grey matter decrease, compared to controls, is most marked in those who use cannabis.

5.7.5 Nicotine

More than 3 decades of research indicates that smoking has both acute and chronic effects on cognition. Difficulty in concentrating is part of nicotine withdrawal and a likely barrier to success in smoking cessation attempts; most of the neuropsychological research using fMRI has focused on cognitive domains generally classified as 'executive functions' including sustained attention and working memory. Studies of smoking abstinence have identified functional brain correlates of increased reactivity to smoking-related cues, and worsening of concentration.

Smoking has not been identified as a cause of dementia, but being a significant cardiovascular risk factor, smoking has been associated with white matter changes on MRI in many population and clinical studies.

Structural MRI studies comparing smokers and non-smokers indicate that smokers have reduced grey matter volume and density in prefrontal cortical areas involved in executive function. Research on the effects of smoking a cigarette confirms that smoking leads to the release of dopamine in brain reward areas and to nicotinic receptor binding. Smokers' and non-smokers' brains have also been shown to differ in the distribution of nicotinic receptors. Among nicotine-naïve individuals, the $\alpha4\beta2$ nicotine receptor has high densities in thalamus, followed by midbrain, pons, cerebellum and cortex. Comparisons of non-smokers and recently abstinent smokers using SPECT/PET show that smokers have higher densities of $\alpha4\beta2$ nicotinic receptors both in cerebral cortex and in striatum. Finally, compared to non-smokers, smokers have lower levels of brain monoamine oxidase (MAO). These lower levels of MAO are thought to be due to the ingestion of MAO-inhibiting compounds in cigarette smoke and may play a role – either independently or in synergy with nicotine – in smoking reinforcement.

Suggested Reading

Alzheimer's Disease

Frisoni GB, Fox NC, Jack CR Jr, Scheltens P, Thompson PM (2010) The clinical use of structural MRI in Alzheimer disease. Nat Rev Neurol 6:67–77

Henneman WJP, Sluimer JD, Barnes J et al (2009) Hippocampal atrophy rates in Alzheimer disease: added value over whole brain volume measures. Neurology 72:999–1007

Jack CR, Knopman DS, Jagust WJ, Shaw LM, Aisen PS, Weiner MW, Petersen RC, Trojanowski JQ (2010) Hypothetical model of dynamic biomarkers of the Alzheimer's pathological cascade. Lancet Neurol 9:119–128

Perrin RJ, Fagan AM, Holtzman DM (2009) Multimodal techniques for diagnosis and prognosis of Alzheimer's disease. Nature 461:916–922

Rinne JO, Brooks DJ, Rossor MN et al (2010) 11C-PiB PET assessment of change in fibrillar amyloid-beta load in patients with Alzheimer's disease treated with bapineuzumab: a phase 2, double-blind, placebo-controlled, ascending-dose study. Lancet Neurol 9(4):363–72

Scheltens P, Fox N, Barkhof F, De Carli C (2002) Structural magnetic resonance imaging in the practical assessment of dementia: beyond exclusion. Lancet Neurol 1(1):13–21

Tolboom N, Yaqub M, van der Flier WM, Boellaard R, Luurtsema G, Windhorst AD, Barkhof F, Scheltens P, Lammertsma AA, van Berckel BN (2009) Detection of Alzheimer pathology in vivo using both 11C-PIB and 18F-FDDNP PET. J Nucl Med 50:191–197

Frontotemporal Lobar Degeneration

Adlam AL, Patterson K, Rogers TT, Nestor PJ, Salmond CH, Acosta-Cabronero J, Hodges JR (2006) Semantic dementia and fluent primary progressive aphasia: two sides of the same coin? Brain 129(Pt 11):3066–80

Cairns NJ, Bigio EH, Mackenzie IR et al (2007) Neuropathologic diagnostic and nosologic criteria for frontotemporal lobar degeneration: consensus of the Consortium for Frontotemporal Lobar Degeneration. Acta Neuropathol (Berl) 114(1):5–22, Epub 2007 Jun 20

Chan D, Fox NC, Scahill RI, Crum WR, Whitwell JL, Leschziner G, Rossor AM, Stevens JM, Cipolotti L, Rossor MN (2001) Patterns of temporal lobe atrophy in semantic dementia and Alzheimer's disease. Ann Neurol 49(4):433–42

Gorno-Tempini ML, Dronkers NF, Rankin KP et al (2004) Cognition and anatomy in three variants of primary progressive aphasia. Ann Neurol 55(3):335–46

Mackenzie IR, Neumann M, Bigio EH et al (2010) Nomenclature and nosology for neuropathologic subtypes of frontotemporal lobar degeneration: an update. Acta Neuropathol 119(1):1–4

McKhann GM, Albert MS, Grossman M, Miller B, Dickson D, Trojanowski JQ, Work Group on Frontotemporal Dementia and Pick's Disease (2001) Clinical and pathological diagnosis of frontotemporal dementia: report of the Work Group on Frontotemporal Dementia and Pick's Disease. Arch Neurol 58(11):1803–9

Neary D, Snowden JS, Gustafson L et al (1998) Frontotemporal lobar degeneration: a consensus on clinical diagnostic criteria. Neurology 51(6):1546–54

Rohrer JD, Guerreiro R, Vandrovcova J et al (2009) The heritability and genetics of frontotemporal lobar degeneration. Neurology 73(18):1451–6

Rascovsky K, Hodges JR, Kipps CM et al (2007) Diagnostic criteria for the behavioral variant of frontotemporal dementia (bvFTD): current limitations and future directions. Alzheimer Dis Assoc Disord 21(4):S14–8

Dementia with Lewy Bodies and Parkinson's Disease Dementia

Aarsland D, Ballard CG, Halliday G (2004) Are Parkinson's disease with dementia and dementia with Lewy bodies the same entity? J Geriatr Psychiatry Neurol 17(3):137–145

McKeith IG, Dickson DW, Lowe J et al (2005) Diagnosis and management of dementia with Lewy bodies: third report of the DLB consortium. Neurology 65:1863–1872

McKeith I, O'Brien J, Walker Z et al (2007) Sensitivity and specificity of dopamine transporter imaging with I-123-FP-CIT SPECT in dementia with Lewy bodies: a phase III, multicentre study. Lancet Neurol 6(4):305–313

O'Brien JT, Paling S, Barber R, Williams ED, Ballard C, McKeith IG, Gholkar A, Crum WR, Rossor MN, Fox NC (2001) Progressive brain atrophy on serial MRI in dementia with Lewy bodies, AD, and vascular dementia. Neurology 56(10):1386–1388

Watson R, Blamire AM, O'Brien JT (2009) Magnetic resonance imaging in lewy body dementias. Dement Geriatr Cogn Disord 28(6):493–506

Whitwell JL, Weigand SD, Shiung MM et al (2007) Focal atrophy in dementia with Lewy bodies on MRI: a distinct pattern from Alzheimer's disease. Brain 130:708–719

Progressive Supranuclear Palsy (PSP)

Williams DR (2005) Characteristics of two distinct clinical phenotypes in pathologically proven progressive supranuclear palsy: Richardson's syndrome and PSP-parkinsonism. Brain 128(6):1247–1258

Schrag A (2000) Differentiation of atypical parkinsonian syndromes with routine MRI. Neurology 54(3):697–702

Oba H (2005) New and reliable MRI diagnosis for progressive supranuclear palsy. Neurology 64(12):2050–2055

Litvan I, Agid Y, Calne D et al (1996) Clinical research criteria for the diagnosis of progressive supranuclear palsy (Steele-Richardson-Olszewski syndrome): report of the NINDS-SPSP international workshop. Neurology 47:1–9

Multiple System Atrophy (MSA)

Ozawa T (2004) The spectrum of pathological involvement of the striatonigral and olivopontocerebellar systems in multiple system atrophy: clinicopathological correlations. Brain 127:2657–2671

Watanabe H (2002) Progression and prognosis in multiple system atrophy: an analysis of 230 Japanese patients. Brain 125:1070–1083

Seppi K (2005) How to diagnose MSA early: the role of magnetic resonance imaging18. J Neural Transm 112: 1625–1634

Huntington's Disease

Aylward EH, Sparks BF, Field KM et al (2004) Onset and rate of striatal atrophy in preclinical Huntington disease. Neurology 63:66–72

Bates G, Harper PS et al (eds) (2002) Huntington's disease. Oxford University Press, New York

Feigin A, Tang C, Ma Y et al (2007) Thalamic metabolism and symptom onset in preclinical Huntington's disease. Brain 130:2858–67

Paulsen JS, Zimbelman JL, Hinton SC et al (2004) fMRI biomarker of early neuronal dysfunction in presymptomatic Huntington's Disease. AJNR Am J Neuroradiol 25: 1715–1721

Rosas HD, Feigin AS et al (2004) Using advances in neuroimaging to detect, understand, and monitor disease progression in Huntington's disease. NeuroRx 1(2):263–72

Wild EJ, Tabrizi SJ (2007) Huntington's disease phenocopy syndromes. Curr Opin Neurol 20(6):681–7

Metal Metabolism Disorders

http://www.nbiadisorders.org/

Hayflick SJ, Westaway SK, Levinson B, Zhou B, Johnson MA, Ching KH, Gitschier J (2003) Genetic, clinical, and radiographic delineation of Hallervorden-Spatz syndrome. N Engl J Med 348:33–40

van Wassenaer-van Hall HN, van den Heuvel AG, Algra A, Hoogenraad TU, Mali WP (1996) Wilson disease: findings at MR imaging and CT of the brain with clinical correlation. Radiology 198(2):531–6

Prashanth LK, Sinha S, Taly AB, Vasudev MK (2010) Do MRI features distinguish Wilson's disease from other early onset extrapyramidal disorders? An analysis of 100 cases. Mov Disord 25(6):672–8

Dentatorubral-Pallidoluysian Atrophy

Koide R, Onodera O, Ikeudu T et al (1997) Atrophy of the cerebellum and brain stem in dentatorubral-pallidoluysian atrophy: influence of GAG repeat size on MR findings. Neurology 49:1605–1612

Tomiyasu H, Yoshii F, Ohnuki I, Ikeda JE, Shinohara Y (1998) The brainstem and thalamic lesions in dentatorubral-pallidoluysian atrophy: an MRI study. Neurology 50: 1887–1890

Prion-Linked Dementias

Collie DA, Sellar RJ, Zeidler M, Colchester AC, Knight R, Will RG (2001) MRI of Creutzfeldt-Jakob disease: imaging features and recommended MRI protocol. Clin Radiol 56(9):726–39

Hyare H, Thornton J, Stevens J et al (2010) High-b-value diffusion MR imaging and basal nuclei apparent diffusion coefficient measurements in variant and sporadic Creutzfeldt-Jakob disease. AJNR Am J Neuroradiol 31:521–6

Wadsworth JD, Hill AF, Beck JA, Collinge J (2003) Molecular and clinical classification of human prion disease. Br Med Bull 66:241–54

Macfarlane RG, Wroe SJ, Collinge J, Yousry TA, Jäger HR (2007) Neuroimaging findings in human prion disease. J Neurol Neurosurg Psychiatry 78:664–70

Tschampa HJ et al (2007) Radiological assessment of CJD. Eur Radiol 17:1200–11

Meissner B, Kallenberg K, Sanchez-Juan P et al (2009) MRI lesion profiles in sporadic Creutzfeldt-Jakob disease. Neurology 72:1994–2001

Ukisu R, Kushihashi T, Tanaka E, Baba M, Usui N, Fujisawa H, Takenaka H (2006) Diffusion-weighted MR imaging of early-stage Creutzfeldt-Jakob disease: typical and atypical manifestations. Radiographics 26(Suppl 1):S191–204

Recreational Drugs and Alcohol

Bartsch AJ, Homola G, Biller A, Smith SM, Weijers HG, Wiesbeck GA, Jenkinson M, De Stefano N, Solymosi L, Bendszus M (2007) Manifestations of early brain recovery associated with abstinence from alcoholism. Brain 130:36–47

Brody AL, Mandelkern MA, Jarvik ME, Lee GS, Smith EC, Huang JC et al (2004) Differences between smokers and nonsmokers in regional gray matter volumes and densities. Biol Psychiatry 55(1):77–84

Ernst M, Matochik JA, Heishman SJ, Van Horn JD, Jons PH, Henningfield JE et al (2001) Effect of nicotine on brain activation during performance of a working memory task. Proc Natl Acad Sci USA 98(8):4728–4733

Fowler JS, Logan J, Wang GJ, Volkow ND (2003) Monoamine oxidase and cigarette smoking. Neurotoxicology 24(1):75–82

Geibprasert S, Gallucci M, Krings T (2009) Alcohol-induced changes in the brain as assessed by MRI and CT. Eur Radiol 20(6):1492–1501, Epub ahead of print

Lingford-Hughes A (2005) Human brain imaging and substance abuse. Curr Opin Pharmacol 42(1):42–6, Review

McClernon FJ (2009) Neuroimaging of Nicotine Dependence: key findings and application to the study of smoking-mental illness comorbidity. J Dual Diagn 2:168–178

Mechtcheriakov S, Brenneis C, Egger K, Koppelstaetter F, Schocke M, Marksteiner JA (2007) Widespread distinct pattern of cerebral atrophy in patients with alcohol addiction revealed by voxel-based morphometry. J Neurol Neurosurg Psychiatry 78:610–614

Mukhin AG, Kimes AS, Chefer SI, Matochik JA, Contoreggi CS, Horti AG et al (2008) Greater nicotinic acetylcholine receptor density in smokers than in nonsmokers: a PET study with 2-18F-FA-85380. J Nucl Med 49(10): 1628–1635

Yücel M, Solowij N, Respondek C, Whittle S, Fornito A, Pantelis C, Lubman DI (2008) Regional brain abnormalities associated with long-term heavy cannabis use. Arch Gen Psychiatry 65(6):694–701

Contents

Synonyms

Vascular dementia (VaD), post-stroke dementia, multi-infarct dementia, strategic infarct dementia, arteriosclerotic dementia, Binswanger's disease, 'état lacunaire', subcortical ischaemic vascular dementia (SIVD)

6.1 Introduction

Vascular dementia (VaD) is the second most common type of dementia after AD, especially in the elderly. Recently, the scope of VaD has been broadened to include also more subtle cognitive dysfunction that can be caused by vascular damage to the brain, designated as 'vascular cognitive impairment'. The term VaD implies the existence of dementia, which is assumed to be secondary to cerebrovascular disease. Not only is the latter difficult to prove, VaD is also a heterogeneous entity. It encompasses a group of conditions related to a variety of pathophysiological vascular mechanisms. The neuropathological changes associated with VaD include multifocal and/or diffuse disease, and also single focal lesions. The range of pathology includes:

- Diffuse confluent age-related white matter changes (ARWMC)
  - Also referred to as subcortical arteriosclerotic encephalopathy (SAE)
- Multi-lacunar state ('état lacunaire')
- Multiple (territorial) infarcts
- Strategic cortical–subcortical or watershed lesions
- Cortical laminar necrosis (granular cortical atrophy)
- Delayed post-ischaemic demyelination (see Sect. 7.5.5)
- Hippocampal sclerosis

F. Barkhof et al., *Neuroimaging in Dementia*,
DOI: 10.1007/978-3-642-00818-4_6, © Springer-Verlag Berlin Heidelberg 2011

Reflecting this heterogeneity, the underlying vascular pathology may involve either the grey or the white matter, the small or the large vessels and have a local or systemic cause. In any case, brain imaging is essential for the diagnosis, as it reveals most of the (macroscopic) cerebrovascular pathology.

Given that VaD has many expressions and even more causes, we attempt to classify it in this chapter by the main underlying pathophysiological mechanism (i.e. large vessel, small vessel, systemic). Within the group of small vessel disease, several homogenous disorders can be discerned from the garden variety; these will be discussed in Sect. 6.4.3. Likewise, within the group of systemic disorders, several specific diseases can be singled out (e.g. vasculitis). Deep venous thrombosis and dural arteriovenous fistulae are also vascular abnormalities that may rarely cause venous hypertensive encephalopathy or bilateral thalamic congestion, and lead to dementia. Since the latter are associated with brain swelling, they will be discussed in Chap. 8.

The relevance of many vascular lesions in patients with dementia is often unclear, given the frequent coexistence of vascular and degenerative pathology, especially in the very old. In addition, cerebrovascular disease and late-onset Alzheimer disease (AD) share common risk factors. The concept of VaD is changing but – regardless of what it may be – it is fair to say that there is a considerable proportion of patients with dementia and substantial cerebrovascular disease. The pathophysiological interplay and overlap between AD and VaD is discussed in Sect. 6.7 in more detail.

6.2　History and Nosology

6.2.1　Binswanger's Disease, 'État lacunaire', Multi-infarct Dementia, Strategic Infarct Dementia – Current Concepts

In 1672, Thomas Willis provided the first accurate clinical observations of patients with post-stroke dementia. In 1894, Otto Ludwig Binswanger described the so-called *encephalitis subcorticalis chronica progressiva* and 'arteriosclerotic cerebral degeneration'. The following year, Aloïs Alzheimer wrote about

'arteriosclerotic atrophy of the brain'. Both of them were probably referring to a clinical condition characterised by slowly progressive dementia and subcortical white matter atrophy. It was assumed that the white matter atrophy was secondary to vascular insufficiency and could result in dementia. In 1896, the term 'arteriosclerotic dementia' was included in the text book *Psychiatrie* by Emil Kraepelin. It became clear that 'arteriosclerotic dementia' was a clinical condition that should be differentiated from the neurosyphilitic general paresis of the insane (dementia *paralytica*), a major cause of dementia and insanity at that time. In 1902, Alzheimer named the clinical condition believed to be secondary to vascular insufficiency (not to focal cerebrovascular lesions) as Binswanger's disease. Currently, Binswanger's disease is considered a synonym for subcortical ischaemic VaD due to diffuse confluent microvascular white matter damage.

The term lacune (derived from the Latin word *lacuna*, plural: *lacunae* – meaning 'depth' or 'loss') was first described by Amédée Dechambre in 1838. Although lacunes are usually considered to result from ischaemic infarcts of the deep perforating small vessels, the term is sometimes (confusingly) also used to encompass microhaemorrhages and enlarged perivascular (Virchow–Robin) spaces, according to a classification proposed by Poirier and Derouesné in 1984. Durand-Fardel described the so-called 'état criblé' in 1842, a denomination that now refers to the occurrence of multiple enlarged Virchow–Robin spaces in the basal ganglia (see Fig. 3.14). In 1901, Pierre Marie introduced the term 'état lacunaire' (lacunar state), still used to describe subcortical ischaemic VaD secondary to multiple lacunar infarcts.

In 1974, the term multi-infarct dementia (MID) was coined by Vladimir Hachinski and colleagues as a cause of mental deterioration in the elderly due to multiple strokes, giving the clinicians a new perspective that (multi) focal cerebrovascular lesions could lead to dementia. The rationale underlying the concept of MID was that multiple infarcts would have a synergistic effect on mental functions, resulting in dementia independently of the specific location (topography) or severity of lesions.

In 1975, Hachinski et al. also proposed a clinical scale to aid in the differentiation between what was then called MID and AD. The scale represents a set of clinical features thought to be characteristic of dementia caused by vascular disease, and characteristics of vascular disease in itself: abrupt onset, stepwise deterioration and fluctuating course, history of strokes and focal neurological signs. However, the scale does not take into account imaging criteria; neither does it provide any evidence for a causal relationship between dementia and vascular disease. Although the corresponding score is still useful as a 'bedside' clinical tool to distinguish between AD and VaD, the scale ignores the fact that both neurodegenerative and vascular pathology frequently coexist, a notion increasingly supported by neuroimaging and pathological findings. Currently, this perhaps represents the major limitation of the Hachinski ischaemic score.

The advent of modern neuroimaging with computed tomography (CT) and magnetic resonance imaging (MRI) allowed for a better diagnosis and reclassification of MID and VaD. In 1993, the National Institute of Neurological Disorders and Stroke (NINDS)–Association Internationale pour la Recherche et l'Enseignement en Neurosciences (AIREN) formulated criteria that recognised structural neuroimaging as a crucial element for the diagnosis of VaD, providing indications on the topography and severity of cerebrovascular lesions thought to be causative of dementia. In 1995, the term strategic infarct dementia was proposed by Tatemichi and colleagues for patients with an abrupt change in behaviour after acute (single) infarcts occurring in functionally critical areas of the brain, as documented by CT or MRI. Currently, both MID and strategic infarct dementia are not very widely used terms anymore, but both can be considered as synonyms of large vessel VaD – the subgroup of VaD secondary to multiple or single strategic large vessel cortical–subcortical or subcortical (e.g. watershed) cerebrovascular lesions.

Neuroimaging also provided added value for the detection of cerebrovascular lesions, as small vessel disease, and so-called silent strokes may occur without noticeable clinical symptoms. Finally, neuroimaging has shown that an overlap may exist between different expressions of cerebrovascular disease in patients with VaD, such as coexistence of large vessel and small vessel disease. Actually, a patient with VaD may either have diffuse confluent ARWMC and/or multiple lacunes (isolated small vessel disease), infarcts located in strategic regions of the brain (isolated large vessel disease) or a combination of both. Furthermore, and as previously mentioned, neuroimaging is increasingly providing evidence (supported by pathological findings) that

cerebrovascular and neurodegenerative pathology frequently coexist.

6.2.2 Current Diagnostic Criteria and Limitations

Vascular pathologies like atherosclerosis and arteriolosclerosis are extremely common in the general (ageing) population. They are a necessary, but not sufficient prerequisite for the diagnosis of VaD. Ideally, criteria for VaD should include criteria for dementia, vascular disease and some sort of evidence for a relation between the two. In practice, the latter may be rather difficult to establish and even falsely assumed, for example. AD may develop in elderly subjects with concomitant atherosclerosis. By contrast, in some clearly vascular cases (e.g. extensive white matter disease), it may be difficult to proof the vascular nature of the lesions, especially when the usual risk factors (e.g. hypertension, diabetes, smoking or hypercholesterolaemia) are absent. The issue is further complicated by the fact that VaD is quite a heterogeneous group of disorders and by the fact that agreement among neuropathologists on the exact criteria and their interpretation is limited – there is not a true gold standard. Therefore, it is important to use strict diagnostic criteria for VaD that will be discussed in the following sections, and not over-interpret the finding of vascular changes in the elderly presenting with cognitive decline, since such abnormalities are ubiquitous.

6.2.2.1 DSM-IV Criteria for Vascular Dementia

In 1994, the DSM-IV criteria for VaD were proposed by the American Psychiatric Association on the basis of a general definition of dementia (Table 6.1). Vascular disease is clinically defined (physical examination) or inferred from laboratory evidence. No criteria are mentioned to establish a causal relationship between dementia and vascular disease ('judged to be aetiologically related').

Note that the DSM-IV criteria for VaD do not really require neuroimaging evidence of cerebrovascular damage. More specific criteria including neuroimaging features for VaD will be discussed in the following sections.

Table 6.1 DSM-IV criteria for vascular dementia

(**a**) Multiple cognitive deficits manifested by both memory impairment and one or more of the following cognitive disturbances: aphasia, apraxia, agnosia, disturbance in executive functioning
(**b**) The cognitive deficits cause significant impairment in social or occupational activities and represent a significant decline from a previous level of functioning
(**c**) Focal neurological signs and symptoms (e.g. exaggeration of deep tendon reflexes, extensor plantar response, pseudobulbar palsy, gait abnormalities, weakness of an extremity) or laboratory evidence indicative of cerebrovascular disease (e.g. multiple infarcts involving the cortex and the underlying white matter) that are judged to be aetiologically related to the disturbance
(**d**) The deficits do not exclusively occur during the course of a delirium

6.2.2.2 NINDS-AIREN Criteria for Vascular Dementia

In comparison to histopathological examination, the most specific diagnostic criteria for VaD are those formulated by NINDS-AIREN. These criteria were developed for research purposes and emphasise the heterogeneity of both clinical syndromes and pathological subtypes of VaD, the need to establish a temporal relation between stroke and the onset of dementia as well as the crucial importance of brain imaging to support clinical findings. The clinical and radiological parts of the NINDS-AIREN criteria were proposed in 1993 – operational definitions for their radiological part were subsequently specified in 2003 and will be discussed in the following section.

6.2.2.3 Operational MRI Definitions for the NINDS-AIREN Criteria

The NINDS-AIREN criteria provide a list of possible vascular lesions considered relevant for the pathogenesis; however, without clearly defining their imaging criteria in terms of topography and severity of lesions. To enhance their clinical implementation, operational definitions for the radiological part of the NINDS-AIREN criteria were subsequently defined (Table 6.2).

6.2.3 Towards More Homogeneous Criteria

By far, most patients with VaD have small vessel rather than large vessel disease. Therefore, research criteria were specifically formulated for subcortical ischaemic VaD, now recognised as the most broad and homogeneous subtype. Recognising that the NINDS-AIREN criteria do not cover the subgroup of patients with small vessel VaD in too much detail, Erkinjuntti and colleagues proposed separate criteria for subcortical ischaemic vascular dementia (SIVD) in 2000. The brain imaging criteria are described in Table 6.3 – these encompass Binswanger's disease and 'état lacunaire'.

6.3 Large Vessel Vascular Dementia

Synonyms

Large vessel vascular dementia (VaD), post-stroke dementia, multi-infarct dementia, strategic infarct dementia

Large vessel VaD results from multiple or single cortical–subcortical or subcortical (e.g. watershed) cerebrovascular lesions involving strategic regions of the brain, such as the hippocampus, paramedian thalamus and the thalamocortical networks.

6.3.1 Aetiology, Pathology and Genetics

The risk factors for VaD are believed to be the same as those for stroke in general. The aetiology and pathology of large vessel VaD may be described considering either the type of brain lesion or the underlying type of vessel abnormality. Brain lesions mostly include: large vessel cortical–subcortical or subcortical infarcts (e.g. watershed infarcts) and haemorrhages. Vessel abnormalities encompass atherosclerosis and embolic sources. Vasculitis is another cause of large vessel disease (see Sects. 6.5 and 7.2).

Table 6.2 Operational imaging definitions for the NINDS–AIREN criteria

---

(**a**) Topography criteria

Large vessel stroke – arterial territorial infarct involving the cortical grey matter
  • Anterior cerebral artery (ACA) infarcts – only bilateral ACA infarcts suffice
  • Posterior cerebral artery (PCA) infarct – involving one of the following regions:
    1. Paramedian thalamus (in contact with the third ventricle)
    2. Inferior medial temporal lobe

  • Association areas – a middle cerebral artery (MCA) infarct involving:
    1. Parieto-temporal cortex (e.g. angular gyrus)
    2. Temporo-occipital cortex

  • Watershed territories – infarcts between MCA and ACA or between MCA and PCA involving:
    1. Superior frontal region
    2. Parietal region

Small vessel disease
  • Ischaemic pathology resulting from occlusion of small perforating arteries:
    1. Extensive white matter lesions (leukoaraïosis), or
    2. Multiple basal ganglia, thalamic and frontal white matter lacunar infarcts:
        i. ≥2 lacunar infarcts in the basal ganglia, thalamus or internal capsule, AND
        ii. ≥2 lacunar infarcts in the frontal white matter, or
    3. Bilateral thalamic lesions

(**b**) Severity criteria
  • Large vessel disease of the dominant hemisphere – in the absence of clinical information, the left hemisphere is considered to be the dominant one
  • Bilateral large vessel hemispheric strokes – only the infarct located in the non-dominant hemisphere should involve an area listed under topography
  • Extensive leukoencephalopathy involving at least ¼ of the total white matter:
    – Confluent lesions – grade 3 on the ARWMC scale – in at least two regions, AND
    – Beginning confluent – grade 2 on the ARWMC scale – in two other regions

(**c**) Fulfilment of radiological criteria for probable VaD
  • Large vessel disease – a lesion must be scored in at least one subsection of both topography and severity (both the topography and severity criteria should be met)
  • Small vessel disease – for white matter lesions, both the topography and severity criteria should be met; for multiple lacunar infarcts and bilateral thalamic lesions, only the topography criterion is sufficient

---

Source: Modified from Stroke 2003;34:1907–1912.

Table 6.3 Neuroimaging criteria for subcortical ischaemic vascular dementia (SIVD)

(**a**) Computed tomography:

Extensive periventricular and deep WM lesions: patchy or diffuse symmetrical low attenuation
  AND
Absence of cortical or cortical–subcortical (non-lacunar) territorial infarcts, watershed infarcts, or haemorrhages indicating large vessel disease. No other cause of WM lesions

(**b**) Magnetic resonance imaging:
1. Extensive periventricular and deep WM lesions, and lacune(s) in the deep grey matter
  OR
2. Multiple lacunes (e.g. >5) in the deep grey matter, and at least moderate WM lesions
  AND
Absence of cortical and/or cortical–subcortical (non-lacunar) territorial infarcts, watershed infarcts, haemorrhages, and other specific causes of white matter lesions

---

Source: Adapted from J Neural Transm Suppl 2000;59:23–30

Infarcts may either be complete or incomplete. Complete infarcts correspond to areas of tissue destruction/liquefaction and are the most frequent pathological findings in large vessel VaD. Incomplete infarcts may only represent demyelination and oedema and are mostly due to small vessel pathology. Therefore, they will be more extensively discussed in Sect. 6.4.

Genetic factors play a role in VaD. Genes conferring susceptibility to vascular cognitive impairment may be grouped into two different classes: genes that confer susceptibility to cerebrovascular disease and genes that determine the brain tissue response to cerebrovascular insults. As to the former, some progress has been made in identifying genes that confer susceptibility to hypertension and stroke during the past few years. In addition, several monogenic forms of cerebrovascular disease have been identified. The best defined examples relate to small vessel VaD (Sect. 6.4.3). Even less is known about genes that determine brain tissue vulnerability to cerebrovascular disease, but they include the ones related with AD, such as the apolipoprotein E genes, which play a role in subjects with ARWMC (see also Sect. 6.7).

6.3.2 Clinical Presentation, Epidemiology and Treatment

A diagnosis of vascular cognitive impairment or dementia is made by demonstrating the presence of cognitive change from medical history and examination, and by showing that the patient had vascular events sufficient to produce the cognitive change; the latter requirement can be difficult, even in large vessel VaD. From the clinical point of view, findings that increase the likelihood of large vessel VaD are a history of transient ischaemic attack (TIA) and/or cerebrovascular accident, 'stepwise decline' in mental status, and the presence of abnormal neurologic signs. In general, executive dysfunction is commonly seen in VaD and memory impairment is usually less severe than in AD.

The reported frequency of large vessel VaD in demented populations is very variable among studies. Up to one third of stroke survivors exhibit dementia within 3 months after their stroke. As with cerebrovascular disease in general, treatment options include control of vascular risk factors.

6.3.3 Neuroimaging Strategy and Key Findings in Large Vessel VaD

In general, T2-weighted MRI sequences (including FLAIR) are far more sensitive to depict cerebrovascular disease than CT, although CT was found to be more specific than MRI in predicting subsequent symptomatic cerebrovascular disease. However, both modalities perform relatively well in depicting the large vessel infarcts that may cause VaD. Marked hypointensity on T1-weighted images usually represents tissue destruction, and is a marker for complete infarcts (Fig. 6.1).

The key neuroimaging features to identify patients with large vessel VaD rely on the assessment of the topography and severity of large vessel disease, according to the NINDS-AIREN criteria (Table 6.2 and Fig. 6.2).

More advanced neuroimaging modalities, such as diffusion-weighted imaging (DWI) and perfusion MR or CT techniques like CT angiography and perfusion may be of use in VaD. In particular, the value of DWI is already well established in clinical practise for the diagnosis of recent onset ischaemia

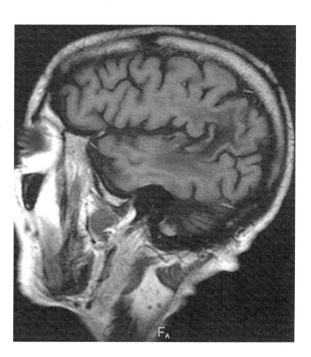

Fig. 6.1 Parasagittal left T1-weighted image showing a cortical–subcortical hypointensity in the parietal-temporal region involving the left angular gyrus, which is part of a critical association area of the brain. The hypointensity on T1 represents tissue destruction – complete infarct

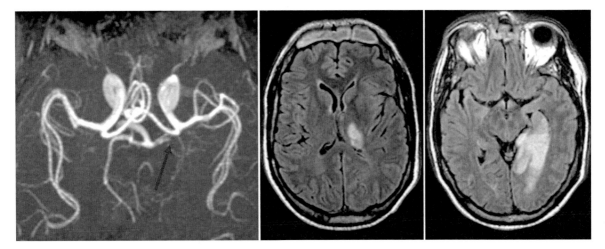

Fig. 6.2 Medial temporal lobe infarct of dominant hemisphere. MRA (*left*) shows an occlusion in the left posterior cerebral artery (*arrow*), and FLAIR (*middle* and *right*) depicts an infarction involving both the left thalamus and the left medial temporal lobe (including the hippocampal body)

(Fig. 3.7). DWI may detect recent infarcts causing the so-called stepwise decline in patients with large vessel VaD.

MR/CT and catheter angiography (DSA) may also be useful to identify occluded vessels (Fig. 6.2), arterial stenoses and extracranial arterial dissections. Doppler sonography may complement the assessment of arterial stenoses and dissections. Angiographic techniques are crucial for the diagnosis of less frequent pathologies that may rarely cause dementia and involve large vessels, such as deep venous thrombosis (Fig. 6.3) and dural arteriovenous fistulae (Chap. 8). Catheter angiography (DSA) can be useful for the interventional therapy of these abnormalities.

6.3.4 Differential Diagnosis

The differential diagnosis of large vessel VaD mostly corresponds to the one applied to cerebrovascular disease involving large vessels in general and includes, for example:

- Cerebral amyloid angiopathies (Sect. 6.4.3.2)
- Vasculitis (Sect. 6.5)
- Sickle cell disease (Sect. 6.6.1)
- Infections; for example syphilis (Sect. 7.2.4)
- Mitochondrial diseases (Sect. 7.4.3)
- Venous thrombosis and dural arteriovenous fistulae (Chap. 8)

Fig. 6.3 Deep venous thrombosis with thalamic congestion. Axial FLAIR (**a**) showing bilateral thalamic hyperintensity (*red arrows*) due to venous congestion caused by a straight sinus thrombosis, confirmed by digital subtraction angiography (**b**), which fails to opacify the deep veins including straight sinus (*dotted line* represents expected course)

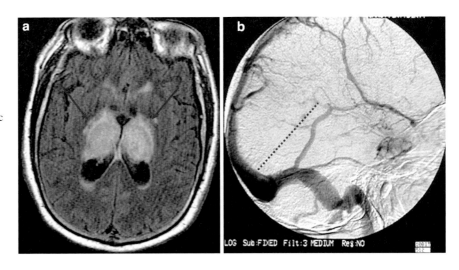

- Moyamoya disease
- Fibromuscular dysplasia
- Cardiac or other source of emboli

6.4 Small Vessel Vascular Dementia

Synonyms

Small vessel vascular dementia (VaD), arteriosclerotic dementia, Binswanger's disease, 'état lacunaire', subcortical ischaemic vascular dementia

6.4.1 Age-Related White Matter Changes, Lacunes, 'État criblé' – What Is Abnormal for Age?

Small vessel VaD may either result from diffuse confluent white matter lesions (Binswanger's disease) or from multiple subcortical lacunar infarcts ('état lacunaire'). According to the NINDS-AIREN criteria, it may result from bilateral thalamic lesions (Fig. 6.4) as well. Cerebral

autosomal dominant arteriopathy with subcortical infarcts and leucoencephalopathy (CADASIL) is also an example of a clinical condition representing small vessel VaD (see Sect. 6.4.3.3). Cerebral amyloid angiopathies (CAA) are additional subtypes of small vessel VaD (see Sect. 6.4.3.2), but they may have associated large vessel pathology, such as deposition of amyloid in the wall of medium-to-large vessels leading to cortical–subcortical (lobar) haemorrhages. Both CADASIL and some forms of CAA have a genetic basis.

Ischaemic pathology resulting from occlusion of small perforating arteries (Fig. 6.5) may become apparent as white matter lesions or as focal infarcts of the deep small vessels (e.g. lacunar infarcts).

As mentioned in Sect. 4.4, white and deep grey matter hyperintensities on T2-weighted images and FLAIR images are generally considered surrogate markers of ischaemic small vessel disease in elderly subjects. Given that white matter hyperintensities progressively accumulate with age, they are usually

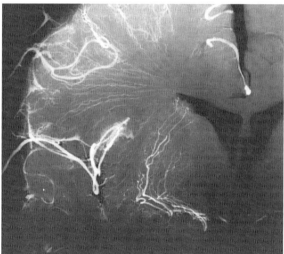

Fig. 6.5 Deep white matter vasculature. Post-mortem radiograph of a normal cerebral hemisphere after the injection of a barium sulphate-gelatin suspension into the right middle cerebral artery. This X-ray demonstrates that the periventricular and deep white matter, and the deep grey matter nuclei are irrigated by long, perforating medullary arteries that are relatively small in diameter (small vessels), exit at right angles from their 'parent' vessels, and give few or no collaterals. As such, the regions supplied by these small vessels constitute a 'watershed territory' susceptible to ischaemia in the presence of decreased cerebral perfusion (Reproduced with permission from Coffey CE, Figiel GS. Neuropsychiatric significance of subcortical encephalomalacia. In: Carrol BJ, Barret JE, editors. Psychopathology and the brain. Raven; 1991)

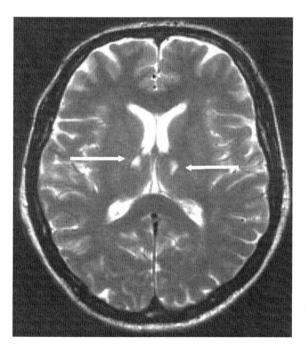

Fig. 6.4 Bilateral thalamic infarcts. Axial T2-weighted image showing bilateral thalamic infarcts (*white arrows*), a feature sufficient to diagnose VaD according to the NINDS-AIREN criteria

referred to as age-related white matter changes, or ARWMC. Moreover, they are associated with vascular risk factors as well as with other types of cerebrovascular lesions. According to the NINDS-AIREN criteria, white matter changes alone may be sufficient to cause dementia when at least ¼ of the white matter is involved (Table 6.2). Although this proportion has been defined arbitrarily, it is in accordance with the finding that only severe white matter disease is associated with cognitive dysfunction.

Diffuse confluent ARWMC are always abnormal for age. In fact, extensive white matter changes predominantly involving the periventricular and deep white matter, but relatively sparing the U-fibres, are the imaging correlate of Binswanger's disease. By contrast, in patients with CADASIL, diffuse white matter signal abnormalities involving the U-fibres predominantly occur in the temporal, temporopolar and frontal regions. Microbleeds (see Sects. 3.2, 3.3 and 4.3.1) are also present in a considerable proportion of patients with CADASIL, as well as in patients with CAA.

The occurrence of multiple enlarged Virchow–Robin spaces in the basal ganglia, a condition referred to as 'etát criblé' (see Sect. 4.3) is also pathological. The association of 'état criblé' with diffuse confluent white matter lesions (Fig. 6.6), and cognitive impairment is a frequent finding.

6.4.2 Subcortical Ischaemic Vascular Dementia

Most patients with a diagnosis of VaD have small vessel rather than large vessel disease. Subcortical ischaemic VaD is now recognised as the most broad and homogeneous subtype of small vessel VaD. It encompasses both Binswanger's disease and 'état lacunaire'.

6.4.2.1 Aetiology, Pathology and Genetics

The risk factors for subcortical ischaemic VaD also include risk factors for stroke, in general, especially those leading to small vessel disease, such as hypertension and diabetes. The aetiology and pathology of subcortical ischaemic small vessel VaD may be classified by either the type of brain lesion or the underlying type of vessel abnormality:

- Brain lesions
  - Small vessel cortical microinfarcts
  - Infarcts of the perforating deep small vessels
  - 'État criblé' (multiple enlarged deep grey matter Virchow–Robin spaces)
  - Microbleeds
  - Diffuse white matter lesions (incomplete infarctions)

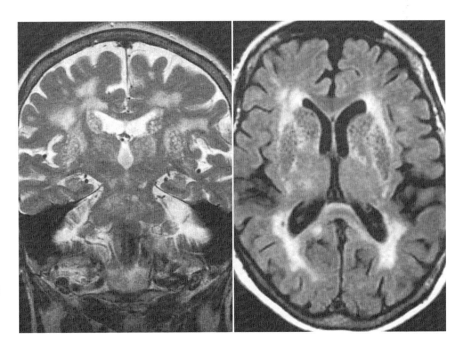

Fig. 6.6 'État criblé' with extensive white matter hyperintensities (WHM). Coronal T2-weighted (*left*) and axial FLAIR (*right*) images showing multiple enlarged perivascular spaces in the basal ganglia ('état criblé') and diffuse confluent WMH highly suggestive of ischaemic small vessel disease

• Vessel abnormalities include:
  – Atherosclerosis
  – Arteriolosclerosis
  – Amyloid angiopathy
  – Embolic sources

The two major pathological expressions of small vessel disease are lacunar infarcts (complete infarction) and white matter changes (incomplete ischaemic changes). Although small vessel disease is more likely to cause white matter changes, it certainly causes grey matter abnormalities as well (e.g. microinfarcts). Microinfarcts are not currently depicted by means of the available conventional neuroimaging modalities (Fig. 6.7).

Lacunar infarcts are complete infarcts of deep small vessels. Some authors also consider this definition dependent on size (from 2–3 to 15–20 mm in diameter). Enlarged Virchow–Robin spaces correspond to extensions of the subarachnoid space around small vessels and may be misclassified as cystic lacunar infarcts, especially when they are large or irregular. Diffuse white matter lesions usually represent incomplete infarcts. From the pathological point of view, they may correspond to diffuse myelin and axonal loss (demyelination), oedema, gliosis, spongiosis (vacuolation) and to breakdown of ependymal lining.

Genes that determine brain vulnerability to cerebrovascular disease seem to play a relevant role in subjects with ARWMC – including those related to AD, for example presenilins, amyloid precursor protein and the apolipoprotein E genes (see also Sect. 6.7).

6.4.2.2 Clinical Presentation, Epidemiology and Treatment

Many individuals with microvascular ischaemic VaD have a significant amount of white matter changes before presenting with a clinically recognised transient

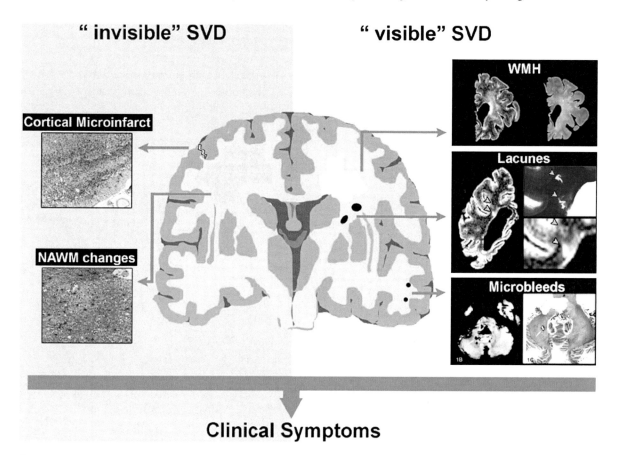

Fig. 6.7 A schematic representation of small vessel disease (SVD). Different expressions of SVD are shown, including post-mortem fluid-attenuated inversion recovery magnetic resonance images and histological sections. Note that cortical microin-farcts and normal-appearing white matter changes are only histopathologically depicted. (Illustration kindly provided by Dr. Alida A. Gouw)

ischaemic attack or stroke. As the disease may be slowly progressive, it usually lacks the so-called stepwise decline more typical of large vessel VaD. Neuroimaging is, therefore, even more relevant for the diagnosis. From the neuropsychological point of view, the most striking characteristics are executive dysfunction, mental slowness, problems with decision making, poor organisational ability, difficulties in adjusting to change (impaired executive functions of the frontal lobe), attention difficulties and apathy. This clinical syndrome has been called subcortical dementia. Whereas present, memory dysfunction is not the principal feature. Gait abnormalities may be a clinical clue.

Vascular disease of the brain, particularly hypertensive small vessel disease, is a more important factor in producing cognitive impairment and dementia than previously thought. Due to the possible concomitance between vascular and neurodegenerative pathology, the true incidence of vascular versus mixed dementia is unknown. However, subcortical ischaemic VaD represents the most broad and homogeneous subtype of patients with dementia attributable to cerebrovascular disease.

The primary treatment for subcortical ischaemic VaD is the control of vascular risk factors. It is expected that with an early identification and treatment of risk factors, vascular cognitive impairment and dementia could decrease, but this still remains uncertain. In addition to classical vascular risk factors (smoking, hypertension, diabetes, cholesterol) it may be worthwhile to exclude less common risk factors, for example vitamin B deficiency, mild hyperhomocysteinaemia and even Fabry disease and clotting disorders.

Once cognitive problems are present, studies have shown that cholinesterase inhibitors are effective. As in AD, the results are modest and only seen in some patients, but they are useful and should be considered when deciding to start treatment.

6.4.2.3 Neuroimaging Strategy and Key Findings

White matter changes on MRI are visible as diffuse hyperintense abnormalities on T2 and FLAIR (Fig. 6.8), but usually not prominently hypointense on T1-weighted images as in true infarction with liquefaction (Fig. 6.1). On CT, white matter changes appear as mildly hypodense areas, sometimes referred to as leukoaraïosis.

Patients with subcortical ischaemic VaD have widespread (diffuse or multifocal) cerebrovascular

supratentorial pathology. In addition, infratentorial abnormalities are also frequent among these patients (Fig. 6.9).

The combination of different types of MR images is important to correctly detect and classify brain abnormalities. As previously mentioned, hypointensity on T1-weighted MR usually represents tissue destruction, a surrogate marker for complete infarcts. Conversely, ischaemic abnormalities, hyperintense on T2 and isointense on T1 (e.g. ARWMC), represent incomplete infarcts. FLAIR images enable identification of cystic lesions, and the combination of FLAIR with T1 may be useful to differentiate the more aggressive lesions from those that might be less likely to cause cognitive impairment.

Lacunar infarcts are focal complete infarcts of deep small vessels. Contrary to focal incomplete infarcts, lacunar infarcts are markedly hypointense on T1 (Fig. 6.9) and FLAIR. Actually, lacunar infarcts may have signal intensity similar to CSF on all MR sequences, although FLAIR may often reveal an irregular rim of hyperintensity around the lacune; see Fig. 3.12. Finally, the sensitivity of T2-weighted images to depict thalamic lesions is superior to FLAIR and, given the well-known clinical relevance of these lesions, FLAIR should not be used as the only T2-weighted sequence to detect thalamic lesions in patients suspected of having VaD (Fig. 3.5).

Perfusion-weighted imaging is an advanced MR technique that constitutes a good alternative to nuclear medicine for the evaluation of microvascular changes in the brain. In particular, arterial spin labelling (ASL) represents a promising MR technique to evaluate brain perfusion in patients with or at risk of having VaD (Sect. 3.4.2).

Neuroimaging modalities other than MR or CT may also play a role in small vessel VaD. Transcranial Doppler sonography is now able to provide valuable information on cerebrovascular resistance, cerebrovascular reserve and cerebral perfusion. Patients with small vessel VaD have a significant increase in vascular resistance and a decrease in vascular reserve.

6.4.2.4 Differential Diagnosis

The causal relation between vascular lesions alone and dementia is not always clear. Such a relation may be expected when patients are young and unlikely to have associated Alzheimer's pathology; when cognitive

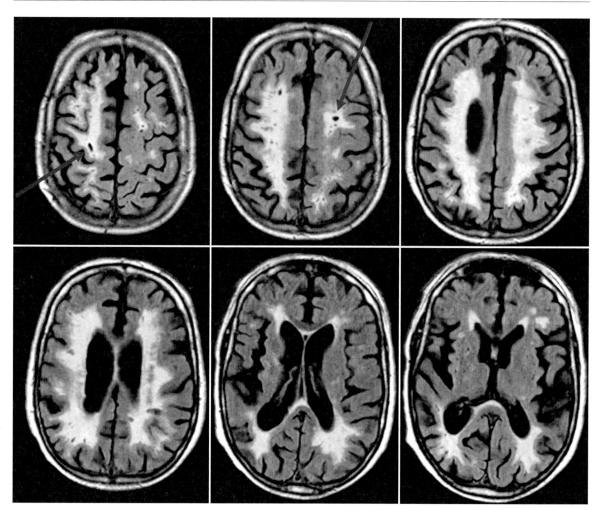

Fig. 6.8 Extensive and diffuse WMH. Axial FLAIR shows white matter hyperintensity (WMH) predominantly involving the deep and periventricular white matter, but relatively sparing the U-fibres, a pattern typical of Binswanger's disease. Scattered lacunar infarcts (*red arrows*) in the white matter are also identifiable

functions are normal before stroke, impaired immediately after, and do not worsen over time; when vascular lesions are located in strategic regions and when well-defined vasculopathies known to cause dementia are proven, for example CADASIL and CAA (see Sect. 6.4.3). In other circumstances, it is possible that both neurodegenerative and vascular pathology may contribute to cognitive impairment. Therefore, the overlap between cerebrovascular and neurodegenerative disease is significant, which makes 'mixed dementia' (see Sect. 6.7) the most relevant differential diagnosis of subcortical ischaemic VaD.

Specific types of microangiopathy will be discussed in Sect. 6.4.3, but the differential diagnoses for small vessel VaD otherwise includes:

- Vasculitis (Sect. 6.5)
- HIV encephalitis (Sect. 7.2.2)
- Progressive multifocal leukoencephalopathy (PML - Sect. 8.3.2)
- Subacute sclerosing panencephalitis (SSPE, Sect. 7.2.2.)
- Acute disseminated encephalomyelitis (ADEM) and MS (Sect. 7.3.2)
- Adult polyglucosan body disease (Sect. 7.4.6)
- Leukodystrophies (see Sect. 7.4)
- Mitochondrial diseases (see Sect. 7.4.3)
- Encephalopathy secondary to cytotoxic drugs; for example methotrexate (Sect. 7.5)
- Carbon monoxide poisoning (see Sect. 7.5.4)
- Post-radiation encephalopathy (Sect. 7.6)

Fig. 6.9 Examples of focal lesions in patients with small vessel vascular dementia. (**a**) Axial T2-weighted image of a 78-year-old patient showing multiple small vessel cerebellar infarcts, some involving the cerebellar cortex (*arrows*). (**b**) Coronal T1-weighted image of an 80-year-old patient showing multiple infratentorial lacunes in the basilar pons (*large arrow*), and supratentorial lacunes occurring in the right basal ganglia region and in both thalami (*small arrows*). (**c, d**) Axial T2-weighted images of a 66-year-old patient showing multiple deep cerebellar and pontine microbleeds (*arrows*). (Reproduced with permission from Stroke 2006;37:105–110)

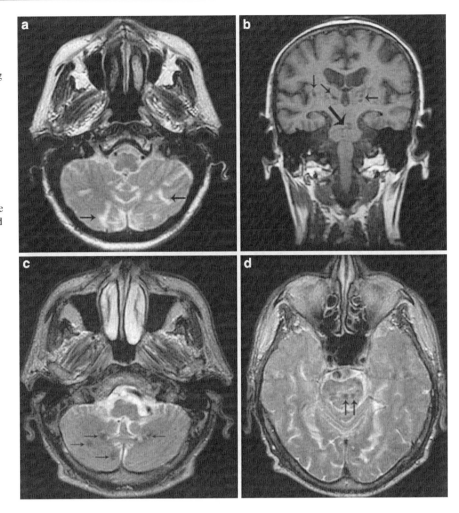

6.4.3 Specific Types of Microangiopathy

6.4.3.1 Introduction and Classification

The majority of cases with small vessel disease will be sporadic or idiopathic; that is the underlying pathophysiology is not attributable to a specific cause, but rather to general vascular risk factors (e.g. hypertension, diabetes) leading to arteriosclerosis, hypoperfusion and to white matter damage. In a small fraction of patients, specific diseases can be discerned with an established aetiology. Recognising those subtypes is clinically meaningful in terms of management, prognosis and genetic counselling. Therefore, in the following sections, we will separately discuss several types of microangiopathy with microhaemorrhages (e.g. amyloid angiopathies), as well as cerebral autosomal dominant arteriopathy with subcortical infarcts and leukoencephalopathy (CADASIL), and Fabry's disease.

6.4.3.2 Microangiopathy with Microhaemorrhages

Introduction

Cerebral microhaemorrhages or microbleeds (MBs) have mostly been identified by radiologists on MRI rather than by histopathological examination. They correspond to small intramural and perivascular haemorrhages. MBs are defined as small foci of low signal intensity on T2* or susceptibility-weighted images not attributable to vessels, calcification or other pathology like cavernomas (see Chap. 3.3.2.5). MBs occur in sporadic small vessel disease (SVD), often in a central distribution involving the basal ganglia. By contrast, their distribution is mostly cortical–subcortical (lobar) in specific disorders such as cerebral amyloid angiopathies (CAA). Amyloid angiopathy may also occur secondarily in prion diseases, such as Creutzfeldt–Jakob disease and Gerstmann–Sträussler–Scheinker syndrome. Table 6.4 provides a radiological differential diagnosis of cerebral MBs.

Sporadic CAA

Aetiology, Genetics and Histopathology

In addition to the hereditary types of CAA discussed in the following sections, sporadic cases of CAA may also present with symptomatic or subclinical (imaging only) intracerebral haemorrhage. Sporadic CAA is a common cerebrovascular pathology of the elderly and is caused by the deposition of β-amyloid in the media and adventitia of small-to-medium sized cerebral arteries. The prevalence of CAA in autopsy specimens increases from 2% at the age of 50 years to 74% in subjects above the age of 90 years. In western countries, CAA is the leading cause of lobar intracerebral haemorrhage as opposed to hypertensive bleeding occurring in the basal

ganglia. In addition, severe CAA is a risk factor for dementia, in general, and is a feature of AD, in particular. Specific amyloid precursor protein (APP) mutations (e.g. APP693) and APP duplications lead to severe CAA. In 'sporadic' CAA, the apolipoprotein E (APOE) ε2 and ε4 alleles are risk factors: the latter leads to a higher propensity for amyloid β (Aβ) 40 to be deposited in vessel walls. Histopathologically, degenerative vascular changes may accompany amyloid deposition, including fibrous thickening with an 'onion skin' appearance of the vessel wall, 'double barrelling', thinning of the degenerated vessel wall (sometimes with microaneurysm formation), fibrinoid necrosis and evidence of blood breakdown products around the affected blood vessels. In severe CAA, the smooth muscle cell layer is completely lost and degenerative changes of the affected vessel walls are often accompanied by evidence of leakage of blood. In some cases, a marked inflammatory response occurs, probably underlying an angiitis-like illness occurring in a subset of patients.

Clinical Presentation, Epidemiology and Management

Sporadic CAA may present as subcortical vascular dementia, stroke or as an acute vasculitis-type of illness.

Table 6.4 Differential diagnosis of cerebral MBs

Disease	MB distribution	Additional MR features	Clinical clues
Sporadic SVD	Basal ganglia	Lacunes, extensive WML	Hypertension, other vascular risk factors
CAA	Cortico-medullary	Subpial blood, macroscopic haematoma	TIA, stroke
HCHWA APP mutations (e.g. APP693) and duplications; and other familial amyloidoses	Cortico-medullary	Frequently macroscopic haematoma, but sometimes just extensive white matter changes	Stroke, seizures, family history
FBD/FDD	Unknown	Rarely macroscopic haematoma	Ataxia
CADASIL	Cortico-medullary	Temporal WML	Migraine, family history
Cavernomas	Random	Popcorn lesions, high signal on T1	Epilepsy
Collagen type 4 mutations	Brainstem and cerebellum	Extensive WML, dilated VRS	Stroke, migraine, retinal artery tortuosity
Radiation vasculopathy	Random	Atrophy and WML	Radiation therapy
DAI	Cortico-medullary, corpus callosum, brainstem	Cortical contusions	History of trauma

For abbreviations, see text and list on page XV

Table 6.5 Boston criteria for CAA

1. Definite CAA. Full post-mortem examination demonstrating:
   * Lobar, cortical, or cortical–subcortical haemorrhage
   * Severe CAA with vasculopathy
   * Absence of other diagnostic lesion

2. Probable CAA with supporting pathology. Clinical data and pathologic tissue (evacuated haematoma or cortical biopsy) demonstrating:
   * Lobar, cortical, or cortical–subcortical haemorrhage
   * Some degree of CAA in specimen
   * Absence of other diagnostic lesion

3. Probable CAA. Clinical data and MRI or CT demonstrating:
   * Multiple haemorrhages restricted to lobar, cortical or cortical–subcortical regions (cerebellar haemorrhage allowed)
   * Age of 55 years and over
   * Absence of other cause of haemorrhage

4. Possible CAA. Clinical data and MRI or CT demonstrating:
   * Single lobar, cortical or cortical–subcortical haemorrhage
   * Age of 55 years and over
   * Absence of other cause of haemorrhage

Source: Neurology 2001;56:537–539

The diagnosis of CAA can be made with varying levels of confidence using the Boston criteria (Table 6.5). In subjects with stroke/haemorrhage, anti-coagulants convey an increased risk for repeated haemorrhage (particularly when multiple MBs are present). Patients with an inflammatory type of CAA may benefit from steroid treatment, certainly when there is marked oedema on MR imaging; see also Sect. 8.9. In subjects with AD, multiple MBs (and presumably severe CAA) convey a worse prognosis.

Neuroimaging Strategy and Findings

Structural imaging is important to assess macroscopic haemorrhage in both symptomatic (Fig. 6.10a) and asymptomatic (Fig. 6.10b) subjects. Demonstration of MBs requires T2*-weighted imaging techniques, such as gradient echo. Susceptibility-weighted imaging (SWI) is a particularly sensitive technique that uses phase information to enhance visibility of MBs. The sensitivity to detect MBs is dependent on many technical factors, including field strength, echo time and slice thickness. Higher field strength leads to many more MBs being detected. Due to this variation in acquisition, generalisable normative data are hard to provide, but as a rule-of-thumb, more than 3 MBs in a lobar distribution at 1.5 T are indicative of CAA. Amyloid PET imaging (e.g. ^{11}C-PIB) shows increased uptake in CAA.

Neuroimaging strategy and findings in CAA
Structural imaging
* Evidence of past lobar haemorrhage on CT or MR
* Multiple MBs at the corticomedullary junction on T2*-images
* Subpial siderosis on T2*-images
* White matter hyperintensities (WMH) and lacunes

PET scanning (unconfirmed)
* Increased binding on amyloid imaging

HCHWA

Synonyms

Hereditary cerebral haemorrhage with amyloidosis (HWHWA), Dutch type (HCHWA-D), icelandic type (HCHWA-I), oculo-leptomeningeal amyloidosis, APP-related CAA

Aetiology, Genetics and Histopathology

Several autosomal dominant single-gene disorders cause cerebral haemorrhage with amyloidosis. The best studied of these involve certain specific point mutations in the APP gene on chromosome 21. Such disorders result in early-onset cerebral amyloid angiopathy and/or cerebral 'parenchymal' amyloidosis: cerebral haemorrhage is a common, but not inevitable, feature in members of these families. The Dutch mutation (at position 693) is the prototypical example (HCHWA-D), but other mutations are well described and all encode changes at a similar position on APP within the sequence of the Aβ peptide (clustered around the alpha secretase site); e.g. Flemish (692), Italian (693), Iowa (694) and Arctic (693) mutations. In all these APP-related CAAs, meningocortical arteries are affected by β-amyloid deposits, leading to aneurysmal dilatation or thinning of the vessel wall and to fibrinoid necrosis. Vessels in the deep hemispheric structures (e.g. thalamus and basal ganglia) and brain stem are relatively spared. Duplications in APP also cause CAA, but cerebral haemorrhage is less frequent.

In the Icelandic-type (HCHWA-I), a mutant of the cysteine protease inhibitor cystatin-c occurs on chromosome 20. In oculo-leptomeningeal amyloidosis, the affected protein is transthyretin (TTR), whose gene is located on chromosome 18. In HCHWA-I, the amyloid angiopathy is more widely distributed, involving arteries in the cerebrum, cerebellum and the brain stem. In oculo-leptomeningeal amyloidosis, deposition of abnormal TTR occurs in the vitreous and the leptomeningeal blood vessels.

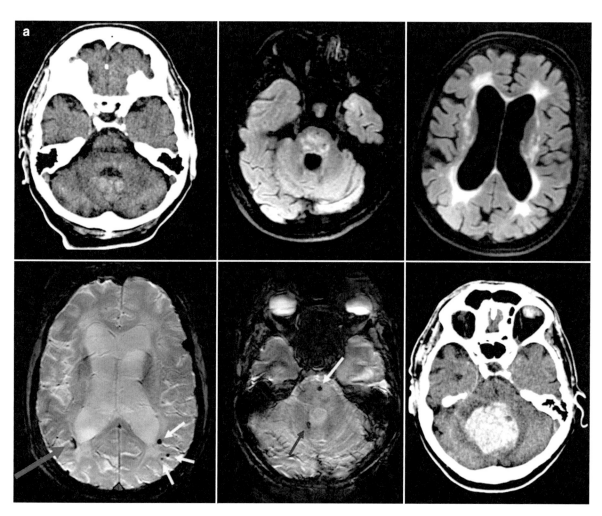

Fig. 6.10 (**a**) Symptomatic CAA. This 62-year-old woman had diabetes type II and a history of cardiac disease, for which she was treated with oral anti-coagulants. She was recently forced to stop working due to forgetfulness. Baseline CT (*upper left*) and MRI showed confluent ischaemic white matter lesions with multiple microbleeds (*light blue arrows*), but also evidence of a silent macroscopic bleeding (*dark blue arrow*) consistent with CAA. A few months later she was admitted with a stroke and the corresponding CT (*lower right*) showed a large cerebellar haematoma, in close proximity to a previous cerebellar microbleed (*red arrow*)

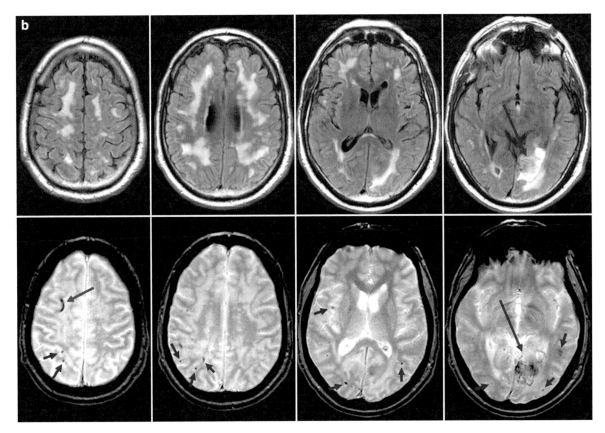

Fig. 6.10 (continued) (**b**) Asymptomatic CAA. This 67-year-old man presented with mild memory disturbances and cognitive slowing clinically diagnosed as MCI (MMSE 30/30). MR revealed confluent ischaemic white matter lesions (Fazekas grade III) on FLAIR (*upper row*) and multiple microbleeds (*blue arrows*) on gradient echo T2*-weighted images (*lower row*). However, there is also evidence of past (*purple arrow*) and recent (*red arrow*) haemorrhages, consistent with a diagnosis of CAA

Clinical Presentation, Epidemiology and Management

HCHWA-D usually presents in the 6th decade, mostly with intracranial haemorrhage or dementia without stroke. Patients with HCHWA-I usually present their first episode of haemorrhage in the 3rd or 4th decade of life. The TTR variants may manifest in many ways, usually with sensorimotor polyneuropathy with or without associated autonomic neuropathy.

Neuroimaging Strategy and Findings

Structural imaging may reveal multiple haemorrhages and multiple MBs in a typical lobar distribution. In advanced cases, the central regions become affected as well. Additional white matter lesions develop as the disease progresses.

Familial British and Danish Dementia with Amyloid Angiopathy

Synonyms

Familial British dementia (FBD), Worster-Drought syndrome, familial cerebral amyloid angiopathy – British type, familial Danish dementia (FDD), heredopathia ophthalmo-oto-encephalica

Aetiology and Genetics

Both FBD and FDD are extremely rare early-onset autosomal dominant disorders characterised by progressive cognitive impairment, spasticity and cerebellar ataxia. Both are associated with a stop codon mutation in the BRI gene located on chromosome 13, resulting in the production of amyloidogenic fragments. Histopathologically, there is hippocampal neurofibrillary degeneration and widespread parenchymal and vascular amyloid deposits.

Symptoms and Course

Both diseases present over the age of 60 years. FBD is characterised by impaired recognition and memory recall progressing to dementia, progressive spastic tetraparesis and cerebellar ataxia. FDD is additionally characterised by cataracts and deafness. In contrast to HCHWA and CAA, cerebral haemorrhage is less common.

Neuroimaging Findings

FBD patients have extensive periventricular white matter hyperintensities mostly affecting the occipital lobe and corpus callosum as well as lacunar infarcts, but no signs of frank intracerebral haemorrhage. No data are available on the occurrence of MBs. The corpus callosum can be severely atrophic as in some forms of hereditary spastic paraparesis. Amyloid imaging may show increased cerebellar rather than cerebral uptake in FBD. Imaging findings in FDD are lacking.

6.4.3.3 CADASIL

Synonyms

Cerebral autosomal dominant arteriopathy with subcortical infarcts and leukoencephalopathy (CADASIL), hereditary multi-infarct dementia, familial Binswanger's disease

Aetiology, Genetics and Histopathology

In 1977, Stevens et al. described a hereditary disorder with recurrent subcortical infarctions and progressive neurological deficits leading to dementia, pseudobulbar palsy and severe disability; similar cases were perhaps also described by van Bogaert in 1955. The acronym CADASIL was coined by Elizabeth Tournier-Lasserve et al. in 1993. CADASIL is caused by a mutation in the NOTCH3 gene. NOTCH3 codes for a transmembrane receptor protein on the surface of smooth muscle cells surrounding arteries. Accumulation of pathologic NOTCH3 receptor protein in small and medium-sized cerebral arteries results in thickening and fibrosis of the walls of small and medium-sized arteries, leading to cerebral infarctions (Fig. 6.11). CADASIL is the most common hereditary disease presenting with stroke.

Clinical Presentation, Epidemiology and Management

The clinical picture is dominated by recurrent subcortical ischaemic infarctions and cognitive decline typically beginning around 40–60 years of age. Other clinical manifestations include migraine with aura and prominent mood disturbance with apathy. The cognitive profile in CADASIL is characteristically subcortical and is similar to that seen in small vessel VaD in general. A definite diagnosis of CADASIL can be made when the clinical findings are supported by a mutation of the NOTCH 3 gene on chromosome 19 or with histopathological confirmation from skin or brain biopsy. Probable CADASIL is suggested by brain imaging findings (Table 6.6). CSF, blood and neurophysiology are typically non-contributory. There is no specific treatment beyond anti-platelet treatment and management of vascular risk factors.

Neuroimaging Strategy and Findings

Whereas CT may reveal leukoaraïosis, MR imaging is the critical imaging modality in CADASIL due to its sensitivity to show the characteristic involvement of the temporal pole (Figs. 6.12 and 6.13), which has a very limited differential diagnosis (multiple sclerosis, myotonic dystrophy and adult polyglucosan body disease). Additional findings include lacunes, MBs and widened Virchow–Robin spaces at the cortico-medullary junction. Diffusion-weighted imaging may show restricted diffusion in recent infarcts, and diffusion tensor imaging (DTI) may show a more generalised involvement of the (partially normal appearing) brain tissue. MR spectroscopy may reveal lactate (Fig. 6.14). The imaging findings are summarised in Table 6.6.

6.4.3.4 Fabry's Disease

Synonyms

Fabry's disease (FD), Anderson-Fabry disease, α-galactosidase A deficiency

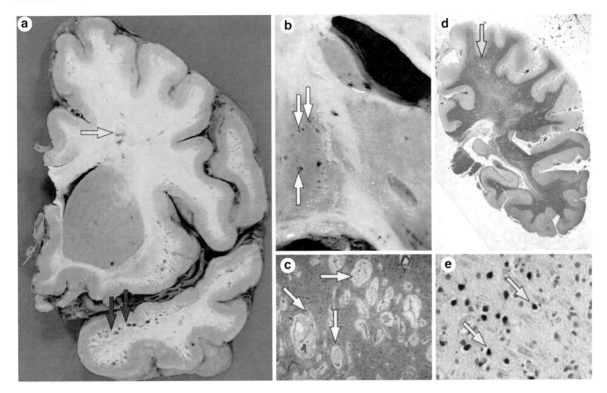

Fig. 6.11 Pathological features of cerebral lesions in CADASIL. (**a**) Coronal section of the left hemisphere at the level of the caudate nucleus. Multiple subcortical lacunes (*small infarcts*) corresponding to dilated perivascular spaces are detected at the cortico–subcortical junction in the temporal lobe and insula (*red arrows*). Other lacunes are present in the centrum semi-ovale (*green arrow*). (**b**) Macroscopic section of the striatum and thalamus showing a notable area of status cribrosus (*arrows*) with (**c**) microscopy (haematoxylin and eosin, ×40) confirming accumulation of dilated perivascular spaces (*arrows*). (**d**) Coronal section of the right hemisphere at the level of the pulvinar nucleus. Klüver-Barrera stain (*luxol cresyl violet*) shows myelin loss in the centrum semi-ovale and small lacunes in the white matter (*arrow*). Note the relative sparing of the cortex by ischaemic lesions. (**e**) Neuronal apoptosis in layer 3 of the occipital cortex (TUNEL technique). Several neurons are positively stained, some of which have pyknotic nucleus (×400; *arrows*). (Reprinted with permission from Lancet Neurology 2009;8: 643–653–and with thanks to Françoise Gray, Department of Pathology, Hôpital Lariboisière, Paris, France)

Table 6.6 Neuroimaging strategy and findings in CADASIL

Structural imaging – MRI is the preferred modality
- Extensive/confluent white matter lesions
  - Temporal pole involvement (one of the earliest features)
  - U-fibre involvement at the vertex
  - Deep white matter and basal ganglia
    External capsule involvement (not specific)
- Lacunes in the basal ganglia, thalamus and deep white matter
- Widened Virchow–Robin spaces
  - 'État criblé' in the basal ganglia
  - Lacune-like lesions at the cortico-medullary junction
- Cerebral MBs–especially in the thalamus and brainstem
- Hypointensity on T2 in the basal ganglia

Additional MR findings
- MR angiography – typically normal
- DWI may show restricted diffusion in recent lesions
- Spectroscopy may reveal lactate

NB – contraindication for catheter angiography (DSA), because of a high complication rate

Aetiology, Genetics and Histopathology

FD is an X-linked recessive disorder resulting from α-galactosidase-A deficiency. Progressive accumulation of globotriaosylceramide in lysosomes leads to pathological lipid storage in many organs, including the vascular endothelium throughout the CNS. Thickening of vessel walls and their obstruction leads to ischaemic infarctions and aneurysm formation.

Clinical Presentation, Epidemiology and Management

The prevalence of FD is probably underestimated; FD accounts for ~1% of young male stroke cases and should be considered when proteinuria is present. Characteristic clinical manifestations of FD include acroparaesthesias,

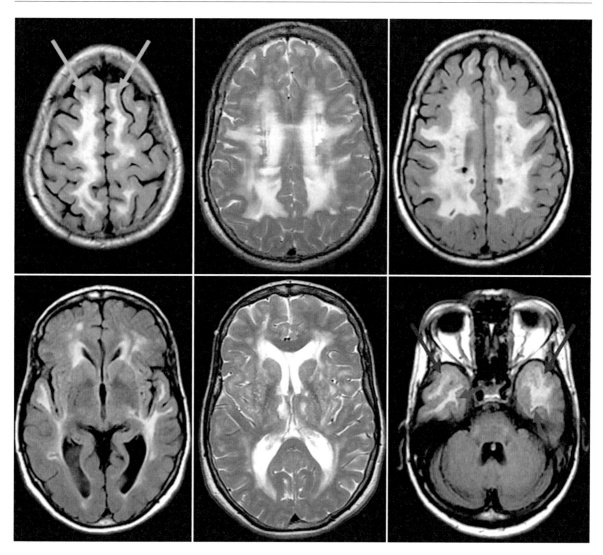

Fig. 6.12 MRI findings in CADASIL. This 57-year-old woman presented with memory disturbances, disorientation and difficulties in walking and speaking. Her MMSE was 26/30. MRI revealed confluent white matter changes on FLAIR and T2 (*middle*), with a characteristic involvement of the U-fibres at the vertex (*green arrows*) and temporal pole (*red arrows*); note also juxtacortical fluid spaces in the temporal pole on FLAIR. Genetic testing revealed a notch-3 mutation

angiokeratoma, corneal opacity, hypohidrosis, gastrointestinal symptoms, renal, and cardiac dysfunction. CNS manifestations include hemiplegia, hemianaesthesia, aphasia and seizures. In men developing stroke below the age of 65 years, FD should be routinely excluded. The diagnosis of FD is established by demonstrating a deficiency of α-galactosidase in plasma, leukocytes, urine or by means of genetic testing. Enzyme replacement therapy has been available since 2001.

Neuroimaging Strategy and Findings

The imaging findings in FD are mostly non-specific small and large vessel pathology on CT or MRI.

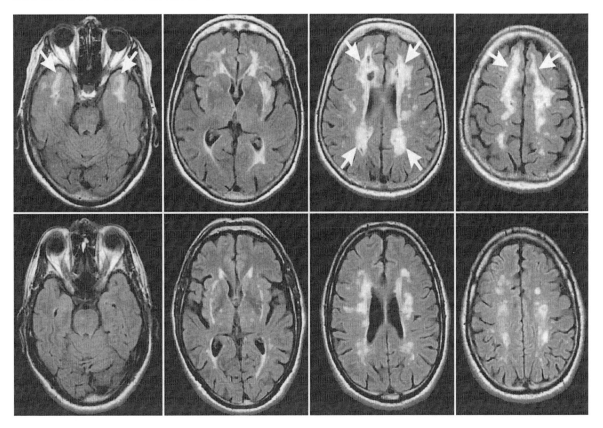

Fig. 6.13 CADASIL versus sporadic small vessel disease. The characteristic involvement of the anterior frontal lobes and the temporal poles in the CADASIL patient (*arrows in the upper row*) is not seen in the patient with sporadic small vessel disease (*lower row*). Note that involvement of the external capsule is not a distinguishing feature. (Reprinted with kind permission from Radiology 2000;218: 443–445)

Dolichoectasia may occur, especially in the vertebrobasilar system. Small vessel disease may manifest as diffuse white matter changes or multiple bilateral lacunar infarcts. Large vessel territorial infarcts may occur as well.

A relatively specific MR imaging finding is signal change in the pulvinar. Twenty to 30% of the patients have high signal intensity in the pulvinar on T1-weighted images, and low signal intensity on T2*-weighted images in the more severe cases. These changes correspond to calcification on CT (Fig. 6.15). The proportion of subjects showing pulvinar changes increases with age. Increased cerebral blood flow in the posterior circulation, particularly in the thalamus, suggests that the dystrophic calcification may be secondary to cerebral hyperperfusion and to a selective vulnerability of the pulvinar and adjacent thalamic nuclei. CT may show more extensive calcium deposits, particularly in more severely affected patients, involving the cerebral cortical–subcortical junction, globus pallidus, pulvinar and the cerebellar cortico-medullary junction – a pattern similar to Fahr's disease.

Beyond focal vascular lesions, quantitative MR techniques may show diffusely abnormal DTI characteristics of the normal appearing WM and reduced NAA/Cr ratio as well as enhanced brain activity on fMRI, probably as a compensatory mechanism.

Fig. 6.14 Proton MR proton spectroscopy in CADASIL. Long-echo spectrum shows reasonably preserved ratios of N-acetylaspartate, creatine and choline. At 1.3 ppm there is a lactate doublet

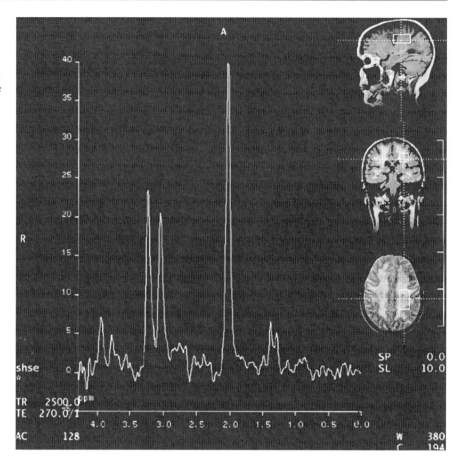

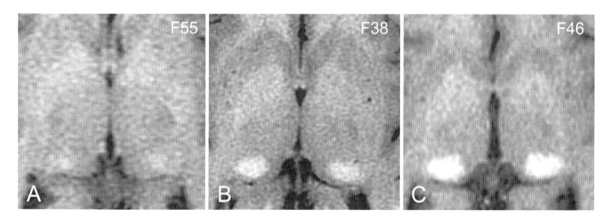

Fig. 6.15 Comparison of CT and MR findings in the posterior thalamus. (**a–c**), T1-weighted images through the thalamus in three patients with mild (**a**), moderate (**b**), and marked (**c**) hyperintensity, respectively. (**d–f**), Corresponding CT scans demonstrate increased attenuation indicating calcification is present only in the moderate and marked cases. (**g–i**), Corresponding gradient-echo T2*-weighted images demonstrate that susceptibility-induced signal intensity loss is seen in the pulvinar in only the moderate and marked cases. Numbers in upper right are patient identifiers. (Reprinted with permission from Am J Neuroradiol 24:1096–1101)

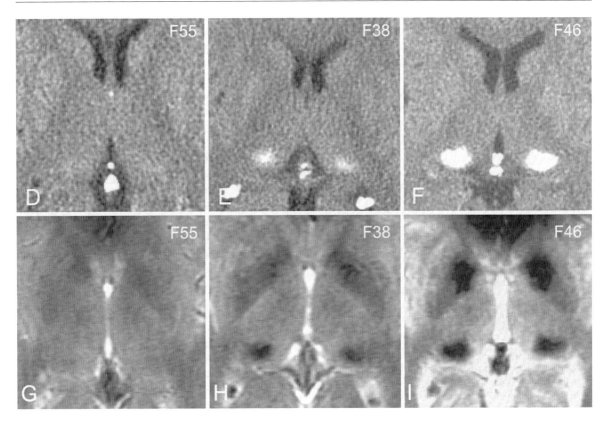

Fig. 6.15 (continued)

6.5 Vasculitis

In this chapter, we discuss primary central nervous system (CNS) vasculitis, and several causes of secondary CNS vasculitis.

6.5.1 Primary CNS Vasculitis

Synonyms

Primary CNS vasculitis, primary angiitis of the CNS (PACNS), granulomatous angiitis of the nervous system, isolated CNS vasculitis

6.5.1.1 Aetiology and Histopathology

Primary angiitis of the CNS is a disease with unknown aetiology with patchy inflammation preferentially affecting small leptomeningeal and parenchymal vessels. Histologically, two subtypes exist: granulomatous angiitis and benign angiopathy.

6.5.1.2 Clinical and Laboratory Findings

Presentation of PACNS is quite variable, and classically involves a triad with headache, organic brain syndrome and multifocal neurological deficits. The course may be fluctuating, multiple sclerosis-like or gradually progressive with subcortical dementia. Being isolated to the CNS, patients with PACNS often lack systemic manifestations. In addition, routine laboratory blood tests, including the erythrocyte sedimentation rate (ESR), may be normal in a variable proportion of cases; brain biopsy (including leptomeninges) is the only definitive proof. As a result, suspicion of vasculitis is a major indication for brain biopsy, since the diagnosis is elusive otherwise and the condition is potentially treatable. Therapeutic options include steroids and cyclophosphamide treatment.

6.5.1.3 Neuroimaging Strategy and Findings

Conventional digital subtraction angiography (DSA) may show beading produced by alternating stenosis and dilation, but this finding is not specific and negative catheter angiography is an invasive procedure that does not rule out vasculitis. MRI is a more sensitive test, being abnormal in perhaps 80% of patients, but often is non-specific by showing hyperintense lesions on T2-weighted or FLAIR images in the white matter and cortex. The pattern of gadolinium enhancement (punctiform in multiple locations) may suggest PACNS, but the differential diagnosis includes other vasculitides, malignancies and infectious disease (Fig. 6.16).

6.5.2 Secondary CNS Involvement

Secondary vasculitis of the CNS is more common than primary CNS vasculitis, and encompasses several systemic illnesses, including generalised autoimmune diseases like systemic lupus erythematosus (SLE), Behçet's disease, Sjögren's syndrome, Susac's syndrome and a variety of systemic vasculitides including Wegener's granulomatosis and polyarteritis nodosa. Secondary CNS vasculitis can be induced by hard drugs like amphetamine, cocaine, but also by over-the-counter drugs like ephedrine.

In this section, only two disorders are discussed in some more detail: SLE, given its high prevalence (of neuropsychiatric involvement), and Susac's syndrome, for its typical imaging appearance. Imaging findings in other secondary vasculitides include non-specific vascular lesions and, occasionally, punctiform gadolinium enhancement.

6.5.3 Systemic Lupus Erythematosus (SLE)

SLE is an autoimmune disease of unknown origin that can be drug induced. The most common cerebral manifestation of SLE is an organic encephalopathy that can be detected by using MRI and PET/SPECT. It may be due to cytokine-mediated effects on vascular endothelium and antiphospholipid antibodies. In addition to small vessel vasculopathy, true vasculitis may occur in large-to-medium sized vessels, which may lead to a stroke-like presentation (that may also be caused by emboli from Libman–Sacks endocarditis). The diagnosis is based on the criteria of the American College of Rheumatology, as well as on laboratory tests including the erythrocyte sedimentation rate (ESR), C-reactive protein (CRP) and anti-nuclear antibodies (ANA).

Neuropsychiatric symptoms are frequently found in SLE, and incidental cases have been described where dementia is the manifesting symptom. While multi-infarct dementia can be caused by SLE, chronic encephalopathy and inflammation due to active vasculitis can affect cognition as well. MRI findings in acute presentations include multiple infarcts, swelling of the basal ganglia, and punctiform contrast enhancement (Fig. 6.17). In chronic cases, the most common correlate of NP-SLE is generalised brain atrophy, including the hippocampus, and non-specific WM lesions. Quantitative MR techniques (MTR, DTI) may show abnormal tissue integrity in neuropsychiatric SLE, and SPECT/PET may show reduced cerebral perfusion and metabolism.

6.5.4 Susac's Syndrome

Susac's syndrome is a microangiopathy caused by endothelial dysfunction affecting the precapillary arterioles of the brain, retina and the inner ear. It mostly affects middle-aged women. Clinical symptoms include encephalopathy, branch retinal artery occlusions and hearing loss. Headache is often the presenting feature, but neuropsychiatric disturbances can occur and include memory loss, confusion and dementia.

Typical (characteristic) MRI findings in Susac's syndrome
* Multiple lesions (sometimes small, but also 'snow-ball'-like)
    – Centred on the corpus callosum

* Enhance with gadolinium in the acute phase
    – May have associated leptomeningeal enhancement

* May have restricted diffusion in acute phase

* Progress to a 'punched-out' appearance on follow up

Although any part of the corpus callosum may be involved (Figs. 6.18 and 6.19), the central fibres are

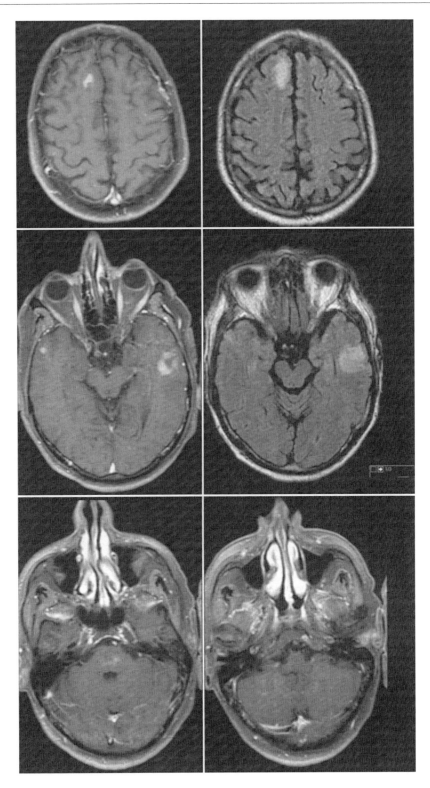

Fig. 6.16 PACNS. This 48-year-old patient presented with multifocal neurological symptoms, a high erythrocyte sedimentation rate and eosinophilia. FLAIR images (*upper-right*) show hyperintense lesions in multiple cortical areas, with patchy punctiform contrast enhancement on T1-weighted images (*left column and bottom-right*), both supra and infratentorially. The major vessels to the brain were normal on MRA (not shown). The patient was diagnosed (and successfully treated) as having a PACNS

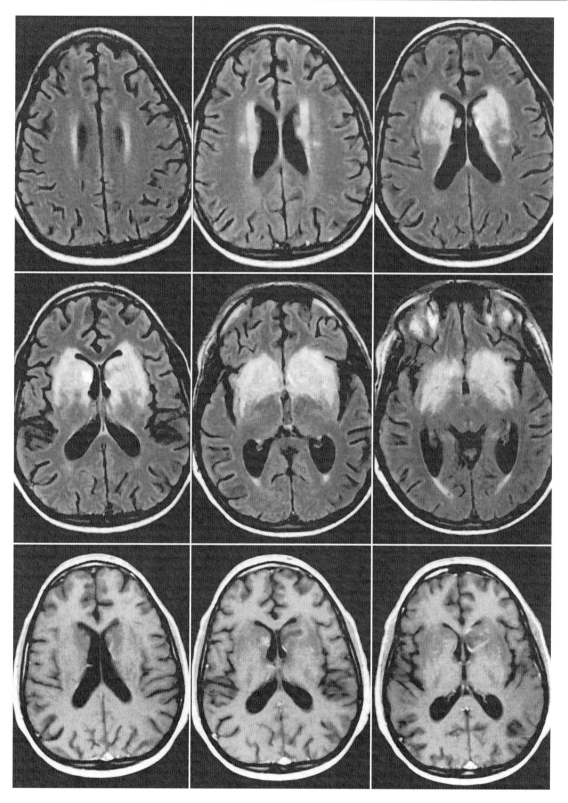

Fig. 6.17 SLE. This 36-year-old woman with SLE presented with confusion. The *upper two rows* represent transverse FLAIR images showing severe involvement of the striatum with swelling and oedema. CT did not show calcification. The *bottom row* shows contrast-enhanced T1-weighted images, illustrating some punctiform enhancement in the caudate nuclei

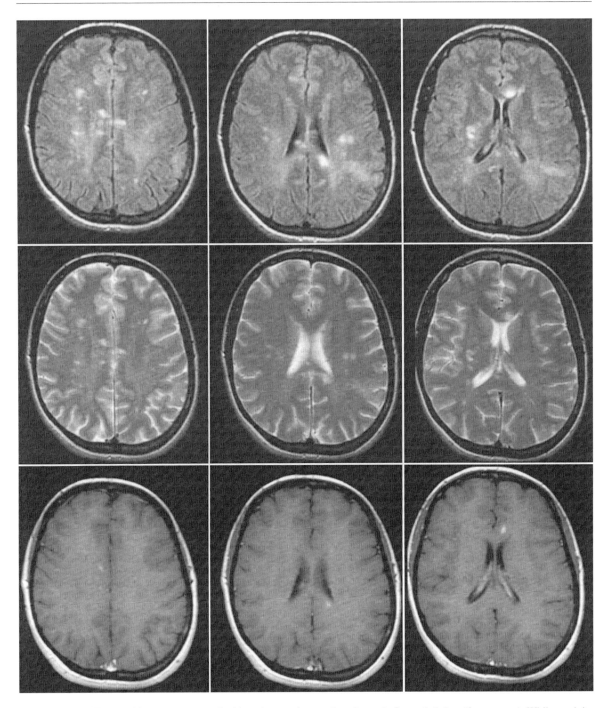

Fig. 6.18 This 33-year-old woman presented with a picture of mental slowing, memory disturbances, headache and drowsiness that developed over a month. MRI showed multiple white matter lesions on FLAIR (*top row*) and T2-weighted (*middle row*) images that mainly involved the corpus callosum, and partly enhanced after gadolinium (*bottom row*). While reminiscent of MS, the 'cotton wool-ball' or 'snow-ball' appearances suggest Susac's syndrome, which was subsequently diagnosed based on auditory loss and retinal vessel abnormalities (Thanks to Alex Rovira, Barcelona, for providing this case)

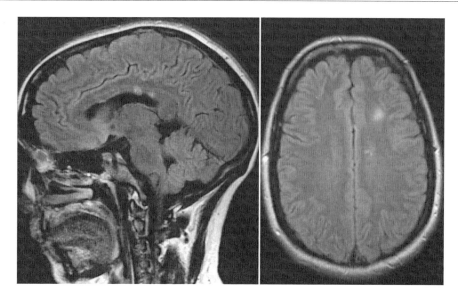

Fig. 6.19 Callosal involvement in Susac's syndrome. Sagittal and axial FLAIR images of a middle-aged woman presenting with behavioural disturbance and cognitive decline that had progressed over several weeks: a psychiatric condition or a rapidly progressive dementia were considered. The sagittal FLAIR image shows two characteristic 'snow-ball' lesions, the smaller one (also displayed on the axial image) predominantly involves the central fibres of the corpus callosum, but relatively spares the peripheral ones. The diagnosis was confirmed by means of retinal examination. A good response was seen with immunosuppression. However, frontal deficits persisted

predominantly involved and there is relative sparing of the peripheral ones. Extensive miliary and leptomeningeal enhancement can also occur. Cerebral angiography findings are almost always normal, because the involved precapillary arterioles can not be depicted, but the diagnosis can be confirmed by fluorescein angiography of the retina that often shows branch retinal artery occlusions, as well as the pathognomonic multifocal fluorescence of the branch arterioles. The pathogenesis of this syndrome is unknown, but it is believed as an autoimmune endotheliopathy. Patients tend to spontaneously improve, but treatment options include immunosuppressants like steroids, cyclophosphamide and immunoglobulin, often combined with aspirin.

6.6 Systemic Hypoxia

In addition to ischaemia caused by vascular wall changes, as discussed in the preceding sections, systemic disorders can also cause ischaemia and lead to cognitive dysfunction and ultimately dementia.

Systemic causes of ischaemia include cellular dysfunction (mitochondrial disease, discussed in Sect. 7.4.3), but also anaemia and hypotension.

6.6.1 Anaemia

Cerebral function and neuronal survival depends on oxygen availability at the cellular level and may be adversely affected by a number of contributing factors including a susceptibility to failure of local energy production (e.g. mitochondrial disorders) exacerbated by hypoxia, regionally impaired oxygen delivery (e.g. atherosclerosis) or general oxygen delivery failure (e.g. due to anaemia; cardiopulmonary arrest or failure).

Anaemia is associated with an increased incidence of cognitive impairment, probably through diffuse anoxic damage to vulnerable (grey matter) regions. Such damage may lower the threshold for the development of neurodegenerative processes, such as AD or may contribute to cognitive deficits through a lowering of cognitive reserve. In fact, vascular dementia and AD share many vascular risk factors, including anaemia. Anaemia is also often associated with other co-morbidity, such as with chronic obstructive pulmonary disease (COPD), which may, in itself, lower the threshold for cognitive decline. The reverse situation (i.e. increased levels of haemoglobin), including polycythaemia, carries the risk of thrombotic vascular events in the brain. Polycythaemia, probably due to a hyperviscosity state, is reversibly associated with subcortical (speed and attention) cognitive deficits.

Macrocytic anaemia is a feature of vitamin B12 deficiency which in itself is associated with (potentially reversible) dementia, mostly secondary to white matter demyelination (see Sect. 7.5.2). Similarly, transient low blood oxygenation, leading to delayed white matter demyelination and cognitive impairment, will be discussed in Sect. 7.5.5.

Special cases of anaemia include sickle cell disease and cerebral malaria.

In sickle cell disease (haemoglobin SS), a low haematocrit per se predisposes to cognitive impairment, probably through diffuse ischaemic damage to grey matter. The occurrence of infarctions (caused by abnormal coagulability) additionally predisposes to cognitive impairment. Infarctions may manifest themselves either as evident strokes or be clinically silent (MRI only); subclinical cognitive decline may occur in childhood.

In cerebral malaria, anaemia also is combined with intravascular occlusion of capillaries, and a complex inflammatory process with alterations in the blood–brain barrier, leading to impaired homeostasis and cerebral oedema. In the short term, this may cause seizures, altered consciousness and coma. In the long term, structural damage and cognitive impairment ensue.

6.6.2 *Transient Global Amnesia*

Synonym

Acute transient amnesia

6.6.2.1 Aetiology and Histopathology

Although first recognised by Morris Bender in 1956, the name transient global amnesia (TGA) was coined by C. Miller Fisher and Raymond Adams in 1958. It occurs in 5–25 of 100,000 individuals. The pathophysiology of TGA remains speculative; suggested causes include epilepsy, a vascular (TIA) event, and migraine-related phenomena. In a proportion (~30%) of cases a precipitating event such as emotional stress or physical exertion can be identified. Venous 'congestion' in the context of a Valsalva manoeuvre has also been suggested as a mechanistic cause and may explain the history of precipitating events.

Although the pathological substrate for TGA is unclear there have been suggestions that damage to CA1 neurons of the cornu ammonis (the so-called Sommer – or vulnerable – sector of the hippocampus) may be responsible. The existence of a watershed area within the CA1 vulnerable sector of Sommer, secondary to the existence of an anastomosis between the upper and lower hippocampal arteries, has been suggested as a predisposing factor. Some support for hippocampal involvement has been provided by diffusion imaging studies discussed below.

6.6.2.2 Clinical and Laboratory Findings

TGA is a clinical syndrome characterised by an abrupt onset of profound anterograde and retrograde amnesia, without further neurological deficits, resolving within 24 h. A typical feature is perseverance (i.e. constantly repeating the same statements or questions). Diagnostic criteria include:

- A reliable witness report of definite loss of recent memory (anterograde amnesia)
- Absence of clouding of consciousness or any other cognitive impairment other than amnesia.
- No focal neurological signs or deficits during or after the attack.
- No features of epilepsy in the past 2 years, and no recent head injury.
- Resolution within 24 h

There are no established laboratory tests. Depending on the clinical features, MRI and/or EEG are indicated; transient epileptic attacks are an important differential. The differential diagnosis otherwise includes basilar artery thrombosis, cardioembolic syndrome, migraine variants, syncope or related paroxysmal spells.

6.6.2.3 Neuroimaging Strategy and Findings

The most informative imaging modality is MRI including DWI. Foci of abnormal high signal intensity with corresponding low apparent diffusion coefficient (ADC) in the Sommer sector of the hippocampus may be seen in the first few days (max 48–72 h) after the onset of symptoms, suggesting a vascular cause (Fig. 6.20). Such abnormalities may be unilateral or bilateral, and may subsequently become visible on high-resolution T2-weighted images. TGA abnormalities do not evolve into cystic lesions, and should be differentiated from benign cavities of the vestigial hippocampal sulcus that occur with normal ageing (Chap. 4).

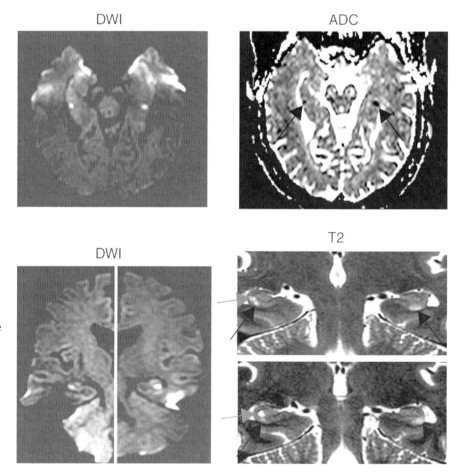

Fig. 6.20 Representative 3T MRI showing typical increased DWI signal and lowered ADC corresponding to lesions on T2-weighted images. On the high-resolution T2-weighted images, note the bilateral lesions in the CA1 sector of the cornu ammonis (*red arrow*) extending over 4–5 mm (slice thickness 2 mm), which are clearly separated from the cavity of the pre-existing vestigial hippocampal sulcus (*green arrow*) more deeply located, between the cornu ammonis and the dentate gyrus. (Reprinted with permission from Brain 2006;129:2874–2884)

6.7 Mixed Dementia and Interrelationship of Vascular Pathology and Alzheimer's Disease

Wiesje van der Flier

6.7.1 Introduction

Traditionally, demented patients are diagnosed with a specific label, on the basis of clinical criteria and ancillary investigations. MRI is an essential part of the diagnostic work-up, and can provide positive evidence for the presence of neurodegenerative (e.g. atrophy) and cerebrovascular disease (e.g. infarcts, white matter hyperintensities, microbleeds). The majority of patients receive a diagnosis of AD. The second most frequently diagnosed type of dementia in the elderly is VaD. Whereas neuropathologically, senile plaques and neurofibrillary tangles are presumed to underlie AD, cerebrovascular pathology, by definition, causes VaD. However, it is increasingly recognised that these two types of neuropathological damage are often concomitantly observed, and their effects may not be easily disentangled. In this section, we will focus on the interrelationship between these two types of pathology, and their manifestation on MRI. We first provide evidence for the role of additional cerebrovascular pathology in AD. Subsequently, we will focus on the opposite, describing neurodegenerative changes in VaD. We will then discuss several hypotheses regarding the relationship between neuropathological and cerebrovascular brain changes. Finally, a future perspective is sketched, where MRI may have a decisive role in the choice of therapy for individual patients, regardless of the specific diagnostic label.

6.7.2 The Contribution of Cerebrovascular Pathology to Alzheimer's Disease

Since the first description of AD at the beginning of the twentieth century, there has been debate about the putative interaction between neurodegenerative and cerebrovascular brain changes in causing AD. For a long time, 'senile dementia' was thought to be caused by intracranial atherosclerosis, while by contrast AD (with young onset, as exemplified by the first patient Auguste D who was only 51 years old) was caused by senile plaques and neurofibrillary tangles. In the course of the twentieth century, senile dementia was discovered to be the same disease as initially described by Alois Alzheimer, caused by plaques and tangles. In this period, atherosclerosis was abandoned as a causative disease mechanism for AD. To date, however, interest is picking up again; as the amyloid cascade hypothesis has not provided definitive answers, and evidence is accumulating that cerebrovascular disease should be taken into account to fully understand the disease.

AD is the most common form of dementia, accounting for more than half of dementia cases. An increasing number of neuropathological studies, however, have shown that, especially in elderly patients over the age of 75 years, mixed disease underlies the clinical syndrome of dementia, whereas pure AD is relatively less common. Persons with multiple neuropathological diagnoses are three times more likely to develop dementia, and most are clinically classified as 'probable' AD. Cerebrovascular disease refers to a variety of abnormalities, including large vessel infarcts, lacunar infarcts and white matter lesions. Intracranial atherosclerosis of the circle of Willis has been related to a diagnosis of AD, and a relationship with the burden of plaques and tangles suggests a direct relationship between the two disease processes. In addition, (cortical) microinfarcts seem to have significant impact on cognitive function. Expressions of cerebrovascular disease on MRI can also be highly diverse. In addition to large vessel infarcts, there are several expressions of small vessel disease that can be appreciated on MRI, including lacunar infarcts, white matter hyperintensities (WMH) and microbleeds. With current methods, cortical microinfarcts cannot be visualised by using MRI.

There is a growing body of literature that illustrates interrelationships between AD and cerebrovascular disease. Not only VaD, but also AD is predicted by a history of stroke. In population studies, clinically silent infarcts predict the subsequent development of dementia, believed to be of the Alzheimer type. Infarcts are, therefore, a risk factor for AD. Having a stroke doubles the risk of developing dementia, often AD. It seems that the stroke may serve to exacerbate pre-existing, subclinical pathology. In line with this, baseline atrophy of the medial temporal lobe, rather than WMH, predicts subsequent cognitive decline in stroke survivors.

WMH are considered to reflect ischaemic, small vessel pathology. Severe WMH may suffice to cause VaD, but WMH are also frequently observed in AD. Subjects with severe WMH have a modestly, but non-significant, increased risk of AD, whereas atrophy of the medial temporal lobe is clearly associated with AD. However, patients in whom both abnormalities are observed on MRI mostly have a clinical diagnosis of AD, implying that the known association between hippocampal atrophy and AD is amplified by the additional presence of WMH (Table 6.7). These results support the view that WMH contribute to the clinical syndrome of AD.

Microbleeds are an expression of cerebrovascular disease that has recently received an increasing amount of attention. Microbleeds have long been considered to be specific for populations such as stroke. In the context of dementia, they are often observed in VaD, cerebral amyloid angiopathy (CAA) and CADASIL. However, it has become clear that microbleeds also frequently occur in patients with AD. One out of five patients with AD has at least one microbleed on T2*-weighted imaging. Of these, the majority has only one microbleed. A small proportion of patients with AD has many (>5) microbleeds; probably reflecting amyloid angiopathy (Fig. 6.21). Amyloid angiopathy typically reflects the combination of Alzheimer's pathology and vascular damage, as it refers to the deposition of amyloid in small-to-medium sized blood vessels of the brain and leptomeninges. The amyloid deposits may cause breakdown of the blood vessel wall, resulting in haemorrhage.

The clinical relevance of microbleeds is not yet clear. In stroke patients without cognitive impairment and in those with subcortical vascular dementia, microbleeds have been associated with cognitive decline. In AD, patients with many microbleeds seem to perform worse on a cognitive screening test than AD patients without any microbleed, regardless of the degree of atrophy, WMH, or disease duration. Such patients also tend to have more abnormal levels of amyloid-beta 1–42 in the CSF than AD patients without any microbleeds, suggesting a direct link with neurodegenerative pathology. In addition, microbleeds convey a more than 2-fold increased risk of mortality, regardless of other expressions of small vessel disease on MRI, or vascular comorbidity (Fig. 6.22). Microbleeds might predispose for future haemorrhages, although the benefits of antithrombotic agents in secondary prevention of vascular diseases seem to outweigh the potential risks.

6.7.3 The Contribution of Neurodegenerative Pathology to VaD

VaD is the second most prevalent type of dementia, certainly over the age of 75 years. Although by definition, cerebrovascular disease underlies dementia in VaD, atrophy is also commonly observed. Neuropathologically, pure VaD is a rare condition, as cerebrovascular disease is often accompanied by neurodegenerative changes. According to the radiological National Institute of Neurological Disorders and Stroke (NINDS)–Association Internationale pour la Recherche et l'Enseignement en Neurosciences (AIREN) criteria, involvement of more

Table 6.7 Odds ratios for AD dependent on medial temporal lobe volume and WMH

Medial temporal lobe	Large	Large	Small	Small
White matter hyperintensities	Mild	Severe	Mild	Severe
AD (number)	9	10	16	23
Control (number)	14	10	4	0
Odds ratio	1.0	1.6	6.2	Infinity
(95% CI)	(Reference)	(0.4–6.2)	(1.3–32.7)	(6.3–inf.)

Data are presented as odds ratios and their exact 95% confidence intervals. Note that, due to the zero cells, the accompanying OR is indeterminably high

Source: Adapted from Neurology 2004;62(10):1862–1864

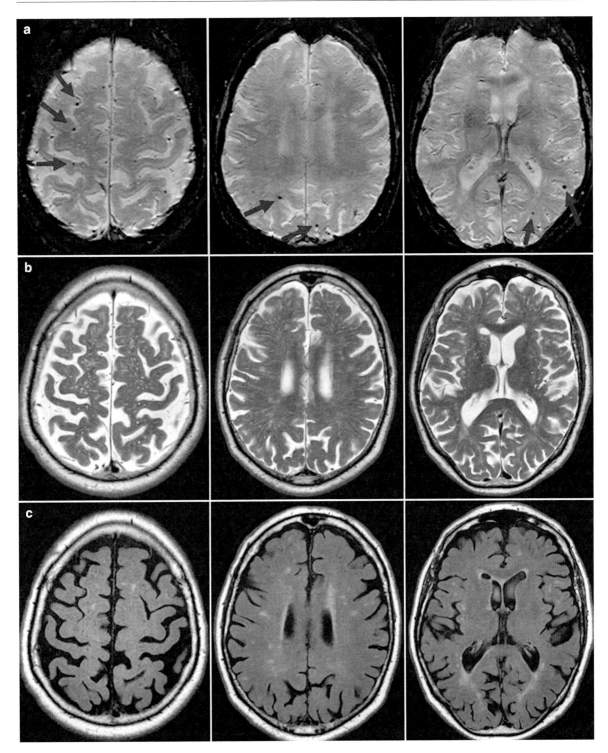

Fig. 6.21 Microbleeds are small, dot-like lesions of low signal intensity in the brain that can be observed on T2*-weighted images. This patient with a clinical syndrome of AD, presented with multiple microbleeds on brain MRI, mostly in the posterior cortex (total count > 50), suggestive of cerebral amyloid angiopathy. Susceptibility-weighted images are presented in **panel A** (examples of microbleeds indicated by *red arrows*). For comparison, matching slices on T2-weighted images (**panel B**) and FLAIR (**panel C**) are shown

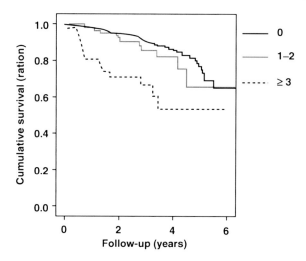

than 25% of the total white matter, multiple lacunes, bilateral thalamic lesions and strategic infarcts can lead to a diagnosis of VaD. In clinical practice, the majority of patients with VaD have small vessel disease, rather than large vessel disease.

In patients fulfilling diagnostic criteria for VaD, atrophy of the medial temporal lobe – generally regarded as an indicator of neurodegenerative (Alzheimer's) pathology – is the strongest predictor of severity of cognitive decline, when compared with measures of cerebrovascular small vessel disease, such as WMH and (lacunar) infarcts. Presumably, the occurrence of medial temporal lobe atrophy (MTA) in patients with VaD is attributable to concomitant Alzheimer's pathology, or even to misdiagnosis (see Fig. 6.23).

Alternatively, MTA may be secondary to vascular pathology, more precisely to small vessel disease and ischaemia. Nonetheless, when there is neuroimaging evidence of mixed pathology (degenerative and

Fig. 6.22 Kaplan-Meier curve illustrating the dose relationship between microbleeds and risk of mortality. Patients with many microbleeds are at increased risk of mortality. Hazard ratio (95% confidence interval): 2.4 (1.4–4.3). (Reproduced with permission from Stroke 2009;40(2):492–498)

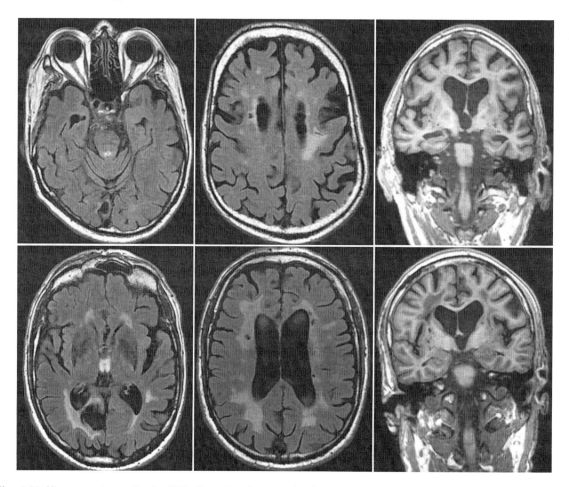

Fig. 6.23 Hippocampal atrophy in VaD. Example of two patients with a diagnosis of vascular dementia, based in both cases on confluent white matter hyperintensities, multiple lacunar infarcts and a large vessel infarct. In the first patient, diagno-sis of vascular dementia is accompanied by clear atrophy of the medial temporal lobe (*upper panel*). In the second patient, there is no evidence of any atrophy of the medial temporal lobe (*lower panel*)

vascular), atrophy seems to predict or correlate better with dementia than small vessel disease. Likewise, approximately 10% of patients fulfilling diagnostic criteria for VaD have midbrain atrophy (Fig. 6.24), which also is associated with cognitive impairment even after correction for abnormalities representing degenerative and vascular supratentorial pathology. It is conceivable that midbrain atrophy may also represent concomitance of degenerative pathology, and that its occurrence in the periaqueductal grey matter may explain the association between midbrain atrophy and cognitive impairment by disruption of mesencephalic connections and neurotransmitter systems.

Correlation of post-mortem findings with ante-mortem MRI measures in a sample of patients with a wide range of cognitive dysfunction has revealed complex relationships of vascular and neurodegenerative pathology with MRI features. During life, volumes of cortical grey matter and hippocampal volumes are the best predictors of cognitive functioning, as opposed to WMH volume and number of lacunes, which are only marginal predictors of cognitive decline. When neuropathological indices are related to MRI findings, cortical grey matter volume on MRI is histopathologically predicted by a combination of

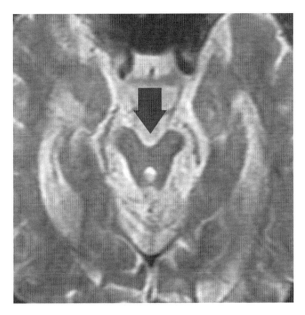

Fig. 6.24 Midbrain atrophy in VaD. Axial T2-weighted image showing midbrain atrophy (*arrow*) in a patient with the clinical diagnosis of vascular dementia. (Reproduced with permission from Stroke 2006;37:105–110)

the degree of Alzheimer's pathology, arteriosclerosis and subcortical infarcts, whereas the hippocampal volume on MRI is best predicted by the degree of Alzheimer's pathology and hippocampal sclerosis. These findings indicate that observations on MRI are often the result of a combination of different underlying neuropathological changes.

6.7.4 How Do Neurodegenerative and Cerebrovascular Changes Interact?

To date, there is dispute about the relation between neurodegenerative and cerebrovascular changes. They may be separate and unrelated processes that co-occur incidentally. The majority of dementias may, therefore, be of the mixed type (Fig. 6.25), but it is also possible that misdiagnoses occur. For example, patients that are diagnosed as VaD, just on the basis of the presence of cerebrovascular damage on MRI, may in fact suffer of dementia mostly due to Alzheimer's pathology. Conversely, it is possible that microinfarcts, which cannot be recognised on MRI, underlie the clinical syndrome of dementia labelled as AD.

A second possibility is that different types of pathology add up, until a threshold is reached and clinical dementia ensues, so that, in the presence of additional cerebrovascular damage, a less severe stage of Alzheimer's pathology is sufficient to cause the clinical syndrome. Third, neurodegenerative and cerebrovascular changes may act in synergy, aggravating each other's effect. This is probably most clearly illustrated by the concept of amyloid angiopathy, where Alzheimer's pathology is deposited in the vessel walls, resulting in vascular damage. A theoretical representation of how neurodegenerative and vascular disease might interplay, via the concept of amyloid angiopathy (visible on MRI as microbleeds) is provided in Fig. 6.26.

The e4 allele of the Apolipoprotein E (APOE) gene is associated with an elevated risk of AD. The mechanism through which APOE e4 contributes to the development of AD has not yet been fully clarified, but it is known that this gene is implicated in lipid transportation, and is associated with both (coronary) atherosclerosis and amyloid angiopathy. These links between APOE e4 and cerebrovascular disease suggest that in APOE e4 carriers, AD may develop as a consequence of impaired vessel walls causing reduced clearance of

Fig. 6.25 Most patients with dementia at old age have mixed disease. Presented are two patients with pronounced atrophy of the medial temporal lobe, and cerebrovascular disease. The upper patient has severe WMH. The lower patient has a thalamic lesion on the right side (*red arrow*). The clinical diagnosis of both patients was possible AD

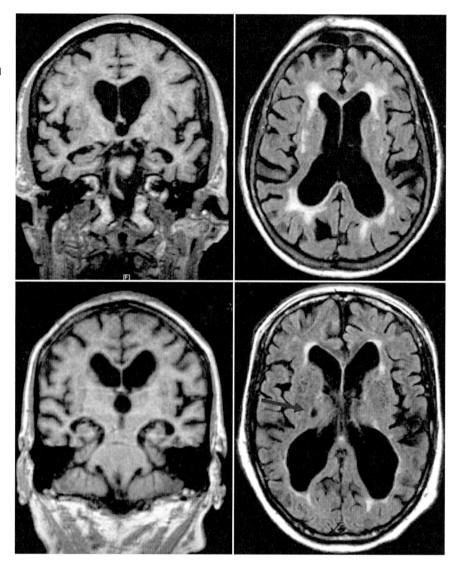

amyloid-beta (red route in Fig. 6.26). Conversely, in APOE e4 non-carriers, the disease would be caused by elevated production of amyloid-beta, resulting in more 'pure' AD. Following this line of reasoning, it is conceivable that mixed disease is especially prevalent among APOE e4 carriers. This is a hypothesis that remains to be proven, however.

There are still many questions left unanswered with respect to the relationships between neurodegenerative and cerebrovascular disease. Not in the least, because it is difficult to study their respective contributions during life. Post-mortem studies by definition are post hoc, and may start too late, as it is always the end stage that is studied. Moreover, there is a time lag between in vivo clinical diagnosis and post-mortem observations of neuropathology, often involving several years. As discussed above, observations on MRI are often not specific for type of neuropathology. For example, global atrophy, often considered to be a measure of neurodegenerative disease, is also predicted by underlying vascular disease. Also, MRI observations may reflect multiple types of underlying neuropathology. Additional studies on the association between neuropathological findings, MRI measures and clinical outcome are certainly needed, to shed light on the role of the respective neuropathological substrates in causing cognitive decline and dementia, and to identify MRI measures that more specifically reflect the underlying neuropathological substrates.

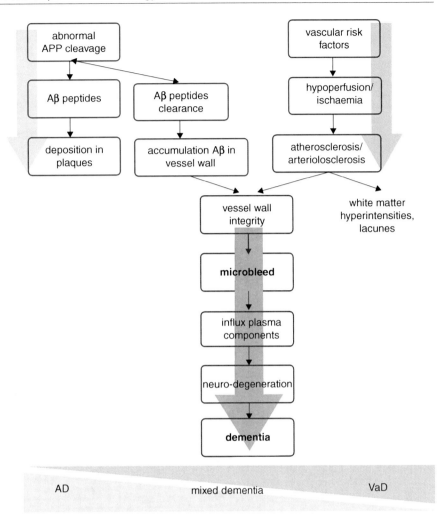

Fig. 6.26 Proposed representation of the interplay between neurodegenerative (*left side, blue representation*; amyloid cascade hypothesis) and vascular (*right side; red representation*; vascular hypothesis) pathological pathways of Alzheimer disease. Amyloid angiopathy (visible on MRI as microbleeds) may play an important role in the interaction between the two. The two routes can both contribute to the clinical syndrome of dementia. Depending on their relative contribution, the diagnosis could be AD, VaD or mixed dementia

6.7.5 Shifting Paradigms

In this section, we have shown accumulating evidence suggesting that patients do not have either AD or VaD; rather, many patients are affected by both types of pathology. Instead of exclusively labelling all patients as either neurodegenerative or cerebrovascular, both disorders could be viewed as a continuum, with purely neurodegenerative disease on one end, and purely cerebrovascular disease on the other end of the spectrum.

Currently, neuroimaging is recommended at least once during the diagnostic work-up of dementia, for the exclusion of surgically treatable disorders, such as tumours or haematomas. In addition, MRI may also add positive evidence for the presence of specific types of neuropathology. It is conceivable that the role of

MRI in dementia will change in the future. We propose a new diagnostic paradigm where – irrespective of specific diagnostic labels – MRI is used both as a starting point for treatment choice, and as a means to monitor disease progression and treatment effect. In this paradigm, where a diagnosis would be more than just a label, treatment and management of the patient would directly follow observations of markers on MRI (Fig. 6.27). For example, observing MTA might be a reason to start cholinesterase inhibitors (currently registered for treatment of AD and dementia with Lewy bodies only, while MTA is also frequently observed in vascular dementia). Evidence of cerebrovascular disease should lead to treatment of vascular risk factors and lifestyle modification (e.g. quit smoking and increase physical activity), even when the clinical diagnosis is probable AD. In this framework, other

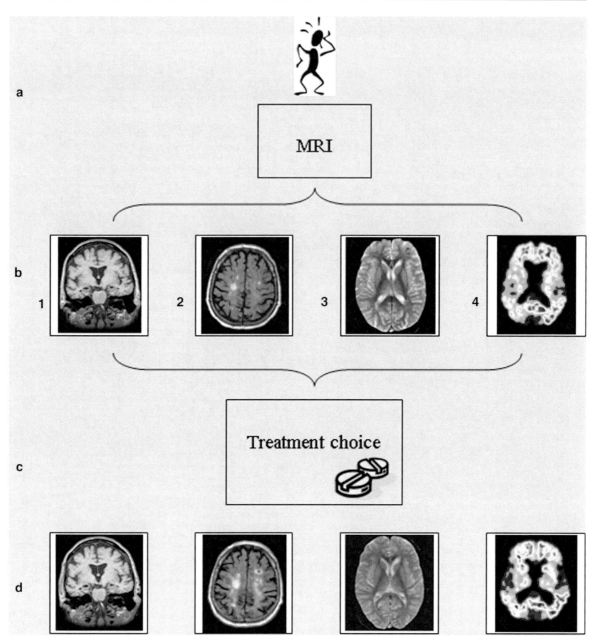

Fig. 6.27 Diagram illustrating the use of MRI as a starting point for treatment choice and monitoring of treatment effect. Please note that this is a representation of a potential future, not an actual recommendation for clinical practise to date. A patient with a clinical syndrome of dementia presents at a memory clinic (**a**). Even when from the specific clinical signs and symptoms it is not perfectly clear which etiological diagnosis best befits this patient, an MRI is made to guide treatment choice. Potential MRI findings with treatment implications could be (**b1**) medial temporal atrophy on coronal T1-weighted images, (**b2**) white matter hyperintensities and/or lacunes on FLAIR images, (**b3**) microbleeds on T2*-weighted images, (**b4**) amyloid load on PIB-PET (as yet, amyloid cannot be demonstrated on MRI, but it is conceivable that this will be the case in the future with ultra-high field MRI). Treatment choice (**c**) will subsequently be based on the observations on MRI, which can be any combination of the observations under (**b**): for example cholinesterase inhibitors in case of atrophy of the medial temporal lobes, treatment of vascular risk factors and explicit advice of lifestyle modification in case of small vessel disease, the observation of microbleeds might prompt carefulness in the use of anticoagulants, and the observation of severe amyloid-load might suggest that vaccination therapy can be useful. Finally, MRI can be used to monitor the effect of the chosen treatment strategy on the evolvement of brain changes (**d**). If necessary, treatment can then be adapted, based on new imaging findings

markers of vascular disease, such as a history of hypertension, diabetes mellitus or hypercholesterolaemia, which have all been related to both the observation of small vessel disease on MRI and to an increased incidence of AD, should be taken into account as well.

By stratification of treatment on the basis of MRI findings, the effectiveness of treatment may be improved. Moreover, with repeated imaging, MRI markers may be used to evaluate treatment effect and to monitor disease progression. Other biomarkers, such as cerebrospinal fluid biomarkers and positron emission tomography, may have added value in combination with the use of MRI.

Suggestions for Further Reading

Bartsch T, Alfke K, Stingele R, Rohr A, Freitag-Wolf S, Jansen O, Deuschl G (Nov 2006) Selective affection of hippocampal CA-1 neurons in patients with transient global amnesia without long-term sequelae. Brain 129(Pt 11):2874–2884

Bastos Leite AJ, van der Flier WM, van Straaten EC et al (2006) Infratentorial abnormalities in vascular dementia. Stroke 37:105–110

Bastos-Leite AJ, van der Flier WM, Van Straaten EC et al (2007) The contribution of medial temporal lobe atrophy and vascular pathology to cognitive impairment in vascular dementia. Stroke 38:3182–3185

Beach TG, Wilson JR, Sue LI, Newell A, Poston M, Cisneros R, Pandya Y, Esh C, Connor DJ, Sabbagh M, Walker DG, Roher AE (Jan 2007) Circle of Willis atherosclerosis: association with Alzheimer's disease, neuritic plaques and neurofibrillary tangles. Acta Neuropathol 113(1):13–21

Birnbaum J, Hellmann DB (2009) Primary angiitis of the central nervous system. Arch Neurol 66:704–709

Buechner S, Moretti M, Burlina AP et al (2008) Central nervous system involvement in Anderson-Fabry disease: a clinical and MRI retrospective study. J Neurol Neurosurg Psychiatry 79:1249–1254

Casserly I, Topol E (2004) Convergence of atherosclerosis and Alzheimer's disease: inflammation, cholesterol, and misfolded proteins. Lancet 363:1139–1146

Chabriat H, Joutel A, Dichgans M, Tournier-Lasserve E, Bousser MG (2009) Cadasil. Lancet Neurol 8:643–653

Chao CP, Kotsenas AL, Broderick DF (2006) Cerebral amyloid angiopathy: CT and MR imaging findings. Radiographics 26:1517–1531, http://www.angiopathy.org/index.html

Cordonnier C, van der Flier WM, Sluimer JD et al (2006) Prevalence and severity of microbleeds in a memory clinic setting. Neurology 66:1356–1360

Erkinjuntti T, Inzitari D, Pantoni L et al (2000) Research criteria for subcortical vascular dementia in clinical trials. J Neural Transm Suppl 59:23–30

Fein G, Di SV, Tanabe J et al (2000) Hippocampal and cortical atrophy predict dementia in subcortical ischemic vascular disease. Neurology 55:1626–1635

Firbank MJ, Burton EJ, Barber R et al (2007) Medial temporal atrophy rather than white matter hyperintensities predict cognitive decline in stroke survivors. Neurobiol Aging 28:1664–1669

Goos JD, Kester MI, Barkhof F, Klein M, Blankenstein MA, Scheltens P et al (2009) Patients with Alzheimer disease with multiple microbleeds: relation with cerebrospinal fluid biomarkers and cognition. Stroke 40(11):3455–3460

Greenberg SM et al (2009) Cerebral microbleeds: a guide to detection and interpretation. Lancet Neurol 8:65–74

Henneman WJ, Sluimer JD, Cordonnier C, Baak MM, Scheltens P, Barkhof F et al (2009) MRI biomarkers of vascular damage and atrophy predicting mortality in a memory clinic population. Stroke 40(2):492–498

Hodges JR, Warlow CP (Oct 1990) Syndromes of transient amnesia: towards a classification. A study of 153 cases. J Neurol Neurosurg Psychiatry 53(10):834–843

Holton JL et al (2002) Familial Danish dementia: a novel form of cerebral amyloidosis associated with deposition of both amyloid-Dan and amyloid-beta. J Neuropathol Exp Neurol 61:254–267

Jagust WJ, Zheng L, Harvey DJ et al (2008) Neuropathological basis of magnetic resonance images in aging and dementia. Ann Neurol 63:72–78

Leys D, Henon H, Mackowiak-Cordoliani MA et al (2005) Poststroke dementia. Lancet Neurol 4:752–759

Mead S et al (2000) Familial British dementia with amyloid angiopathy: early clinical, neuropsychological and imaging findings. Brain 123:975–991

Merino JG, Hachinski VC (2009) Historical perspective. In: Festa JR, Lazar RM (eds) Neurovascular neuropsychology. Springer, pp 1–6

Moore DF, Frank Y, Schiffmann R et al (2003) Increased signal intensity in the pulvinar on T1-weighted images: a pathognomonic MR Imaging sign of Fabry disease. AJNR Am J Neuroradiol 24:1096–1101

Mungas D, Reed BR, Jagust WJ et al (2002) Volumetric MRI predicts rate of cognitive decline related to AD and cerebrovascular disease. Neurology 59:867–873

Neuropathology Group of the Medical Research Council Cognitive Function and Ageing Study (MRC CFAS) (2001) Pathological correlates of late-onset dementia in a multi-centre, community-based population in England and Wales. Lancet 357:169–175

Pantoni L (2010) Cerebral small vessel disease: from pathogenesis and clinical characteristics to therapeutic challenges. Lancet Neurol 9:689–701

Pantoni L, Pescini F, Nannucci S, Sarti C, Bianchi S, Dotti MT, Federico A, Inzitari D (2010) Comparison of clinical, familial, and MRI features of CADASIL and NOTCH3-negative patients. Neurology 74:57–63

Peters R, Burch L, Warner J, Beckett N, Poulter R, Bulpitt C (8 Aug 2008) Haemoglobin, anaemia, dementia and cognitive decline in the elderly, a systematic review. BMC Geriatr 8:18

Peterson PL, Axford JS, Isenberg D (2005) Imaging in CNS lupus. Best Pract Res Clin Rheumatol 19:727–739

Revesz T et al (2009) Genetics and molecular pathogenesis of sporadic and hereditary cerebral amyloid angiopathies. Acta Neuropathol 118:115–130

Schneider JA, Arvanitakis Z, Bang W et al (2007) Mixed brain pathologies account for most dementia cases in community-dwelling older persons. Neurology 69:2197–2204

Snowdon DA, Greiner LH, Mortimer JA et al (1997) Brain infarction and the clinical expression of Alzheimer disease. The Nun Study. JAMA 277:813–817

Steen RG, Miles MA, Helton KJ, Strawn S, Wang W, Xiong X, Mulhern RK (Mar 2003) Cognitive impairment in children with hemoglobin SS sickle cell disease: relationship to MR imaging findings and hematocrit. AJNR Am J Neuroradiol 24(3):382–389

Susac JO et al (2003) MRI findings in Susac's syndrome. Neurology 23(61):1783–1787

van den Boom R, Lesnik Oberstein SA, Ferrari MD, Haan J, van Buchem MA (2003) Cerebral autosomal dominant arteriopathy with subcortical infarcts and leukoencephalopathy: MR imaging findings at different ages-3rd–6th decades. Radiology 229:683–690

van der Flier WM, Middelkoop HA, Weverling-Rijnsburger AW et al (2004) Interaction of medial temporal lobe atrophy and white matter hyperintensities in AD. Neurology 62:1862–1864

van Straaten EC, Scheltens P, Knol DL et al (2003) Operational definitions for the NINDS-AIREN criteria for vascular dementia: an interobserver study. Stroke 34:1907–1912

Vermeer SE, Prins ND, den Heijer T et al (2003) Silent brain infarcts and the risk of dementia and cognitive decline. N Engl J Med 348:1215–1222

Wu CC, Mungas D, Petkov CI et al (2002) Brain structure and cognition in a community sample of elderly Latinos. Neurology 59:383–391

Contents

7.1 Introduction – Focus on White Matter Involvement

Many brain diseases are accompanied by white mater (WM) hyperintensity on T2-weighted (including FLAIR) magnetic resonance imaging (MRI) and by low attenuation on computed tomography (CT) scans. In some diseases, hyperintense lesions in the WM may accompany certain primary neurodegenerative (mostly grey matter) disorders, such as multisystem atrophy (MSA) or adult polyglucosan body disease (APBD). A special category is constituted by vascular disease, in which both grey matter and WM are affected (see chapter 6).

The current chapter focuses on disorders primarily affecting WM, encompassing classic WM diseases like inherited leukodystrophies, but also others with a predominant WM involvement, like some infections, inflammatory demyelinating diseases, and, finally, toxic and traumatic leukoencephalopathies. Obviously, the grey matter can be involved as well in some protypical WM diseases (e.g. multiple sclerosis [MS]). Extensive grey matter involvement in MS accounts, in part, for the pattern of cognitive impairment.

F. Barkhof et al., *Neuroimaging in Dementia*,
DOI: 10.1007/978-3-642-00818-4_7, © Springer-Verlag Berlin Heidelberg 2011

7.2 Infections

7.2.1 Introduction

Numerous infections affect the central nervous system (CNS) leading to meningitis, encephalitis, cerebritis or brain abscess. In most cases, the relationship with systemic infections is obvious, and secondary cerebral involvement manifests without cognitive decline as the lead symptom. This chapter restricts itself to those viral, bacterial and spirochetal infections that may first present with cognitive impairment or dementia and have no associated brain swelling. Those infections that do have associated brain swelling (e.g. Herpes simplex encephalitis and progressive multifocal leukoencephalopathy [PML]) will be discussed in Chap. 8. The findings in syphilis overlap with certain types of vasculitis due to involvement of the vascular wall.

7.2.2 Viral Infections Leading to Dementia

There are several viral infections that can lead to dementia. Some cases of acute viral encephalitis present with cognitive decline, but patients with chronic viral encephalitis more typically have cognitive impairment or dementia as the presenting symptom. Apart from herpes simplex encephalitis, most acute encephalitides are very uncommon (e.g. varicella zoster virus, human herpes virus, Japanese encephalitis and tick-borne encephalitis) and are beyond the scope of this book. The current section on viral encephalitis with dementia therefore concentrates on chronic viral encephalitides leading to dementia, that is HIV-1 encephalitis (HIVE) and subacute sclerosing panencephalitis (SSPE).

7.2.2.1 HIV Encephalitis

Synonyms
Human immunodeficiency virus (HIV) encephalitis (HIVE), HIV encephalopathy, AIDS dementia complex.

Aetiology and Histopathology

The human immunodeficiency virus (HIV-1) is a lentivirus that was discovered in 1983, shortly after the first reports of acquired immune deficiency syndrome (AIDS). When CD4 counts fall below 400/mm^3, CNS infection occurs, leading to HIV-1 encephalitis (HIVE); at autopsy, more than 80% of patients have CNS abnormalities. The resultant dementia syndrome, called HIV encephalopathy or AIDS dementia complex, occurs in 25% of patients with AIDS. In certain cases, dementia is the first presentation of AIDS, and serves as an indicator disease for the diagnosis of AIDS. With the advent of highly active antiretroviral therapy (HAART), opportunistic infections have become less common, survival extended, but there is some evidence that HIVE might have become relatively more common. More recently, HIV-associated neurocognitive dysfunction (HAND) syndrome has been recognised in patients with HIV treated during long periods of time with anti-retroviral therapy – so far, neither the pathophysiological mechanism is known, nor there is an established imaging correlate.

Macroscopically, HIVE brains show cerebral atrophy, white matter pallor and, occasionally, calcifications of the subcortical nuclei. Microscopically, pathognomonic multinucleated giant cells (MGC) are found, resulting from the fusion of infiltrating macrophages. In addition, there are abundant macrophages, as well as microglial nodules. These abnormalities are first found in the deep grey matter structures and surrounding white matter. In later stages, there is spreading to the cortex. Typically, inflammatory cells are absent in this immunodeficient disorder. Viral proteins can be indirectly detected by using immunocytochemistry or an enzyme-linked immunosorbent assay (ELISA), whereas viral nucleic acid can be detected using PCR or in situ hybridisation, even in patients with dementia without MGCs. How established HIVE leads to dementia is unclear, since neuronal damage is not always extensive. Speculation exists about excitotoxicity of viral products, or neuroglial infection. Only in a percentage of patients with HIVE feature accompanying damage to CNS tissue, including myelin pallor. This constellation is sometimes referred to as HIV encephalopathy, a term originally used to describe the neuropsychological findings of AIDS dementia.

Clinical and Laboratory Findings

The essential features of ADC are insidious progressively disabling cognitive impairment accompanied by motor dysfunction, speech problems and behavioural change. Cognitive impairment is characterised by mental slowness, memory disturbances and poor concentration. Focal cortical symptoms have also been described as the sole presenting feature (apraxia, aphasia and agnosia). Motor symptoms include a loss of fine motor control leading to clumsiness, poor balance and tremor. Behavioural changes include apathy, lethargy and diminished emotional responses and spontaneity.

Neuroimaging Strategy and Findings

HIVE is best demonstrated on heavily T2-weighted or FLAIR images (Fig. 7.1) The findings include a mild to moderate increase in signal intensity throughout the periventricular white matter, with no typical predilection – the subcortical U-fibres are characteristically spared. Diffuse cerebral atrophy is evident in advanced stages. In some cases, the white matter abnormalities are not diffuse, but rather focal, and may be indistinguishable from other related pathology, including CMV, lymphoma and PML. Administration of contrast material is essential in patients with AIDS and a differential diagnosis of HIVE. Whereas most opportunistic infections will show enhancement (either in brain parenchyma, the ependyma or meninges), this is typically absent in HIVE. The absence of enhancement in HIVE does of course not rule out PML, but the place and type of lesions usually readily distinguishes the two. PML lesions usually extend into the subcortical U-fibres (and even cortex), with markedly increased signal intensity and some swelling, whereas HIVE results in a more subtle abnormality with less involvement of the U-fibres and cortex, and no swelling. Following therapy, an immune-reconstitution inflammatory syndrome (IRIS) may occur with enlargement of lesions and contrast enhancement. On precontrast T1-weighted images, the signal intensity in HIVE lesions is normal, which helps to differentiate HIVE from PML, in which markedly reduced signal intensity is found. Similarly, the magnetisation transfer ratio is much lower in PML than in HIVE lesions. MR spectroscopy is more sensitive than MR imaging in the detection of HIV infection of the brain, showing increased choline and myo-inositol in the early (asymptomatic) stages. Lowering of N-acetylaspartate (NAA) occurs in cases of HIV encephalopathy, and becomes more obvious in the case of dementia. There are no indications for single photon emission computed tomography (SPECT) or positron emission tomography (PET).

7.2.2.2 Subacute Sclerosing Panencephalitis

Synonyms
Subacute sclerosing panencephalitis (SSPE), inclusion body encephalitis

History, Aetiology and Histopathology

Whereas measles infection can cause an acute meningoencephalitis, in rare cases – for example in the immunocompromised host – a subacute encephalitis follows. More relevant in the context of dementia is an entity called subacute sclerosing panencephalitis (SSPE), which develops 6–8 years after an asymptomatic phase of acute measles infection that occurred during early childhood. Dawson first described the disorder in 1934 under the name 'inclusion body encephalitis', whereas the name SSPE was coined by van Bogaert. This disorder has become extremely rare in the Western world following effective vaccination programmes, but it is still relatively common in Eastern Europe and Asia (esp. India). SSPE is probably determined by a slow infection caused by the hypermutated measles virus with defective M-protein.

Histopathological changes involve the cerebral cortex and white matter, as well as the brainstem, with sparing of the cerebellum. Nerve cell destruction, neuronophagia and perivenous cuffing are found. In the white matter, myelin and axons degenerate, and fibrous gliosis occurs; characteristic eosinophilic inclusions are found in the cytoplasm and nuclei of neurons and glial cells.

Clinical and Laboratory Findings

The onset of SSPE typically occurs in childhood or adolescence; however, cases have been reported with a

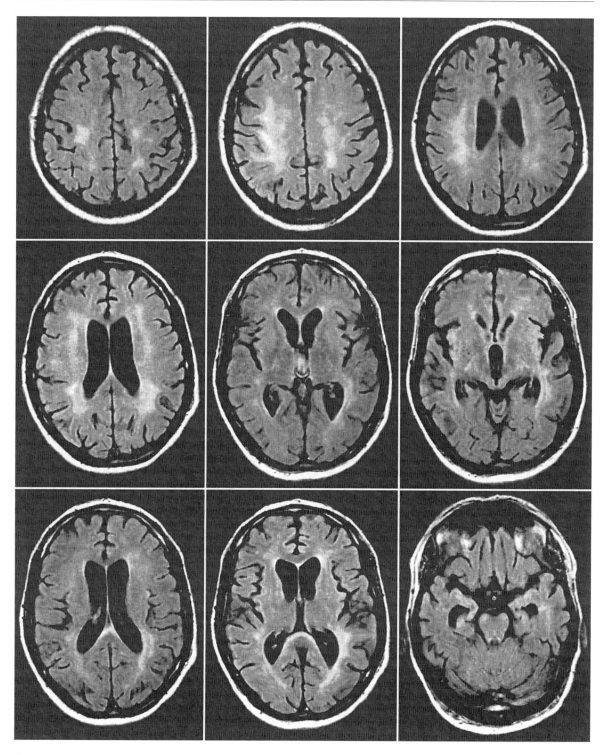

Fig. 7.1 HIV Encephalitis (HIVE). This 43-year-old man presented with subcortical dementia, gait disorders and incontinence. Axial FLAIR images showed ill-defined areas of increased signal intensity throughout the white matter (no enhancement was seen after gadolinium injection). Note also the prominent atrophy, especially in the temporal regions. The subtle and widespread involvement suggested HIV-encephalitis, and subsequent testing revealed HIV antibodies, leading to the diagnosis of AIDS

later onset between 20 and 35 years. SSPE typically has an insidious onset, with subtle slow cognitive decline and behavioural change. A typical feature is the occurrence of rhythmic attacks of myoclonus, seizures and ataxia, finally leading to complete neurological deterioration, with rigidity, unresponsiveness and autonomic dysfunction. Characteristic periodic slow-wave complexes with high voltage are observed on EEG, that can even occur in the asymptomatic stage (Rademecker complex). Ocular and visual manifestations are reported in 10–50% of patients, which include cortical blindness, chorioretinitis and optic atrophy – visual manifestations are relatively more common in adult-onset disease. Death occurs after several months to years, but in a small number of cases (5%) remission occurs. The diagnosis is based on the typical EEG, strongly increased CSF IgG (with normal protein), and strongly raised measles antibody titres (higher in CSF than blood). If recognised early, treatment with interferon and antiviral medication may be beneficial, but inevitably the disease is fatal.

Neuroimaging Strategy and Findings

In the early stages, MRI displays non-specific periventricular white matter lesions, or may even be normal. After several months, lesion sites observed on MRI include the parietal/ occipital grey matter, the adjacent white matter and sometimes the basal ganglia. With time these lesions may become less conspicuous, and lead to marked atrophy (Fig. 7.2). In the later stages, migratory lesions are found in the striatum, with hyperintensity on T2-weighted images starting in the putamen, followed by the caudate, but with initial sparing of the globus pallidus and thalamus. These can be

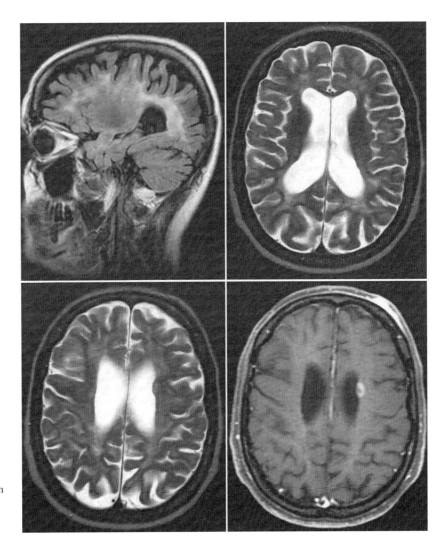

Fig. 7.2 Subacute sclerosing panencephalitis (SSPE). This 35-year-old lawyer presented memory and concentration problems for several months before developing a subacute deterioration with reduced consciousness and epilepsy. MRI revealed white matter abnormalities on FLAIR (*upper left panel*) and T2, partly diffuse, but also partly focal, with one lesion enhancing after gadolinium (*lower right panel*). Note the marked ventricular dilatation secondary to white matter loss

followed by lesions in the substantia nigra and dentate nucleus, clinically evidenced by the occurrence of parkinsonian features.

7.2.3 Bacterial Infections: Whipple's Disease

Aetiology and Histopathology

The disease that carries his name was first described by George Hoyt Whipple in 1907, and it took until about the end of the XXth century before the gram-positive actinobacteria *Tropheryma whipplei* (*T. whipplei*) was finally cultured. Primary Whipple's disease of the brain is extremely rare, but cerebral involvement may also secondarily occur in the course of a systemic infection.

Clinical Presentation and Treatment

Although the primary manifestations are gastrointestinal (e.g. malabsorption), a variety of neurological features occurs in about one third of patients with systemic disease: cognitive changes and altered consciousness, ophthalmoplegia, ataxia, and myoclonus. Preceding systemic manifestations include weight loss, fever, abdominal pain, joint pain and lymphadenopathy. In cerebral Whipple's disease, the most common manifestations are hemiparesis, cognitive dysfunction, seizures and abnormal eye movements. Investigations include duodenal biopsy looking for PAS positive inclusions. The diagnosis of cerebral Whipple's disease is suggested by PCR for *T. whipplei* in CSF. Treatment requires a combination of antibiotics for years, and may be difficult in patients with neurological involvement.

Neuroimaging Strategy and Findings

MRI may provide a diagnostic clue in patients with primary cerebral infection by showing multifocal lesions with enhancing nodules, but may be normal, especially in secondary cerebral involvement. MRI abnormalities more frequently occur than CT abnormalities, but are still only found in 50% of cases.

Lesions show T2 hyperintensity, usually no mass effect, and are located in the medial part of the temporal lobes, the hypothalamic region, or in the pons. Lesions sometimes enhance with gadolinium (Fig. 7.3). Moderate to severe atrophy is a frequent finding. In some cases, multiple mass lesions are found. Spectroscopic and PET/SPECT findings are largely unknown.

7.2.4 Spirochetal Infections and Dementia

7.2.4.1 Neuroborreliosis

Synonyms
Cerebral Lyme Disease, Neuroborreliosis

Aetiology and Histopathology

The name is derived from the city of Lyme in Connecticut, where this infection was first recognised in 1975. It is caused by *Borrelia burgdorfferi*, a spirochetal bacteria living in infected ticks in woods. Through the bloodstream, spirochetes may reach the CNS and infect astrocytes, and induce cytokine release with neurotoxic effects. Alternative pathophysiological mechanisms postulated are an immune response or vasculitis induced by the infection.

Clinical Presentation, Diagnosis and Treatment

The annular erythematous skin reaction that follows the bite of an infected tick (erythema migrans) occurs in 80% of cases, but may be overlooked. In the subacute and chronic phase, neurological symptoms occur in approximately 5% of patients, and include meningitis, radiculoneuritis and even encephalitis. In the late persistent phase, chronic encephalomyelitis occurs in rare cases, which may induce cognitive impairment, weakness in the legs, awkward gait, facial palsy, bladder problems, vertigo and back pain. A variety of psychiatric symptoms can also occur. Diagnosis can be confirmed by serological testing (ELISA followed by Western blot) and treatment with antibiotics like doxycycline.

Fig. 7.3 MRI findings in active cerebral Whipple disease. (**a, b**) Axial FLAIR and (**c**) coronal FLAIR images reveal atrophy of the right mesial temporal lobe with persisting signal abnormality suggesting gliosis. Oedema is present in the left mesial temporal lobe, caudate heads and right putamen. (**d**) Coronal contrast enhanced T1-weighted image shows patchy nodular enhancement involving the mesial temporal lobes, insular cortex and lenticulostriate vessels. (Reprinted with permission from Panegyres PK, *Pract Neurol* 2008;8:311–317)

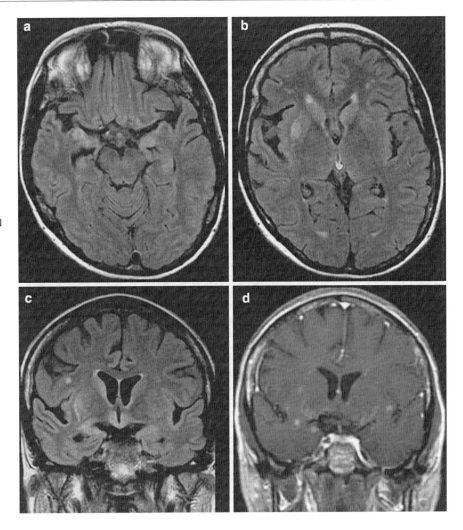

Neuroimaging Strategy and Findings

Many subjects will have normal structural imaging findings. In those with abnormal imaging, enhancement of leptomeninges and (cranial) nerves may be seen with gadolinium. In the chronic phase, nonspecific scattered white matter lesions can be found, indistinguishable from age-related white matter lesions, but sometimes also mimicking inflammatory disorders like MS. Rarely, isolated or multiple mass lesions are found, sometimes with enhancement. SPECT and PET may reveal patchy hypoperfusion, especially in the white matter and subcortical grey matter.

7.2.4.2 Neurosyphilis

Synonyms

Neurosyphilis, generalised paresis of the insane, dementia paralytica *dementia paralytica*.

Aetiology and Histopathology

Infection with *Treponema pallidum* may cause neurosyphilis many (6–50) years after primary infection. The incidence of syphilis is rising in Eastern Europe and co-infection with HIV may interfere with treatment. Like neuroborreliosis, the first stage of neurosyphilis is meningitis, with or without cranial nerve palsy, which can be followed after many years by meningovascular and late vascular stages.

Histopathologically, meningovascular syphilis features endarteritis with perivascular inflammation. The resulting inflammation and fibrosis may induce aneurysmal dilation and luminal narrowing, leading to cerebrovascular thrombosis, ischaemia and infarction. Leptomeningeal granulomas, called gummas, are well-circumscribed masses of granulation tissue (avascular), caused by a cell-mediated immune response to *T. pallidum*. They are often located in the dura, secondarily involving the underlying cortex.

Clinical Presentation, Diagnosis and Treatment

About 50% of cases with neurosyphilis present with psychiatric disturbances including psychosis, cognitive decline and dementia. The remainder present with stroke, probably secondary to vasculitis, cranial nerve palsy, epilepsy (due to focal irritation by gumma) and spinal cord syndromes (including tabes dorsalis). The full-blown picture of neurosyphilis (also referred to as 'general paresis of the insane') includes dysarthria, myoclonus, hyperreflexia and Argyll-Robertson pupil. The diagnosis can be difficult, especially when VDRL serology is negative. Treatment is with antibiotics (e.g. penicillin G), which may be less effective in patients with HIV.

Neuroimaging Strategy and Findings

CT and MR may show cerebral infarction (territorial or lacunar). Arteritis can be revealed by means of catheter angiography (DSA) or MRA (Fig. 7.4). MRI may show non-specific white matter lesions that sometimes involve the basal ganglia and the medial temporal lobe. Gadolinium injection may reveal granulomatous meningitis (thickened, enhancing meninges), sometimes with accompanying hydrocephalus. Gummas are enhancing focal masses in the dura that are predominantly hyperintense on T2-weighted images, but may also present foci of low signal intensity. Generalised atrophy may develop, and the medial temporal lobe may be atrophied, especially in patients with general paresis. In cases with normal MRI, SPECT may reveal hypoperfusion.

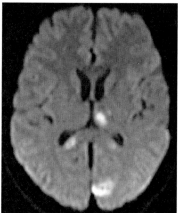 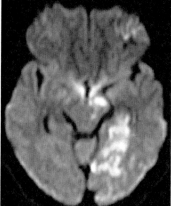 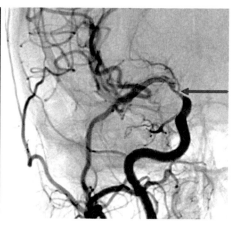

Fig. 7.4 Imaging findings in neurovascular syphilis. This 29-year-old woman was found in a confused state. She was noted to be both alert and cooperative, but disorientated in time and place. There was significant memory impairment involving episodic and prospective memory, and both immediate and delayed recall of visual and spatial information. Reading, writing and naming were normal. Retrograde memory was intact. DWI (*left*) demonstrated multiple areas of ischaemia, including in the left hippocampal region and the left thalamus. Catheter angiography (*right*) revealed stenosis of the distal supra-clinoid right internal carotid artery (*red arrow*). CSF analysis revealed signs of meningitis and positive VDRL assay (1:80), which was confirmed by blood serology that also revealed HIV-1 infection. The patient was treated for large-vessel CNS vasculitis secondary to meningovascular neurosyphilis with intravenous penicillin G for 14 days alongside initiation of highly active anti-retroviral therapy (HAART). After 3 months, the patient was orientated in person, place and time. Immediate recall had normalised but profound difficulties with delayed recall remained. (Reprinted with kind permission from Killian O'Rourke, *J Neurol* 2010;257:669–671)

7.3 Inflammatory Disorders

7.3.1 Introduction

This section deals with inflammatory disorders that affect the CNS and cause cognitive impairment or dementia. The prototypical primary inflammatory CNS disorder is multiple sclerosis and its monophasic variant, ADEM (Sect. 7.3.2). Secondary CNS involvement occurs in a variety of systemic inflammatory disorders. In most cases of secondary CNS involvement, cognitive decline is very uncommon, for example in Behcet's disease; these will not be considered in any detail. In others, CNS involvement occurs more frequently, and can even be the presenting symptom, such as in neurosarcoidosis (Sect. 7.3.3). In terms of pathogenesis and symptomatology, overlap exists with vasculitis (see Sect. 6.5.2).

7.3.2 Multiple Sclerosis and ADEM

Synonyms
Multiple sclerosis (MS), acute disseminated encephalomyelitis (ADEM)

Introduction and History

Already at the time of its first description by Charcot in 1868, it was recognised that MS is not simply a white matter disease, but also affects grey matter. Whereas traditionally regarded as a neurological disorder impacting mainly spinal cord function, the cognitive dysfunction associated with MS and ADEM is belatedly receiving more attention. Cognitive dysfunction is an important cause of disability in these demyelinating disorders with profound consequences for patients and their families. MRI studies have revealed the extent of cortical lesions and grey matter atrophy, in agreement with histopathological studies. Although the cognitive deficits in MS and ADEM arise from both white and grey mater destruction, this chapter will concentrate on the grey matter and neurodegenerative aspects of these disorders.

Aetiology, Histopathology and Genetics

The cause of MS being unknown, one can only speculate about the aetiology of (cortical) MS lesions. Like their white matter counterparts, cortical lesions are defined by focal demyelination although this can only be detected with immunohistochemical stains. There is relatively little inflammation in the cortical MS lesions; damage to neurons, reduction of synapses and dendritic pathology, however, may explain the cortical thinning observed in later phases of the disease. Of special interest are the type III cortical lesions with a subpial configuration, which can be quite extensive in progressive cases, occupying most of the neocortex. They have fuelled speculations that subpial lesions are due to some diffusible factor arising from the meninges or CSF – perhaps antibody and B-cell mediated – a hypothesis so far unconfirmed. Several susceptibility genes for MS have been identified, mostly in the HLA region, but none are specifically related to grey matter involvement or cognition. APOE4 may be a risk factor for more rapid progression in MS and there have been reports that a polymorphism of the APOE regulatory region is associated with cognitive impairment in patients with MS.

Cortical involvement can be limited in early cases – perhaps with a preponderance of cingulated/frontotemporal involvement. In progressive phases, cortical involvement can be quite extensive, and even outweigh white matter damage (Fig. 7.5). Of special interest is the involvement of deep grey matter structures including the hippocampus in MS. Imaging and immunohistochemical studies both reveal frequent and extensive involvement of the hippocampus in MS (Fig. 7.6), associated with memory impairment. Involvement of the thalamus has been related to fatigue. The thalamus is also frequently involved in ADEM, a disease which otherwise has a higher propensity for (cortical) grey matter involvement than MS.

Clinical Presentation, Epidemiology and Treatment

MS often presents with focal neurological deficits caused by lesions in an eloquent long tract, like optic neuritis or a spinal cord syndrome. At the time of presentation, most patients have clinically silent dissemination in space on their MRI scan (Table 7.1), illustrating that clinically silent lesions are common, including the juxtacortical

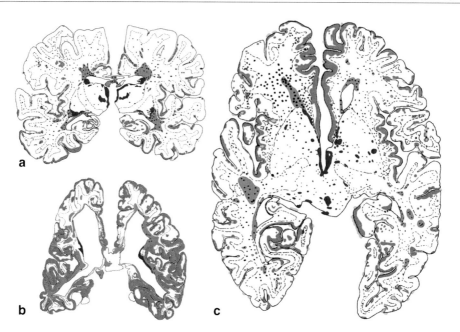

Fig. 7.5 Camera Lucida drawings showing distribution of MS lesions at post-mortem. Focal WM plaques are depicted in *green*, cortical lesions in *orange* and deep GM lesions in *blue*; *black dots* represent perivascular inflammatory infiltrates. (**a**) Secondary progressive MS; male patient, aged 43 years, 16 years of disease duration; moderately affected brain with focal periventricular WM lesions and cortical lesions mainly affecting the cingulated cortex, the insular cortex and the basal temporal cortex. (**b**) Secondary progressive MS, female, aged 46 years, 16 years disease duration; extremely severely affected brain with massive atrophy, extensive cortical demyelination and widespread periventricular demyelination in the white matter. (**c**) Primary progressive MS; aged 55 years, 5 years disease duration; relatively few and small focal white matter lesions, but extensive cortical demyelination, mainly in the cingulated cortex, the insular cortex and the basal temporal cortex. In addition, there is massive diffuse inflammation in the normal-appearing white matter. (Reprinted with kind permission by Kutzelnigg and Lassmann from Handbook of Neurology 2008)

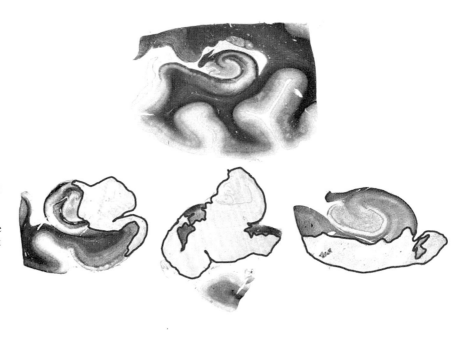

Fig. 7.6 Hippocampal demyelination (Kindly provided by Jeroen Geurts). Myelin staining shows normal myelin density in the hippocampus in a control subject (*top panel*). Note varying degrees of demyelination (outlined in *red*) in the hippocampi of three different MS patients (*bottom panel*), showing almost complete demyelination of the hippocampus and surrounding tissue in one of the cases

Table 7.1 MRI criteria for MS

- Clinical presentation suggestive of MS
  - Other diagnoses ruled out by appropriate tests
    - Consider MRI 'red flags' (Charil, *Lancet Neurol* 2006)

- Dissemination in space (DIS)
  - Clinically: poly-symptomatic onset
  - MRI: 3 or more modified Barkhof criteria
    - ≥1 juxtacortical lesion
    - ≥1 gadolinium enhancement (or ≥9 T2 lesions)
    - ≥1 infratentorial lesion
    - ≥3 periventricular lesions
  - If CSF abnormal: 2 brain lesions sufficient

- Dissemination in time (DIT)
  - New clinical symptoms
  - Repeat MRI: Gad-enhancement or new T2 lesions

Source: (*Ann Neurol* 2005;58(6):840–846)

lesions typical of MS. True cortical syndromes like aphasia, however, are extremely uncommon in MS. There is an increased frequency of epilepsy in MS, attributable to cases with relative high numbers of (juxta) cortical lesions. In established cases of MS, neuropsychological testing reveals cognitive impairment in up to 40% of cases: typically, the deficits are subcortical with declines in attention, concentration, memory and executive function, and worsen with disease progression.

Occasionally, cognitive deficits (or frank dementia) may be the presenting feature of MS and the diagnosis is often only considered after MRI (Fig. 7.7). CSF oligoclonal bands provide supportive evidence. Disease-modifying therapy for MS is clearly reducing relapses and development of (white matter) lesions on MRI, and is likely to reduce progression of cognitive deficits; some studies suggest some symptomatic benefit with cholinesterase inhibitors, but good evidence from large controlled trials is lacking.

Whereas MS, by definition, is disseminated in time, ADEM usually is a monophasic acute demyelinating disorder with a spatial dissemination pattern quite similar to MS. The clinical presentation of ADEM is more often encephalopathic than MS, including drowsiness, disorientation and epilepsy, reflecting its greater propensity for grey matter involvement. Relapsing forms of ADEM can occur, often triggered by the same type of initiating event (e.g. infections or vaccination).

Neuroimaging Strategy and Findings

MRI is the imaging modality of choice (Table 7.2). In addition to the standard protocol with T2 and T1-weighted images (with gadolinium), additional sequences like FLAIR and double-inversion recovery (DIR) are useful. Compared to FLAIR, DIR employs a second inversion pulse, thereby not only nulling CSF signal as in FLAIR, but also white matter signal.

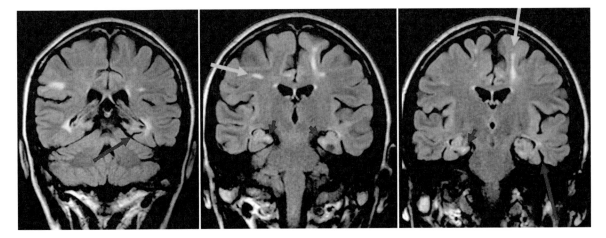

Fig. 7.7 Cognitive presentation in MS. This 25-year-old female patient with MS presented with sudden and rapid onset of memory and concentration deficits, leading to social and psychological invalidity within 3 years. Coronal FLAIR images show multiple typical MS lesions with an ovoid shape in the deep and periventricular white matter (*green arrows*). Intriguing is the extensive amount of cortical involvement in the hippocampus (*short red arrows*) which has led to hippocampal atrophy, as evidenced by widening of the collateral sulcus (*long red arrows*)

Table 7.2 Neuroimaging protocol and findings

MRI Sequences
- 3D sagittal FLAIR isotropic
- 2D transverse proton-density/T2-weighted
- 2D transverse post-gadolinium T1-weighted
- Optional: double inversion recovery (DIR)
  - MRS, DWI not useful, except for pseudotumoural MS

Brain MRI findings in MS
- White matter lesions
  - Periventricular and juxtacortical
  - Ovoid shape due to perivenular origin
  - Temporal lobe typically affected as well
  - Infratentorial lesions: brainstem, cerebellum, spinal cord
  - Gadolinium-enhancement and T1-black holes 'Open-ring sign' in pseudotumoural MS
  - Corpus callosum involvement and 'Dawson fingers'
- Focal cortical lesions
  - Cortical lesions enhance less frequently than WM lesions
  - Truly intracortical lesions are best seen using DIR
  - Juxtacortical lesions (U-fibres) more easily detected using FLAIR
  - Hippocampal lesions best seen on coronal images
- Cortical atrophy
  - Cortical thinning best seen using advanced post-processing
  - Focal cortical atrophy related to (juxta)cortical lesions

Brain MRI findings in ADEM
- Larger, more oedematous lesions
- Variable pattern of enhancement (all lesions, some, or none), mostly depending on the stage of the lesions
- More frequent grey matter involvement
  - Including thalamus and basal ganglia
- Frequent infratentorial lesions
- No new lesions at follow-up

Differential diagnosis of MS and main differentiating imaging findings	
ADEM	Larger and more fuzzy lesions
	Marked thalamic involvement
Leukodystrophies	Symmetric white matter involvement
	• More extensive cortical lesions in mitochondrial encephalomyopathy with lactic acidosis and stroke-like episodes (MELAS)
Vascular disease	Lacunes, infarcts, microbleeds
Infections	Asymmetrical/discontinuous lesions in PML

7.3.3 Neurosarcoidosis

Synonyms

Neurosarcoid; CNS sarcoid; Cerebral Besnier–Boeck–Schaumann disease

Aetiology and Histopathology

Sarcoidosis is a granulomatous inflammation of unknown aetiology that may involve all organs, including the CNS. The histopathological hallmarks include epithelioid granulomas without staining for infectious agents and usually without caseation. These granulomas often incorporate multinucleated giant cells and lymphocytes and are found in the leptomeninges, for example around the cranial nerves and pituitary stalk. Secondary demyelination may occur especially around Virchow–Robin spaces.

Clinical and Laboratory Findings

Whereas sarcoidosis is a relatively common disorder, only ~5% of the cases have clinical CNS involvement, giving an estimated prevalence of ~1:100,000 for neurosarcoidosis. Subclinical involvement probably occurs in 15–20% of cases. The most common manifestation is cranial nerve palsy, followed by headache, epilepsy, pituitary insufficiency and myelitis. In a small number of cases, cognitive impairment is observed. Incidentally, psychosis and dementia have been reported as the presenting feature. A definitive

The resulting images (Fig. 7.8) are grey matter only images and display lesions with even brighter signal.

Advanced MRI techniques are being used in a research setting, but have little diagnostic value. For example, abnormal MTR or FA have been reported in the cortex of patients with MS, and reduced NAA and increased MI/Cr ratios in deep grey matter structures like the thalamus and hippocampus. High field imaging may offer better visualisation of cortical involvement (Fig. 7.9) in the future. Cortical thickness reduction can be better detected using advanced image processing techniques (Fig. 7.10).

There is no established role for nuclear imaging in individual patients, although reduced metabolism on FDG-PET is found on a group-level, whereas increased PK11195 binding (a glial marker) has been reported in grey matter.

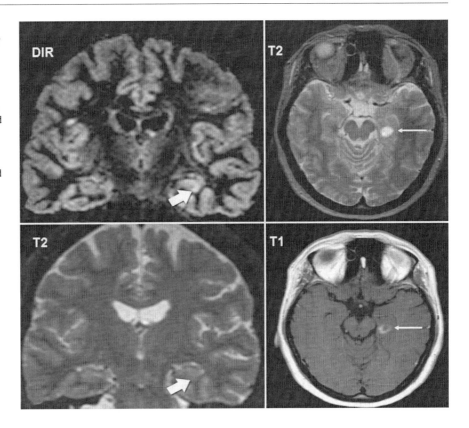

Fig. 7.8 Hippocampal involvement on MRI (Kindly provided by Jeroen Geurts). Coronal DIR image (*upper left*) shows lesion with high signal intensity in the left hippocampus of patient with MS (*short arrow*) that is hard to see on conventional T2-weighted image below. On the right side, transverse T2 and gadolinium-enhanced T1-weighted images show an active lesion in the left hippocampus (*long arrows*) in another patient

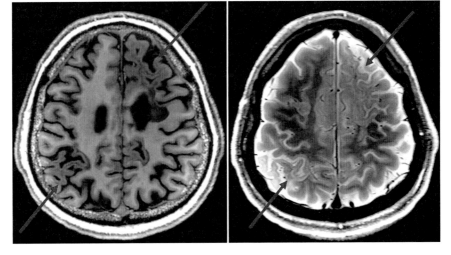

Fig. 7.9 Cortical involvement in MS at 7T (kindly provided by Wolter de Graaf). Note extensive involvement of the juxtacortical white matter on heavily T1-weighted (*left*) and T2-weighted (*right*) MR images with extension into the cortex and accompanying atrophy (*arrows*). (Massimiliano Calabrese, *Neurology* 2010;74:321–328)

diagnosis requires histopathological confirmation. The diagnostic process should confirm CNS involvement and then provide supportive evidence for the underlying disease; in the absence of a positive tissue biopsy, the most useful diagnostic tests are gadolinium-enhanced MRI of the brain and CSF analysis, although both are non-specific. The Kveim test is no longer used because of the risk of transmitting infection. High resolution CT of the chest can be virtually diagnostic in the correct context. Gallium scanning may show patchy uptake in the mediastinum and parotids.

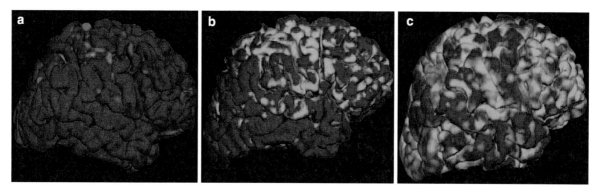

Fig. 7.10 Cortical thickness (CTh) maps in MS. Lateral views of a 3D representation of the brain with CTh maps overlaid in a *red/green* colour scale on the typical (mean) case for each group: (**a**) a healthy 35-year-old man without neurologic problems, mean CTh 2.53 mm; (**b**) a 36-year-old cognitively normal man with 5 years of disease duration of relapsing remitting MS (RRMS), mean CTh 2.32 mm; (**c**) a 34-year-old man also with 5 years of disease duration of RRMS but with cognitive impairment, mean CTh mm 2.05. Cortical thinning below 2.0 mm is displayed in *green*. Normal cortical areas thicker than 2.0 mm are displayed in *red*. (Reprinted with permission from *Neurology* 2010;74;321–328)

Neuroimaging Strategy and Findings

MRI is the imaging method of choice and administration of contrast mandatory to show the typical leptomeningeal enhancement that may extend into the Virchow–Robin spaces or along the cranial nerves (Fig. 7.11). The differential diagnosis includes primary CNS angiitis, lymphoma, tuberculosis and histiocytosis. In cases with extensive granulomatous inflammation, CSF flow may be obstructed leading to hydrocephalus. Non-specific periventricular hyperintense lesions can be seen on T2-weighted images that may strongly resemble multiple sclerosis. Dural lesions may mimic meningioma.

7.3.4 Coeliac Disease

Synonyms

C(o)eliac sprue, non-tropical sprue, endemic sprue, gluten enteropathy or gluten-sensitive enteropathy, and gluten intolerance

Aetiology and Histopathology

Coeliac disease is a multiorgan systemic disease that most commonly affects the gut but also affects other organs, especially the skin. It probably is an autoimmune disease caused by gluten. Up to 10% of patients with coeliac disease and gastrointestinal symptoms have otherwise unexplained neurological symptoms, most often ataxia and peripheral neuropathy. In a small number of subjects with coeliac disease, cognitive decline and even dementia occur, sometimes associated with increased celiac disease activity. The cause of CNS involvement in coeliac disease remains elusive – theories include immune cross-reactivity with CNS antigens and metabolic impairment due to malabsorption.

Clinical and Laboratory Findings

Cognitive dysfunction may feature memory impairment, acalculia, confusion and personality change. In some cases, cognitive decline may be rapid, and lead to suspicion of Creutzfeld–Jacob disease. The diagnosis of celiac disease is based on small bowel biopsy, detection of antigliadin and anti-endomysial antibodies, tissue transglutaminase immunoglobulin A positivity and response to gluten-free diet. Diminution of CNS symptoms by such a diet suggests causality, but does not provide definitive evidence.

Neuroimaging Strategy and Findings

MRI may show generalised cerebral atrophy and hyperintense lesions on T2-weighted and FLAIR images. The latter may include confluent periventricular changes

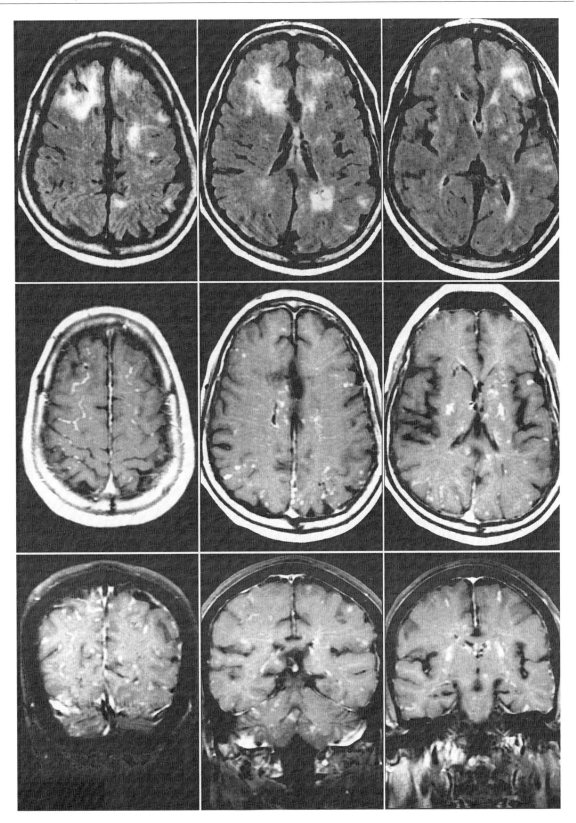

Fig. 7.11 Neurosarcoidosis. A 35-year-old man with auditory problems and neurosarcoidosis developed cognitive disturbances, characterised by a frontal syndrome. The FLAIR images show multifocal white matter lesions, in the periventricular and subcortical regions. The middle row (transverse T1) and bottom row (coronal T1 with fat suppression) show extensive enhancement of the leptomeninges and Virchow–Robin spaces

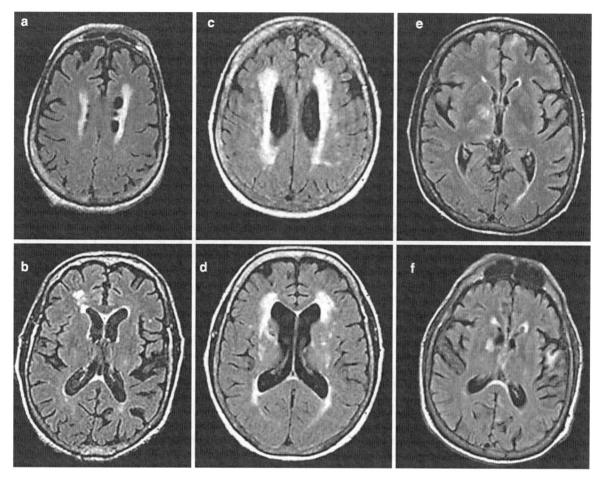

Fig. 7.12 MRI findings in celiac disease with cognitive impairment. FLAIR abnormalities vary between patients, and include confluent areas of periventricular hyperintensities (**a**), scattered hyperintense foci involving the subcortical and deep white matter (**b**), or a combination of both (**c, d**). The interval change over a period of four months in one patient reveals increased signal intensity in the right thalamic lesion and the appearance of a new subcortical lesion in the left temporal lobe (**e, f**) (Reprinted with permission from Keith Josephs, *Arch Neurol* 2006;63: 1440–1446)

(Fig. 7.12a), patchy cortical and subcortical areas of hyperintensity (Fig. 7.12b), or both (Fig. 7.12c and d).

7.4   Inborn Errors of Metabolism

7.4.1   Introduction

A large number of inborn errors of metabolism produce white matter demyelination, referred to collectively as leukodystrophies. Many individual disorders are extremely rare; however, as a group, the leukodystrophies are not uncommon. Most of the leukodystrophies present early in life and many are fatal in childhood. Although dementia may be a later feature of some of these aggressive early-onset diseases, these are beyond the scope of this book. By contrast, a number of leukodystrophies present later in life and do have cognitive decline as a characteristic feature – in this chapter, we will consider some of the more common ones that do so. In addition to the known disorders discussed, there are still a number of unclassified leukodystrophies (e.g. orthochromatic leukodystrophies) that escape current classification systems.

The cardinal imaging feature of the leukodystrophies is involvement of the white matter, with hyperintensity on T2-weighted and FLAIR images. Involvement usually occurs in a strictly symmetric fashion (except for some mitochondrial disorders). The corticospinal tracts and cerebellum are frequently involved structures. In most diseases, there is little in the way of enhancement (except in adrenomyeloneuropathy), which differentiates them from inflammatory disorders like MS and vasculitis. Important diagnostic clues can be obtained from the family history, and from additional involvement of the peripheral nervous system, the eyes, adrenal glands, tendons or other organs. Extensive laboratory and genetic testing is very important in the diagnostic work-up.

7.4.2 Fragile X-Associated Tremor/Ataxia Syndrome (FXTAS)

Synonyms
Fragile-X premutation carrier syndrome

Aetiology and Genetics

Fragile X syndrome is caused by a full-blown trinucleotide repeat mutation, (large >200 CGC expansion) in a gene called FMR1 (fragile X mental retardation gene 1) on the X-chromosome, leading to defective/absent protein production. It is the major genetic cause of mental retardation and autism. Patients with FXTAS have smaller, so-called 'premutation', expansions (50–200 CGC repeats) in the FMR1 gene leading to *excess production* of the FMR1 protein with a later and more subtle onset of symptoms: progressive cerebellar and cognitive dysfunction.

Clinical Presentation and Work-Up

FXTAS usually develops between the ages of 50–80, mostly in men. Symptoms include intention tremor, ataxia, neuropathy, mood instability, irritability, changes in personality in association with memory loss and intellectual decline. The diagnosis (Table 7.3) is based on three factors:

1. Positive carrier testing for the FMR1 premutation
2. Typical findings on neurological examination
3. MRI suggestive of FXTAS

Individuals with FXTAS are often misdiagnosed with other conditions including Parkinson's disease or MSA or considered to have an idiopathic cause of cerebellar degeneration or peripheral neuropathy.

Neuroimaging Findings

MRI can play a crucial role in the diagnosis of FXTAS, by showing characteristic symmetric lesions hyperintense on T2-weighted images in the cerebellar hemispheres close to the dentate nucleus and both middle cerebellar peduncles (MCP sign, Fig. 7.13). In addition, WM lesions can be seen in the cerebral hemispheres hemispheres. Generalised atrophy is an associated feature, and cerebellar atrophy additionally supports the diagnosis (Table 7.3).

The differential diagnosis of the MCP sign includes paraneoplastic syndromes, multiple sclerosis, and postradiation therapy, MSA, dentatorubral-pallidoluysian atrophy (DRPLA) and Wilson's disease.

Table 7.3 Diagnostic criteria for FXTAS

Definite FXTAS	Probable FXTAS	Possible FXTAS
Intention tremor *or* gait ataxia *and* MCP sign on MRI *or* intranuclear inclusions on post-mortem	Intention tremor *and* gait ataxia *or* MCP sign on MRI *and* a minor clinical feature: parkinsonism, executive function deficits, moderate short-term memory deficiency	Intention tremor *or* gait ataxia *and* WM lesions on MRI in the cerebrum *or* moderate generalised brain atrophy

The differential diagnosis of the MCP sign includes paraneoplastic syndromes, multiple sclerosis (MS), and post-radiation therapy, MSA, DRPLA and Wilson's disease

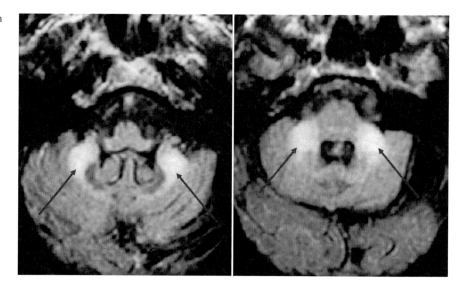

Fig. 7.13 The 'MCP sign' in FXTAS. This 61-year-old man with fragile X premutation showed hyperintense signal (*red arrows*) in the middle cerebellar peduncles (MCP) and adjacent inferior cerebellar white matter on FLAIR images. (Reprinted with permission from Brunberg – *AJNR* 2002;23:1757–1766)

7.4.3 Mitochondrial Dementia

Synonyms

Mitochondrial disorders; respiratory chain disease, mitochondrial myopathy (and numerous specific syndromes – see below)

Aetiology and Genetics

Mitochondrial disorders are caused by acquired or inherited mutations in one of the genes encoding the respiratory chain proteins within mitochondria. These can either be mutations in mitochondrial DNA (with a strictly maternal inheritance), as in mitochondrial encephalomyopathy with lactic acidosis and stroke-like episodes (MELAS) and Kearns–Sayre syndrome (KSS), or a chromosomal mutation, as found in Leigh syndrome.

Clinical Presentation and Work-Up

Mitochondrial disorders are typically multi-system in nature: depending on the gene and protein involved, many organs of the body can be affected, but the brain is particularly vulnerable due to its high energy demand. Systemic involvement (e.g. myopathy, cardiac or gastrointestinal symptoms or haematologic disorders) may precede CNS manifestations. The same mutation may cause a different type of organ involvement in different individuals. Cerebral involvement can present with psychiatric symptoms (altered consciousness, personality change, psychosis) or cognitive decline (verbal, executive, visuo-spatial). A feature of some mitochondrial disorders with cognitive involvement may be an episodic encephalopathy. Brain MRI and MR spectroscopy are important diagnostic pointers as well as tests of serum and CSF pyruvate and lactate. Diagnostic confirmation may be provided by histopathological examination of affected tissue (e.g. muscle biopsy) and/or genetic testing for (mitochondrial) mutations.

Neuroimaging Findings

Plain CT scans of the brain may show calcification; whereas MRI shows hyperintense lesions on T2-weighted and FLAIR images, often involving the deep grey matter or the cortex. Structures frequently involved are the lentiform nucleus, periaqueductal grey matter and dentate nucleus (e.g. in Leigh syndrome). In MELAS, multifocal cortical lesions can be found (Figs. 7.14 and 7.15). In the acute phase, lesions may have associated swelling and show restricted diffusion on DWI. In the chronic phase, tissue loss ensues. Despite its lower energy demand, white matter is also frequently involved and sometimes even in isolation (e.g. diffuse U-fibre involvement in KSS). In mitochondrial neurogastrointestinal encephalopathy (MNGIE) a diffuse symmetrical

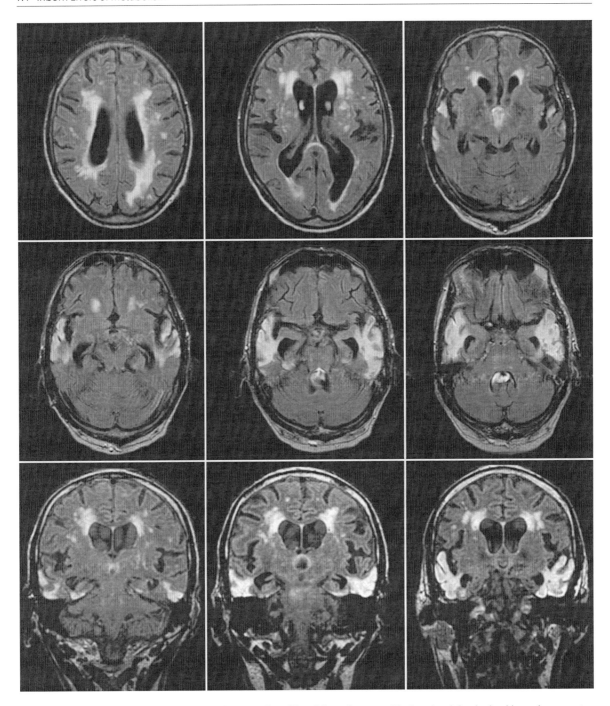

Fig. 7.14 Adult-onset mitochondrial encephalomyopathy with lactic acidosis and stroke-like episodes (MELAS). A previously healthy 60-year-old man presented with cognitive impairment. MRI showed scattered white mater lesions, but also extensive bilateral temporal lesions involving both white and grey mater, consistent with non-territorial ischaemia. Laboratory work-up revealed MELAS. (Case kindly provided by Debbie Duyndam)

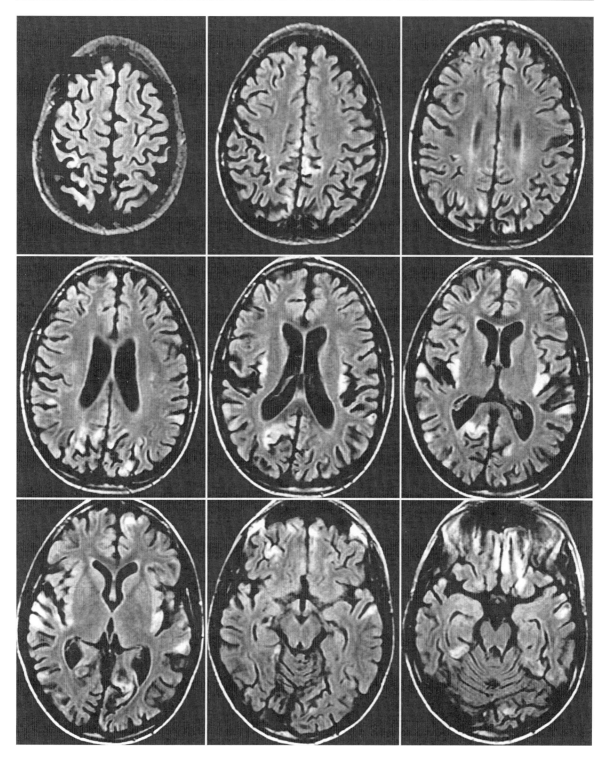

Fig. 7.15 Cognitive impairment in a child with mitochondrial encephalomyopathy with lactic acidosis and stroke-like episodes (MELAS). Axial FLAIR images from a 9-year-old boy with MELAS who developed cognitive impairment and then progressive dementia. Note multiple hyperintense lesions in the cortex with a random distribution, some with associated atrophy, and others (apparently more acute) with associated swelling

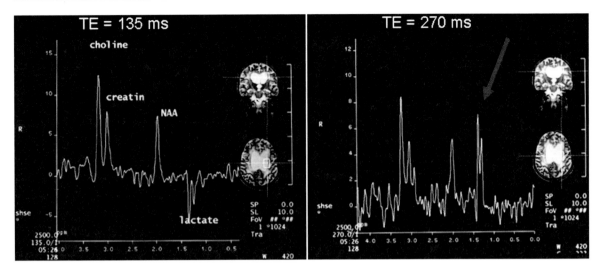

Fig. 7.16 Lactate on MR spectroscopy. Single-voxel spectroscopy revealing a lactate doublet resonating at 1.35 ppm. Note that at TE = 135 ms the phase is negative, but positive at TE = 270 ms due to J-coupling

leukoencephalopathy sparing the corpus callosum is seen. MR spectroscopy can be very useful by demonstrating lactate (a doublet resonating at 1.35 ppm) – see Fig. 7.16.

Differential diagnosis includes hypoxic insult, moyamoya disease, vasculitis, ADEM, vascular dementia (including CADASIL) and prion disease. In addition to examining the pattern of lesions, MR angiography and gadolinium-enhancement can be useful in the work-up. Lactate found by spectroscopy occurs in all ischaemic disorders, but when found throughout the brain and beyond periods of clinical deterioration should raise suspicion of a mitochondrial disorder.

7.4.4 Diffuse Neurofibrillary Tangles with Calcification (DNTC)

Synonym
Kosaka–Shibayama disease

Introduction

Kosaka proposed the term 'DNTC' for a rare form of presenile dementia in 1994 when he added two cases

to the first one described in 1973. Since then, 30 more cases have been reported. Most patients are Japanese; only one Caucasian has been described.

As the name suggests, DNTC is characterised by its imaging and histopathological features: imaging shows cerebral calcification and frontotemporal atrophy, whereas autopsy findings reveal neurofibrillary tangles and neuropil threads with an absence of amyloid plaques. In the early stages of the disease, cognition may be preserved. The disorder may be underrecognised due to similarities with Fahr's syndrome.

Clinical Presentation

Women are four times more commonly affected than men. The youngest patient reported was 48-years-old, but most patients present with symptoms in their sixth decade of life. Generally, the first symptoms are amnestic problems with early dysexecutive and frontal behavioural changes. Hallucinations, delusions and depression may accompany the neurological deterioration and parkinsonism is present in about half of the patients reported. A variant of the disorder with agnosia and apraxia as well as biparietal atrophy on MRI mimicking Alzheimer's disease has been reported. The duration of illness ranges from 3 to 24 years.

Differential Diagnosis

Fahr's syndrome (synonyms Chavany–Brunhes syndrome and Fritsche's syndrome) is a hereditary disorder characterised by movement disorders (such as rigidity, hypokinesia, tremor, choreoathetosis and ataxia), behavioural and mood disturbance, and dementia. In contrast to DNTC, the latter is never the presenting symptom. Imaging in both disorders shows a characteristic pattern of calcification in the basal ganglia and the cerebellum. A similar pattern of calcification, without asymmetric frontotemporal atrophy, can be seen in hyperparathyroidism with abnormal serum calcium and phosphorus, pseudoparathyroidism, mitochondrial encephalopathy, and lead poisoning. The cognitive and behavioural features may mimic FTLD or AD. However, imaging features of calcification combined with atrophy of the superior temporal gyrus and hippocampus point towards a diagnosis of DNTC.

Neuroimaging Findings

Essential imaging features are calcification in the cerebellum, basal ganglia and sometimes periventricular white matter, best seen on CT or with susceptibility-weighted MRI. Asymmetric atrophy of the temporal and frontal lobes is present with involvement of the superior temporal gyrus, parahippocampal gyrus and hippocampus. Several authors have added the finding of focal lesions hyperintense on T2-weighted MRI in the centrum semi-ovale, temporal and frontal lobes, probably associated with arteriosclerotic changes in arteries (please note that none of these patients suffers from hypertension). SPECT and PET imaging may show decreased blood flow or hypometabolism in the temporal and frontal lobes. Notably, with FDOPA-PET, no abnormalities in the basal ganglia have been discovered (Fig. 7.17).

7.4.5 Cerebrotendinous Xanthomatosis (CTX)

Synonyms

Cerebrotendinous xanthomatosis (CTX), van Bogaert–Scherer–Epstein syndrome

Aetiology, Genetics and Histopathology

CTX is a rare metabolic disorder (MIM #213700) with an autosomal recessive mode of inheritance first described by van Bogaert in 1937. In patients with CTX, bile acid synthesis is impaired due to a defect in the activity of the hepatic mitochondrial enzyme sterol 27-hydroxylase of the cytochrome P450 family. Excessive cholestanol accumulates in many tissues (tendons, brain and lung). An indirect effect is a low level of HDL that may lead to the accumulation of tissue sterols in atheromas and xanthomas by reduction of reverse sterol transport. The gene CYP27A1 has been mapped to chromosome 2q: the gene has been cloned, its structure determined and a number of different mutations reported.

Neurological dysfunction results from the deposition of cholesterol and cholestanol in the nervous system and the replacement of cholesterol by cholestanol in myelin layers, affecting the stability of the myelin sheath. Macroscopic findings include generalised brain atrophy with multiple yellowish deposits in the choroid plexus and white matter. Microscopically, extensive demyelination and gliosis is observed in the cerebellar white matter and the pyramidal tracts with multiple dispersed lipid crystal clefts. In addition, there is accumulation of foamy cells and homogeneous myelin-like material, especially around the vessels. Since accumulation of cholesterol and cholestanol may be neurotoxic, there is axonal damage and loss as well.

Clinical Presentation and Treatment

Most patients are of borderline or low intelligence and their school performance can be poor. The more specific clinical manifestations usually appear in late childhood or early adolescence, or even later. Early manifestations of the disease include cataracts, diarrhoea and (Achilles) tendon xanthomas. During the second and third decades, neurological problems gradually become manifest with signs of cerebellar ataxia, spastic paraparesis and tetraparesis, signs of dysfunction of the posterior columns and of a peripheral neuropathy. About 40% of patients develop epilepsy. In the third decade, a decline in intellectual functions usually occurs, but there is a wide range in the rate of cognitive decline. Changes in personality and

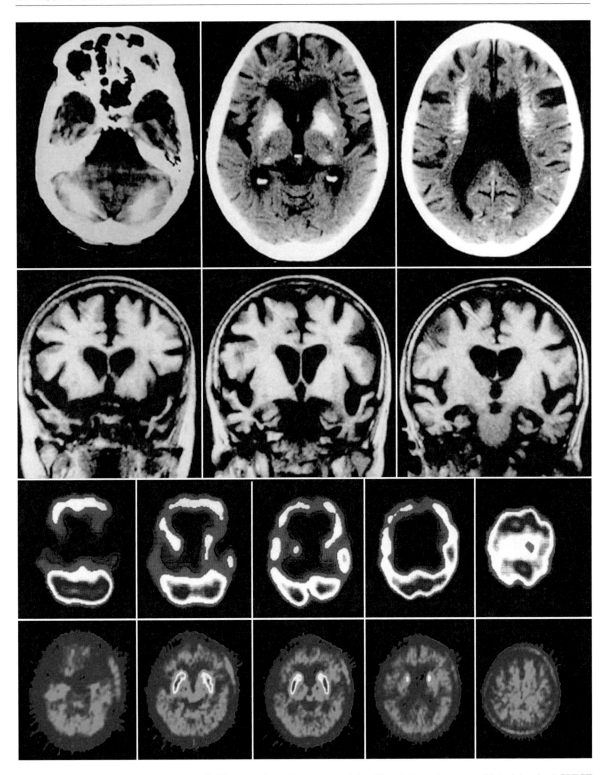

Fig. 7.17 Neuroimaging in diffuse neurofibrillary tangles with calcification (DNTC). This 76-year-old woman presented with altered behaviour, memory impairment, and the MMSE score was 17/30. Cranial computed tomography (CT) (*top row*) shows symmetrical calcification of the cerebellum, basal ganglia, pulvinar, and of the deep and periventricular white matter. Coronal T1-weighted MRI (*second row*) shows symmetrical atrophy of the temporal and frontal lobes. Atrophy is particularly distinct at the temporal tip, although the entire temporal lobe is involved. SPECT (*third row*) shows a marked reduction in blood flow mostly in the temporal lobes. Blood flow and metabolism in the basal ganglia, parietal and occipital lobes, and cerebellum were normal. A PET scan (*bottom row*) using [18F]6-fluoro-L-DOPA shows normal uptake in the basal ganglia. (Reprinted with permission from *J Neurol Sci* 2003;209:105–109)

psychiatric symptoms may be present. Many other signs and symptoms may also be present, including optic atrophy, muscle wasting, parkinsonism, impaired lung function, osteoporosis and endocrine dysfunction. There are no obligatory signs (tendon xanthomas can be absent) and there is a marked variety in nature and degree of progression, even within one family. The diagnosis, therefore, can easily be missed, perhaps particularly when cognitive decline is the presenting symptom. Biochemical diagnosis is made by determining the excessive urinary excretion of bile alcohols and the serum cholestanol levels. The diagnosis can be confirmed by demonstrating lack of sterol 27-hydroxylase activity in fibroblasts. CTX is a treatable disorder. Treatment with chenodeoxycholic acid interrupts the vicious circle of defective endogenous bile acid synthesis leading to absence of negative feedback, and increased cholesterol. The progression of CNS damage can be halted or delayed, with a reversal of cognitive symptoms and improved motor function.

Neuroimaging Strategy and Findings

Structural imaging using CT or MRI may reveal cerebral and cerebellar atrophy. Parenchymal abnormalities are best appreciated on MRI, and include ill-defined hyperintense lesions in the periventricular region on T2-weighted images, combined, in some patients, with focal lesions in the deep white matter. In some patients, there are prominent perivascular spaces. More characteristic lesions are found in the basal ganglia and posterior fossa (Fig. 7.18). Hyperintense lesions are found in the medial part of the globus pallidus in the majority of patients, and involvement of the corticospinal tracts is frequently observed. The most characteristic and frequent finding, however, is cerebellar involvement, starting with lesions in the dentate nucleus, extending towards the corpus medullare. Such lesions show high signal intensity on T2-weighted images and are isointense or, when extensive, hypointense on T1-weighted images. In patients with long-standing disease, areas of decreased signal intensity can be seen on T2-weighted images in the dentate nucleus (hyperdense on CT) (Fig. 7.18). Additional lesions can be found in the inferior olive and the descending tracts in the spinal cord. Technetium-99m brain SPECT may reveal severe cerebellar hypoperfusion.

7.4.6 Adult Polyglucosan Body Disease (APBD)

Contribution by Bert-Jan Kerklaan

Synonyms
Glycogen storage disease type IV, glycogenosis type IV, glycogen branching enzyme type IV deficiency, amylopectinosis, Andersen's disease.

Introduction

Adult polyglucosan body disease (APBD) is a rare metabolic disorder characterised by accumulation of carbohydrate inclusion bodies in the brain and other organs, such as, muscles and nerves. So far, at least 50 cases have been described. The first patient reported in 1971 was a 59-year-old man with dementia, walking difficulties, urinary incontinence and deficits of the peripheral and central nervous system. Since then, progress has been made in unravelling the biochemical and genetic background of this disorder and exploring of the clinical spectrum. Certain imaging features are highly specific; however, APBD appears to be underdiagnosed.

Clinic Features, Diagnosis and Therapy

The clinical picture is characterised by progressive gait difficulties with spasticity and peripheral sensory loss, urinary incontinence and progressive cognitive deficits. Symptoms most commonly appear in the fifth and sixth decades of life. Patients complain of gait and voiding problems, but mild cognitive deficits arefrequently already present and the majority develop dementia during follow-up. The speed of cognitive deterioration is highly variable and survival ranges from 2 to 25 years. The pattern of cognitive dysfunction is also variable, ranging from memory problems, executive dysfunction to a more pure frontal dementia.

As the disorder is defined by the finding of polyglucosan bodies, the clinical spectrum has broadened over recent years. Extrapyramidal and autonomic symptoms, epilepsy, cardiomyopathy and muscle spasms have been reported to accompany the typical clinical profile of APBD, possibly as part of a concomitant disorder. A myopathy or demyelinating neuropathy

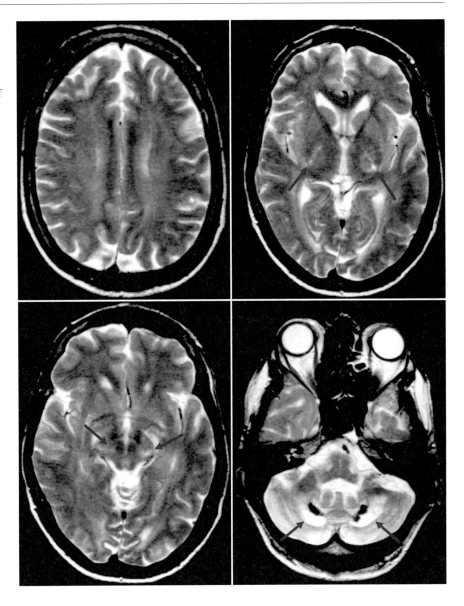

Fig. 7.18 A 51-year-old man with CTX. Transverse T2-weighted images show normal ventricles, a normal cortical pattern, and symmetrical hyperintensity of the supratentorial white matter, most markedly in the optic radiations, posterior limb of the internal capsules, and the corticospinal tracts in the mesencephalon (*arrows*). In the posterior fossa, high signal intensity is seen in the corpus medullare of the cerebellum (*arrows*), in which sharply demarcated and irregular hypointense structures are embedded. These dark areas represent cholestanol deposits and haemosiderin in the dentate nucleus

may also occur without involvement of the central nervous system.

The diagnosis of APBD is confirmed by demonstrating glycogen branching enzyme deficiency in skin fibroblasts, nerve or muscle. When a deficiency is demonstrated, genetic testing should follow, although a known mutation may not be found. A typical clinical profile with a normal enzyme level is an indication for biopsy of the sural nerve or axillary sweat glands, looking for a certain amount and specific distribution of polyglucosan bodies. Therapy is symptomatic.

Differential Diagnosis

The differential diagnosis of the combination of cognitive deficits, gait difficulties with spasticity, incontinence and polyneuropathy in adults is broad and encompasses hereditary disorders including CADASIL and the leukodystrophies, metabolic disorders such as B12 deficiency, toxic encephalopathy (e.g. secondary to chemotherapy or organophosphate), infectious disorders such as syphilis, neuroborreliosis and HTLV-1, paraneoplastic conditions and inflammatory disorders such as neuro-Behçet, SLE and neurosarcoidosis. CADASIL, leukodystrophies, vitamin B12 deficiency, SLE and neurosarcoidosis generally have a distinctive pattern of white and grey matter involvement on MRI.

Pathophysiology

APBD is clinically differentiated from the ordinary gly-cogen storage disorder (GSD) type IV by the age of onset and the organs involved. GSD type IV is a rare autosomal recessive disorder typically affecting the heart and liver of young children with a very poor prognosis. The disorders have in common the accumulation of polyglucosan bodies – inclusion bodies composed of glucose polymers – in affected organs. Polyglucosan bodies have been associated with numerous CNS disorders with different distributions and morphologies. Polyglucosan bodies in APBD are located in neuronal processes throughout the brain and spinal cord and in axons of motor and sensory nerves, but can be found in other organs as well.

APBD is most common in Ashkenazi Jews, with a limited number of mutations in the glycogen branching enzyme gene on chromosome 3 described. Different mutations in the same gene have been detected in non-Jewish people with clinical and pathological APBD. Patients heterozygous for the mutation have only a partial biochemical effect, suggesting an autosomal recessive mode of transmission.

Neuroimaging Findings

MRI demonstrates non-enhancing lesions hyperintense on T2-weighted images progressing towards confluent and symmetric involvement of the periventricular WM, including the internal and external capsules, but with initial sparing of the corpus callosum and subcortical U-fibres. Involvement of the white matter may spread into the region around the fourth ventricle and the temporal poles (see Fig. 7.19). Typically, hyperintensity on T2-weighted images is found in the cervicomedullary region extending to the medulla and pons, with relative sparing of the centre of the pons except for the long tracts. These changes are accompanied by cerebral and spinal atrophy. MRS of the brain shows increased levels of lactate with decreased levels of NAA, in accord with tissue damage involving both axons and myelin.

7.4.7 Neuroaxonal Leukodystrophy (NAL)

Synonyms
Neuroaxonal leukodystrophy with spheroids; Leukoencephalopathy with neuroaxonal spheroids; hereditary diffuse leukoencephalopathy with spheroids (HDLS)

History and Histopathology

Hereditary diffuse leukoencephalopathy with spheroids is an autosomal dominantly inherited progressive condition, first described by Axelsson in 1984. Additional cases have included individuals without a family history (probably *de novo* mutations). The disease is characterised by a loss of myelin and axons, with lipid laden macrophages and gliosis. The central hallmark of neuroaxonal leukodystrophy (NAL) is axonal spheroids – these eosinophilic axonal swellings are observed in the grey matter in a number of conditions including degenerative and metabolic diseases, intoxication, trauma, tumours and neuroaxonal dystrophies, including adult neuroaxonal dystrophies, neurodegeneration with brain iron accumulation (NBIA) and Nasu–Hakola disease (PLOSL). When axonal spheroids are confined to the white matter, the most important disorder to consider in adults is NAL.

Clinical Presentation, Epidemiology and Treatment

There is a large variation in symptomatology and presentation: subtle changes in personality, disorganisation and forgetfulness as well as progression to severe impairment in multiple domains of cognition, with basic neurological functions becoming impaired only late in the course of the disease. The clinical features are caused by early destruction in the deep white matter of the prefrontal cortex, followed by the white matter deep to the association areas of the parietal and temporal lobes and by secondary degeneration in the

Fig. 7.19 This 59-year-old man was referred with cognitive decline of unknown origin. MRI revealed a diffuse leukoencephalopathy. Contrary to ischaemic white matter disease, the temporal lobes were affected as well (*red arrows*), narrowing the differential diagnosis to CADASIL, multiple sclerosis (MS), myotonic dystrophy and APBD. Subsequent laboratory testing confirmed the diagnosis of APBD

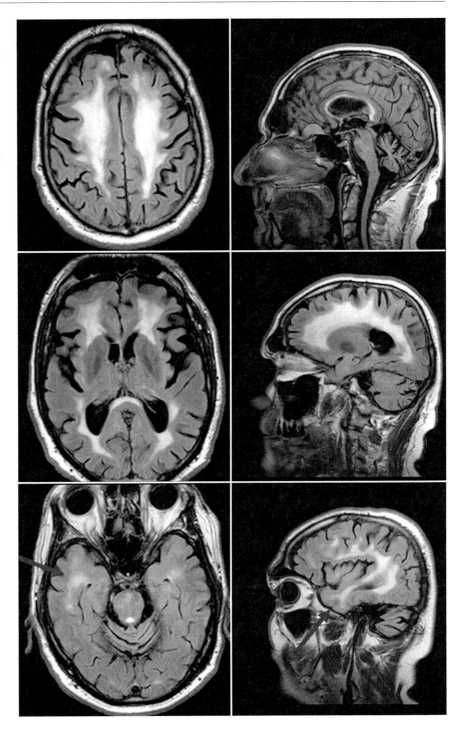

Box 7.1 Neuroimaging findings in NAL and differential diagnosis

MRI (and to a lesser extent CT) demonstrates selective white matter involvement and atrophy of the frontal lobe consisting of:

- Selective frontal subcortical atrophy
  - Ventricular enlargement, especially of the frontal horns
  - Corpus callosum atrophy, especially of the genu
- Bilateral, symmetrical, usually confluent (but may be patchy) signal abnormalities in the white matter of the frontal lobes
  - T2 hyperintense and T1 hypointense on MRI, hypodense on CT
  - Involving the periventricular and subcortical white matter
  - Sparing the U-fibres
- Additional involvement of the parietal white matter later in disease
- Signal abnormalities may extend downwards through the posterior limb of the internal capsule into the brain stem and pyramidal tracts
- Lack of contrast enhancement

Differential diagnosis and main differentiating imaging finding

• FTLD	More cortical atrophy • Less white matter signal change
• Other leukodystrophies	Additional white matter findings • Posterior fossa lesions in ALD • Enhancement in Alexander • Liquefaction in VWM
• Post-traumatic gliosis	Cortical gliosis, focal atrophy, MBs
• Vascular disease	Temporal lobe involvement in CADASIL
• Infections	Asymmetrical/discontinuous lesions in PML
• Inflammation	Discontinuous focal lesions in ADEM/MS

thalamus. Individuals may be thought to have MS or FTD or another leukodystrophy (see Box 7.1). Exact prevalence is unknown. There seems to be

overrepresentation of women. Diagnosis can be confirmed only by biopsy or at autopsy – no genetic or laboratory marker is known. Survival after diagnosis ranges between 2 and 20 years in published cases. Treatment is symptomatic.

Neuroimaging Strategy and Findings

Brain MRI is the imaging modality of choice; there is no established role for nuclear imaging or advanced MR techniques. The main findings are summarised in Box 7.1, along with clues for the differential diagnosis of NAL. A typical example of NAL is presented in Fig. 7.20.

7.4.8 Late-Onset Metachromatic Leukodystrophy

Synonyms
Metachromatic leukodystrophy (MLD); leukodystrophy with arylsulphatase deficiency.

Aetiology, Genetics and Histopathology

Metachromatic leukodystrophy (MLD) is an autosomal recessive lysosomal storage disease characterised by demyelinating white matter damage and an accumulation of sulphatides in brain and other tissues. Incidence is approximately 1/40,000. The majority of MLD cases are caused by deficient activity of arylsulphatase A (ASA) encoded on chromosome 22. Rarely, MLD is due to saposin-B deficiency – a sphingolipid activator. Metachromatic material (cerebroside sulphate) accumulates in brain oligodendrocytes and microglia, peripheral nerve Schwann cells and kidneys. Due to an unknown mechanism, accumulation of lysosomal sulphatide leads to demyelination starting in the periventricular white matter without accompanying inflammation.

Clinical Presentation and Therapy

The three clinical forms – late-infantile, juvenile and adult – are due to different levels of residual enzyme activity. The adult form may present as late as 60 years of age. Adult onset MLD usually first presents

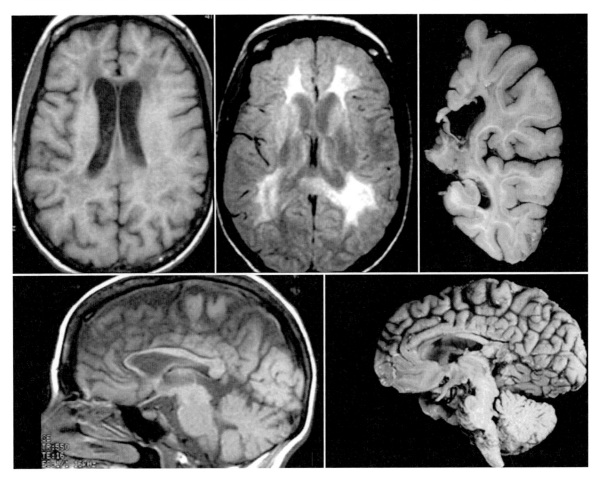

Fig. 7.20 MRI and post-mortem findings in NAL. This 40-year-old woman had a 3-year history of cognitive and behavioural decline followed by more rapid clinical deterioration which included cortical blindness. MRI demonstrates abnormal signal in the white matter underlying the prefrontal cortex and mildly enlarged lateral ventricles, as well as atrophy of the anterior part of the corpus callosum. She died at the age of 45 years and autopsy confirmed leukodystrophy with neuroaxonal spheroids. Note extensive thinning and brown discolouration of the corpus callosum and enlarged lateral ventricle on the mid-sagittal section. Coronal section of the parietal lobe (in a different patient) demonstrates grey discolouration of the white matter with U-fibre preservation. The parietal lobe is more severely affected than the temporal lobe white matter. (Modified with permission from *Brain Pathol* 2009;19:39–47)

with psychiatric symptoms and cognitive decline. The phenotype may resemble schizophrenia or a frontal dementia. Progressive paresis of arms and legs develops subsequently, followed by cerebellar ataxia and extrapyramidal symptoms. The diagnosis is based on measurement of arylsulphatase-A in blood or urine. Therapeutic options include bone marrow transplantation, stem cell therapy and enzyme replacement.

Neuroimaging Strategy

The earliest imaging feature (which may be present in asymptomatic subjects) is the development of initially faint sheet-like changes in periventricular white matter with relative sparing of U-fibres (Fig. 7.21). These are best seen on T2-weighted MRI. There may be a frontal predominance – frontal and parietal/occipital regions are most affected in mild disease. The corpus callosum is involved relatively early but not to the same extent as in ALD or globoid cell leukodystrophy (GLD). As the disease becomes more severe, radial stripes appear peripherally. This tigroid pattern reflects zones of spared myelin in a perivascular distribution within affected white matter. Involvement of the cerebellar white matter and brainstem and basal ganglia may all occur but are usually late findings. After injection of contrast there is no enhancement. Central atrophy develops in later stages. Proton MR

Fig. 7.21 MRI findings in Metachromatic leukodystrophy (MLD). This 30-year-old women presented with memory complaints. Changes in character had been noted by family members and she was thought to be depressed. MRI shows confluent periventricular high signal with severe corpus callosum atrophy. Subsequent laboratory testing revealed abnormally low arylsulphatase-A in leukocytes and the diagnosis of MLD was made. (Case kindly provided by Henry C. Weinstein)

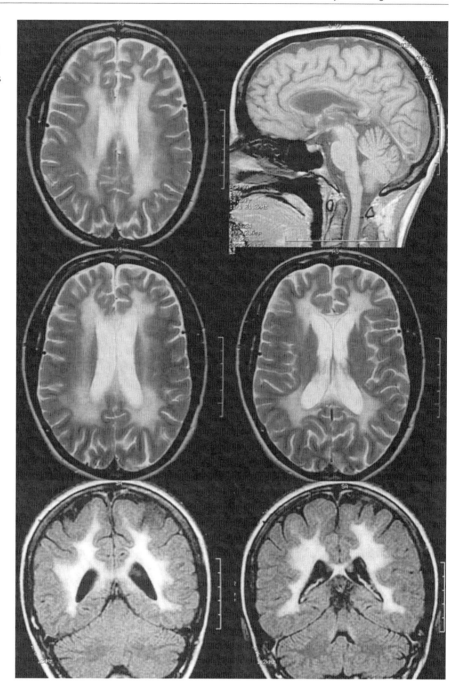

spectroscopy shows a decrease in NAA and high choline, reflecting axonal damage and myelin breakdown – increased myo-inositol probably reflects gliosis. PET or SPECT may reveal thalamic and frontal hypometabolism.

7.4.9 *Adrenomyeloneuropathy (AMN)*

Synonyms

Adrenomyeloneuropathy (AMN), myeloneuropathy variant of ALD; adult-onset adrenoleukodystrophy

Aetiology, Genetics and Histopathology

AMN is the adult-onset form of X-ALD, and has a similar X-linked recessive inheritance. X-ALD and AMN share an impaired capacity to degrade very-long-chain fatty acids (VLCFA), caused by a defect in peroxisomal oxidation. The disease is caused by mutations in the ABCD1 gene that causes a loss of function of a peroxisomal membrane protein whose role is thought to be related to transport of VLCFAs. Other (unknown) genetic or environmental factors (e.g. head injury) may perhaps contribute to explain how individuals in the same family may have different phenotypes. Histopathologically, axonal degeneration with demyelination is found in descending spinal cord tracts. In the cerebral form, demyelination occurs in the cerebellar and cerebral white matter (especially in the splenium of the corpus callosum and the posterior limb of the internal capsule); the active edges of demyelination feature inflammation.

Clinical Presentation and Laboratory Findings

AMN usually presents in the third or fourth decade of life, but occasionally as late as at the age of 60 years. In the typical spinal form, affected males show a slowly progressive paraparesis, urinary incontinence of the urgency type, cerebellar ataxia and signs of a peripheral neuropathy (sensory and autonomic). About half of the patients show signs of cerebral involvement with slowly progressive mild cognitive dysfunction. Visual memory is usually the most affected. Emotional disturbances and depression may occur. AMN should be considered in the differential diagnosis of young-onset dementia, certainly when presenting with spastic paraparesis and urinary incontinence in men.

AMN should also be considered in women, since 10–15% of women heterozygous for X-ALD develop AMN of variable severity around the age of 40 years. Brain MRI may be normal or show involvement of the pyramidal tracts, corpus callosum, and the periventricular white matter with a posterior predilection. In contrast to men, VLCFA may yield false-negative results in heterozygous women and DNA analysis is the recommended test in women.

About two thirds of patients with AMN have overt or biochemical signs of adrenocortical insufficiency, which may precede or follow neurological dysfunction. The disease is slowly progressive over decades, and no effective therapy exists.

Neuroimaging Strategy and Findings

About 50% of symptomatic patients with AMN have abnormal brain MRI scans at presentation, and the extent of cerebral involvement is much more variable than in X-ALD. Presenting findings include signal changes in the posterior limb of the internal capsule and splenium of the corpus callosum (Fig. 7.22). Subsequently, abnormalities become visible in the cerebral peduncles and cerebellar white matter. After contrast injection, enhancement is seen in the most recently affected regions. MR of the spinal cord shows atrophy. MR spectroscopy in cases of cerebral involvement may show decreased NAA and increased choline, reflecting axonal damage/loss and demyelination, respectively.

7.4.10 Vanishing White Matter Disease (VWM)

Synonyms
Childhood ataxia with central nervous system hypomyelination (CACH), Cree leukoencephalopathy, vanishing white matter (VWM), leukodystrophy with ovarian failure, ovarioleukodystrophy

Clinical Presentation and Aetiology

The name 'vanishing white matter disease' (VWM, OMIM #306896) was coined by van der Knaap in 1997. Previously, this disease has also been described as 'childhood ataxia with central hypomyelination' (CACH). It is caused by mutations in any of the five genes encoding subunits of the eukaryote translation initiation factor EIF2B, located on chromosome 3q. Whereas initially thought to be a homogenous disease affecting young children, VWM has a wide phenotypic variation and may present at all ages. In adult-onset cases, the initial presentation may include seizures, psychiatric symptoms, motor deterioration or dementia. The characteristic clinical features of VWM in children – episodes of rapid and major neurological deterioration provoked by fever or minor head trauma – may be absent in adult-onset cases.

Fig. 7.22 MRI findings in AMN with cerebral involvement. This 58-year-old man was wheelchair-bound at the time of the MR examination and had lost most of his cognitive abilities. Proton-density-weighted MRI shows involvement of the splenium of the corpus callosum, the internal capsule and cerebellar white matter. Note enhancement of the most rostral active zones with gadolinium (*upper right*)

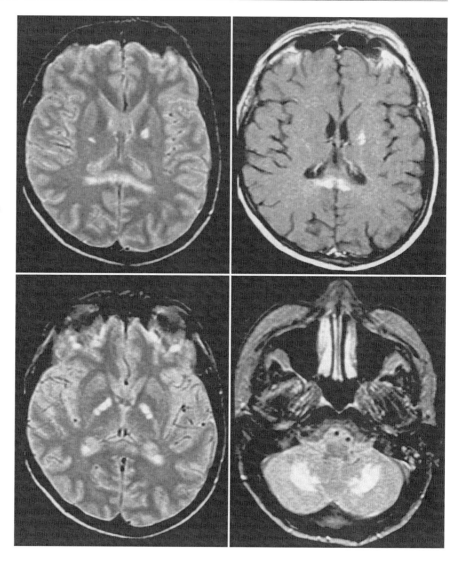

Histopathology

The hallmark pathology of VWM is rarefaction of the deep and periventricular white matter in the frontal and parietal lobes, with relative sparing of the temporal lobes, the corpus callosum, the visual pathways, the anterior commissure and internal capsule. Microscopically, the affected white matter shows myelin pallor, thin myelin sheaths, vacuolation, myelin loss, cystic changes and, rarely, active demyelination. Axonal loss is complete in areas with cavitation, and may be spared in other regions. The radiating stripes as seen on MRI appear to correlate with preserved blood vessels accompanied by reactive astrocytes.

Neuroimaging Findings

MRI of the brain can be diagnostic in full-blown cases of VWM, showing diffuse and homogeneous white matter hyperintensity on T2-weighted images, with remarkably low central signal intensity on T1-weighted images, proton-density and FLAIR images, reflecting severe tissue rarefaction not seen in any other leukodystrophies (Fig. 7.23). In early cases, diagnosis may be more difficult, certainly when tissue rarefaction is not yet apparent. Proton MRS shows a spectrum similar to that of CSF with some lactate and glucose and no or minor 'normal' resonances, but may be unremarkable in earlier stages.

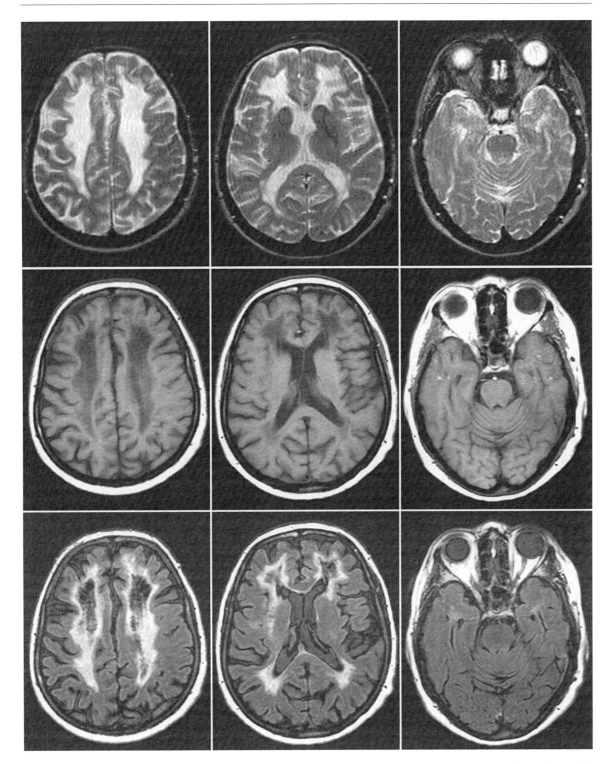

Fig. 7.23 MRI findings in adult-onset VWM. This 55-year-old woman had difficulties at work in the preceding months. The family noted a progressive neglect in personal hygiene, reduced variety in cooking, memory deficits and repetitive speech. MRI revealed a diffuse leukoencephalopathy on T2-weighted images (*upper row*) with characteristic low signal intensity on T1-weighted and FLAIR images (*middle* and *lower rows*) with a central "dot-and-stripe" pattern of preserved tissue. (Based on a case reported by Gascon-Bayarri in *JNNP* 2009, with special thanks to Marjo van der Knaap)

7.4.11 Globoid Cell Leukodystrophy (Krabbe's Disease)

Synonyms

Globoid Leukodystrophy (GLD); adult-onset GLD; Krabbe's disease

Aetiology, Genetics and Histopathology

GLD is a rare autosomal recessive lysosomal storage disorder caused by mutations in the galactocerebrosidase (GALC) gene – over 70 different pathogenic mutations are already known. Deficiency of GALC leads to accumulation of galactosylceramide within multinucleated macrophages of the white matter, forming characteristic 'globoid' cells. A metabolite of galactosylceramide accumulates which is toxic to oligodendroglia and results in damage (demyelination) of the peripheral and central nervous systems. The disease may be subdivided into three types based on age of onset: the most common is the infantile form (with a severe/rapidly progressive phenotype); a rare juvenile form (between 2 and 10 years) and an even rarer adult form, with onset after 10 years, also exist.

Clinical Presentation

Adult forms of GLD have a milder but variable phenotype and a slower rate of progression. Symptoms and signs include spasticity, dementia, ataxia, peripheral neuropathy, and loss of vision. Occasionally, individuals may present with spastic paraparesis and only mild cognitive problems. Nerve conduction can be normal or only mildly affected. Age of onset and survival widely vary even within families. Diagnosis is confirmed by markedly reduced GALC activity in leucocytes. Stem cell treatment appears promising in infantile GLD.

Neuroimaging Findings

MRI shows periventricular white matter abnormalities with relative sparing of the arcuate fibres. The posterior limb of the internal capsule and the corticospinal tracts in the brainstem are involved in most cases. A number of case reports suggest that the earliest and occasionally only site of involvement is the corticospinal tracts (Fig. 7.24). White matter hyperintensities on FLAIR and T2-weighted imaging are common in the more posterior brain regions, with frequent involvement of the splenium of the corpus callosum – a pattern similar to adrenoleukodystrophy (ALD). Atrophy is more conspicuous than in MLD. There is no enhancement after contrast injection. MR spectroscopy shows a decrease in *N*-acetylaspartate and sometimes (not always) high choline.

7.5 Toxic Leukoencephalopathy and Dementia

Mike P. Wattjes

7.5.1 Introduction and Classification

This chapter deals with acquired cognitive decline due to toxic encephalopathy secondary to a selected number of causes. In general, toxic encephalopathy associated with cognitive impairment can be caused by:

- Exogenous/endogenous deprivation of metabolic factors
  - e.g. abnormal intake/uptake/metabolism of vitamins
- Exogenous toxic substances due to substance abuse or inadvertent exposure
  - e.g. drugs, organic solvents
- Endogenous accumulation of toxic substances due to abnormal metabolism
  - e.g. inherited leukodystrophies Sect. 7.4)
- Secondary changes due to vascular pathology or hypoxia

Certain disease entities are based on a combination and/or overlap of the above categories (e.g. dialysis dementia). Toxic encephalopathies are quite heterogeneous in terms of the underlying epidemiology and pathophysiology. However, they all share a specific histopathological feature: demyelination. This can occur in a direct toxic way or be secondary to vascular

Fig. 7.24 MRI findings in GLD. *Upper row* shows images of an 18-year-old patient. Axial FLAIR image on the left shows symmetrically increased signal intensity within the pyramidal tracts (*straight arrows*) and parieto-occipital white matter (*arrowheads*) extending into the corpus callosum (curved arrows on T2-weighted image on the right). The lower row is from his18-year-old monozygotic twin brother presenting with behaviour changes and cognitive decline. Note T2-hyperintensity in the left and right pyramidal tracts (*straight arrows*), and bilateral increased signal intensity in the optic radiation (*curved arrows*). (Courtesy of D.J. Loes, M.D.)

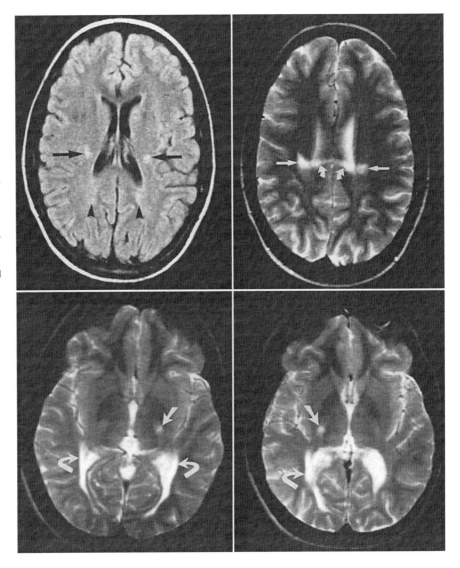

damage and hypoxia. Demyelination can occur diffusely in certain metabolic disorders and toxic diseases, but may also more selectively affect targets such as the midbrain, like in Wernicke's encephalopathy (WE). The histopathological hallmark of demyelination cross-links to other forms of toxic encephalopathy associated with cognitive decline, such as iatrogenic (treatment associated) pathology, like in post-radiation and chemotherapy-related encephalopathies (see Sect. 7.6).

A full description of all toxic encephalopathies is beyond the scope of this book. The current chapter focuses on a certain selection of disorders, which are relatively prevalent or present with typical imaging vignettes. In fact, many other toxic agents such as heavy metals (lead, mercury, tin, arsenic, manganese), drugs (e.g. opiates, cocaine) as well as metabolic factors including hormones (e.g. testosterone, oestrogens, thyroid hormone), and vitamins (vitamin D deficiency) can also lead to toxic encephalopathy. Some of them are discussed in other chapters and others are not associated with discernible neuroimaging findings. Therefore, they are rather diagnosed on clinical and laboratory findings. However, many need to be considered in the differential diagnosis of the diseases discussed in the following chapters.

7.5.2 Vitamin B Deficiencies

7.5.2.1 Vitamin B1 Deficiency

Synonyms
Wernicke's encephalopathy (WE), Korsakoff's syndrome, thiamine deficiency, beriberi

History

WE was first described by Carl Wernicke in 1881 as acute superior haemorrhagic polioencephalitis in two patients with alcohol abuse and in another one with persistent vomiting due to pyloric stenosis after the ingestion of sulphuric acid. The association of WE with malnutrition and thiamine deficiency as an underlying causative factor has been discovered by Campbell and Russell in 1941.

Epidemiology, Aetiology, Clinical Presentation and Treatment

The prevalence of WE in autopsy studies (0.8–2.8%) is substantially higher than clinically reported (0.04–0.13%) and occurs especially in patients with alcohol abuse. WE is associated with many clinical conditions impairing the absorption of thiamine:

- Chronic alcohol abuse
- Malnutrition/unbalanced nutrition
- Chronic gastro-intestinal diseases, recurrent and prolonged vomiting and diarrhoea
- Thyrotoxicosis, magnesium depletion
- Genetic factors
- Haemodialysis, chemotherapy
- Chronic infectious disease (HIV/AIDS), malignancies

Although WE is associated with alcohol misuse, alcohol in itself does not directly cause thiamine deficiency. However, chronic alcohol misuse is associated with unbalanced nutrition and alcohol raises the demand for thiamine. WE is more common in men than in women (male-to-female ratio 1.7:1). A mortality rate of 17% has been reported.

The clinical manifestations of thiamine deficiency in the nervous system can be subdivided into severe short-term deficiency with CNS symptoms (encephalopathy) and mild- to long-term deficiency leading to damage of the peripheral nerves. The early clinical signs of WE are non-specific (headache, fatigue, irritability) which makes an early diagnosis of WE rather difficult. The classical clinical presentation is characterised by oculomotor symptoms, altered consciousness and ataxia. However, this classical triad initially described by Wernicke is only present in 16–38% of the patients.

Recently introduced diagnostic criteria require two of the following criteria for WE:

- Dietary deficiency
- Oculomotor abnormalities
- Cerebellar dysfunction
- Altered mental status/mild memory impairment

Mental status changes occur in the vast majority of WE patients ranging from confusion, apathy, concentration deficits to hallucinations, coma and death. Some patients develop Korsakoff's syndrome, which is characterised by a disproportionate decline of working memory, whereas other cognitive domains are relatively spared. Such patients show severe anterograde amnesia with disorientation in time. Occasionally, emotional changes (euphoria, blandness etc.) may occur.

WE and Korsakoff's syndrome should be treated with a minimum of 500 mg of thiamine hydrochloride intravenously, three times daily, for 2–3 days. In the case of an effective response, treatment should be continued with 250 mg of thiamine daily (intravenously or intramuscularly) for 3–5 days. Dosages of thiamine between 100 and 250 mg are not sufficient to adequately restore the vitamin status.

Aetiology and Histopathology

Thiamine is absorbed in the duodenum and transported through the blood-brain barrier by active and passive mechanisms. In neurons and glia, thiamine is converted to thiamine pyrophosphate, which is involved in several biochemical pathways including the intermediate carbohydrate metabolism (ATP production), lipid metabolism (myelin sheaths, cell membranes) and synthesis of amino acids and neurotransmitters. In addition, thiamine is involved in synaptic transmission (acetylcholinergic, serotonergic).

Due to the normal storage capacity of thiamine in the body, thiamine deficiency leads to neuropathological changes only after 2–3 weeks. After that period, the function of certain thiamine pyrophosphate-dependent enzyme pathways (tricarboxylic acid cycle, pentose-phosphate pathway) starts to decline. This leads to an impaired use of glucose and to a deficient cellular energy metabolism in the brain, which in the end results in cellular death.

The pathological changes symmetrically affect certain areas of the brain obviously very much dependent on a high rate of thiamine-related glucose and oxidative metabolism, such as the periaqueductal grey matter, thalamus, superior cerebellar vermis and mamillary bodies. Acute lesions show microhaemorrhages with spongiosis, but no interstitial infiltrations of macrophages or any capillary proliferation. Chronic lesions feature swelling of astrocytes, neuronal loss, activated microglia, demyelination, astrogliosis and vessel hyperplasia.

Neuroimaging Strategy and Findings

CT might be used in the acute clinical setting, but MRI is much more sensitive and the modality of choice, with a sensitivity of 53% and a specificity of 93% (Box 7.2). Due to the rather distinct involvement pattern, additional advanced MR techniques (e.g. ^{1}H-MR spectroscopy) are not necessary in a clinical routine setting (Fig. 7.25).

7.5.2.2 Vitamin B12 Deficiency

Synonyms
Cobalamin deficiency, subacute combined degeneration

History

The association between vitamin B12 deficiency and neuropsychiatric syndromes has been recognised since the first description of pernicious anaemia in 1849. In 1900, the term *subacute combined degeneration of the spinal cord* was introduced by Russell et al. Vitamin B12 (cobalamin) itself was discovered by George Whipple in 1920, who later shared the Nobel Prize with Minot and Murphy in 1934 for this discovery.

Epidemiology, Clinical Presentation and Treatment

The prevalence of B12 deficiency is probably underrated and might range up to 16% in developed countries. Reduced levels of vitamin B12 become more

Box 7.2 Neuroimaging findings in Wernicke encephalopathy (WE)

Typical MRI findings in WE may include:

- Increased signal on T2/FLAIR in characteristic areas
  - Medial thalamus
  - Mamillary bodies
  - Periaqueductal GM and colliculi
- Increased signal on T1 may occur, due to microhaemorrhages
- Contrast enhancement can be seen in the acute phase
  - May be absent in non-alcohol WE
- DWI may show reduced ADC in the acute phase of the disease

Atypical manifestations include symmetric changes in

- Cerebellum (hemispheres, vermis)
- Other deep GM nuclei: caudate, red nucleus, dentate and cranial nerve nuclei
- Splenium of the corpus callosum
- Cerebral cortex

^{1}H-MR spectroscopy

- Decreased levels of NAA in the thalami and cerebellum indicating axonal damage
- Increased lactate in the cerebellum indicating necrosis

Differential diagnosis

- Mitochondrial disease (e.g. Leigh syndrome)
- Other toxic disorders (hypophosphataemia, methyl bromide intoxication, bromvalerylurea intoxication)
- Posterior circulation ischaemia (e.g. top-of-the-basilar artery syndrome)
- Deep venous thrombosis
- CJD
- Paraneoplastic disease
- Metronidazole treatment (atypical distribution pattern)
- Neuromyelitis optica (NMO) due to aquaporin-4 antibodies

Fig. 7.25 Wernicke encephalopathy. FLAIR images show the characteristic distribution of high signal lesions (*red arrows*) around the third ventricle, affecting the medial thalamus and periaqueductal grey matter

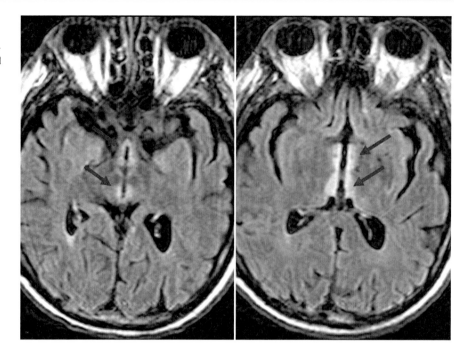

prevalent with ageing, rising to 20% in subjects older than 70 years.

Due to the body storage of 2–5 mg in the liver, B12 deficiency from malabsorption only develops after several years. The early signs of B12 deficiency are often vague and non-specific, such as concentration deficits, memory complaints, irritability and depression. Besides haematological (megaloblastic anaemia) and gastrointestinal manifestations (atrophic glossitis, stomatitis, diarrhoea), chronic vitamin B12 deficiency results in a complex clinical syndrome affecting the peripheral (neuropathy with numbness and altered proprioception) and the central nervous system. CNS symptoms include progressive paraparesis, ataxia, visual changes (due to optic nerve involvement), sleep disturbances, depression, personality changes and memory impairment. The association of lower vitamin B12 with cognitive impairment (adjusted for other risk factors such as age, sex and smoking) is firmly established.

Vitamin B12 metabolism is linked to the metabolism of homocysteine. B-vitamins (B12, B6) are important cofactors for homocysteine catabolism. For example, the enzyme methionine synthase and its cofactor methylcobalamin mediate the remethylation of homocysteine to methionine. Therefore, hyperhomocysteinaemia may indicate B-vitamin deficiency. Similar to B12 deficiency, hyperhomocysteinaemia is more common in the elderly and is associated with an increased risk of cognitive dysfunction. Homocysteine plasma concentrations can predict memory decline with age in the elderly. In addition, high homocysteine plasma levels are an independent risk factor for vascular pathology and vascular dementia.

Supplementation of vitamin B12 (parenteral or oral cobalamin) at an efficient dose for several weeks (or even lifelong) is the only treatment option. Haematological and neurological symptoms can recover, as well as the MRI abnormalities. However, whether adequate cobalamin supplementation or treatment of hyperhomocysteinaemia also positively affects cognition is difficult to predict.

Aetiology and Pathophysiology

Vitamin B12 has an essential role in the functioning of the CNS, being involved in the cell metabolism in terms of DNA synthesis and regulation, as well as in the fatty acid synthesis and energy production. Vitamin B12 is absorbed in the terminal ileum. An 'intrinsic factor' produced by the gastric parietal cells is also needed for its adsorption. Vitamin B12 deficiency can be caused by an impairment of absorption: surgical interventions, chronic gastro-intestinal diseases (e.g. atrophic gastritis, celiac disease), achlorhydria, bacterial overgrowth (e.g. blind loop syndrome), medication (e.g. metformin), chronic infections (e.g. giardiasis or other parasitic infections), medications (e.g. colchicine, neomycin); impaired intake: dietary imbalance (e.g. vegan diet); toxic (nitrous oxide exposure/ nitrous oxide anaesthesia) or hereditary causes such as MTHFR deficiency, homocystinuria and transcobalamin deficiency.

Cobalamin deficiency may lead to diffuse demyelination within the CNS, even though the pathophysiological mechanism remains poorly understood. Histopathologically, vitamin B12 deficiency is characterised by multifocal demyelination in the white matter of the brain and spinal cord. In the spinal cord, multiple areas of demyelination and vacuolisation with myelin breakdown, infiltration of foamy macrophages and axonal pathology can be observed in the posterior and lateral columns – the so-called *subacute combined degeneration*. Demyelination commences in the central parts of the posterior columns of the thoracic cord spreading laterally and cranially to the corticospinal tracts of the cervical spinal cord and medulla oblongata. The histopathological findings in the brain, in terms of demyelination, do not substantially differ from those observed in the spinal cord. In addition, degeneration of the optic nerves can be typically observed.

Neuroimaging Strategy and Findings

In general, CT is not recommended in the detection of vitamin B12 deficiency-related changes due to its low sensitivity. In particular, spinal cord abnormalities cannot be detected by means of CT. Therefore, multi-sequence MRI (including T2-weighted sequences) of the brain and spinal cord is the imaging modality of choice. Spinal cord T2-weighted MRI should be performed in sagittal and transverse orientations, in order to detect typical spinal cord manifestations in the posterior columns ranging from the thoracic spinal cord to the brain stem (Fig. 7.26). Administration of intravenous contrast media is not necessary to establish the

---

Neuroimaging findings in Vitamin B12 deficiency

Brain MRI findings

- Diffuse supratentorial areas of T2 hyperintensity in the periventricular and deep white matter
  - Usually bilateral, but not necessarily symmetrical
  - Large lesions may involve subcortical U-fibres

- Brain stem and the cerebellar white matter are relatively spared

- Blood–brain barrier disruption may occasionally be observed

- Broadening of the gyri due to swelling can be occasionally observed.
  - Atrophy might be present, but can be reversible during/ after treatment

- In patients with additional hyperhomocysteinaemia, silent infarctions may be observed.

- In infants/young children, delayed myelination or dysmyelination instead of demyelination

Spinal cord MRI may show a mild enlargement of the spinal cord and shows signal abnormalities (T2 hyperintense, T1 hypointense) in the dorsal and lateral columns. Contrast enhancement is not observed.

Differential diagnosis:

In general, the differential diagnosis includes toxic, metabolic, infectious and paraneoplastic conditions associated with diffuse demyelination of the central nervous system. Due to the involvement of the spinal cord, the differential diagnosis should particularly include HIV encephalomyelopathy and multiple sclerosis.

---

diagnosis. However, it might be helpful for narrowing/ excluding differential diagnoses.

7.5.3 Chronic Kidney Disease and Dialysis

Synonyms

Chronic kidney disease (CKD), dialysis dementia, dialysis encephalopathy, dialysis dyspraxia, uremic encephalopathy, dialysis disequilibrium syndrome

History

The first reports linking neuropsychiatric symptoms and chronic renal disease were published in the late 1960s. In 1972 the first patients were reported with

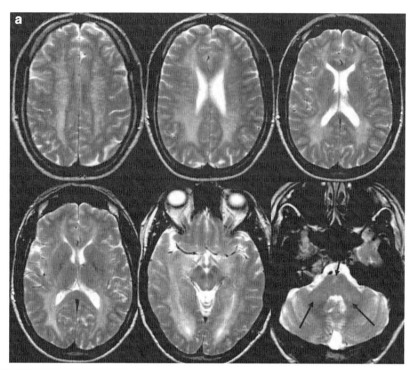

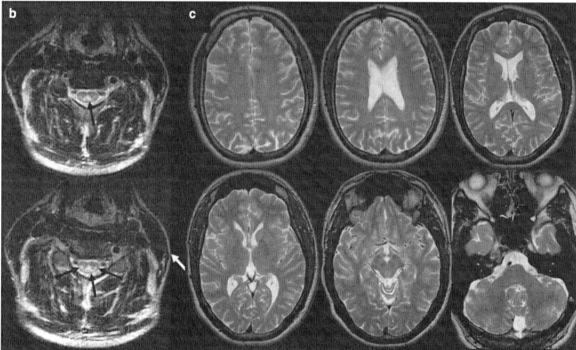

Fig. 7.26 CNS demyelination in vitamin B12 deficiency: A 39-year-old male who presented at a neurological outpatient clinic with progressive neurological complaints including ataxia, paraparesis and numbness of the legs, concentration deficits and memory complaints. The laboratory tests revealed a moderate vitamin B12 deficiency. Initial brain MRI (**a**): transverse T2-weighted images show diffuse confluent hyperintense areas, bilaterally, in the periventricular and deep white matter reaching (but not completely affecting) the U-fibres. The corpus callosum and the infratentorial white matter are relatively spared. However, subtle diffuse white matter changes can also be observed in the pons and the middle cerebellar peduncles (*arrows*). Spinal cord MRI (**b**) shows the typical involvement of *subacute combined degeneration* affecting the posterior and lateral columns (*arrows*) on transverse T2-weighted MR images. The 2-year follow-up MRI (**c**) after adequate supplementation of vitamin B12 shows a complete regression of the diffuse demyelination in the white matter

transient neurological symptoms, such as seizures and speech anomalies during haemodialysis. Subsequent observations confirmed even more fatal encephalopathic complications related to dialysis. A potential role of aluminium as a neurotoxic agent located in phosphorus binders and dialysis water related to chronic cognitive decline in dialysis patients was suggested. During the past decennia it became obvious that neurological complications, in general, and cognitive impairment, in particular, in patients with CKD are an important medical and socio-economic issue that needs further investigation.

Aetiology and Histopathology

Retention of metabolic and toxic waste products (e.g. urea, potassium) and metabolic imbalance reflected by insulin resistance, vitamin D deficiency and hyperparathyroidism are all considered to contribute to cerebral pathology. In particular, hyperparathyroidism might play a major role, since elevated parathyroid hormone levels have neurotoxic effects and secondarily lead to higher brain calcium concentrations, influencing neurotransmission in the CNS. In addition, CKD patients often have substantial (micro) vascular co-morbidity which contributes unfavourably to the clinical outcome as discussed below.

In patients with end-stage CKD undergoing dialysis treatment, rapid shifts in urea concentrations and fluid levels often lead to cerebral oedema during and/or immediate after the dialysis sessions. This pathophysiological process has been referred to as *'dialysis disequilibrium syndrome'*, which was more frequently observed in the early days of dialysis treatment – currently rarer with newer dialysis regimens. However, acute metabolic disturbances (e.g. hypocalcaemia, hypophosphataemia, hyponatraemia), and fluid changes possibly leading to temporary cerebral hypoperfusion during and/or shortly after the dialysis sessions, are still causing acute cognitive disturbances occurring during the dialysis session. The current model of the complex pathophysiology of cognitive impairment in haemodialysis patients is based on a combination and interaction of vascular pathology (stroke, cerebral hypoperfusion), systemic factors (toxic metabolic, genetic, inflammatory etc.) and acute

factors (during the dialysis process) leading to chronic cognitive impairment.

At the time that phosphorus binders and dialysis water contained aluminium, patients who underwent dialysis showed significantly higher concentrations of aluminium in the CSF and cerebral grey matter than non-dialysed subjects. The concentrations of iron can even be (ten times) elevated in patients who developed dementia. However, at present, the discussion about the association between aluminium concentrations and dementia is getting controversial again. The general association between dialysis and cognitive decline has been referred to as *'dialysis dementia'*, a term still used to date.

Clinical Presentation, Epidemiology and Treatment

The prevalence of cognitive impairment in end-stage renal disease is more than double that of the general population. There is both a cross-sectional and a longitudinal relationship between renal function, as measured by the glomerular filtration rate (GFR), and cognitive function. In fact, the prevalence of cognitive decline rises with the severity/grade of kidney dysfunction. CKD reflects an independent risk factor for dementia and acts independently from accompanying risk factors: most patients with CKD have substantial cardiovascular co-morbidity sharing several cardiovascular risk factors including diabetes, hypertension, hyperlipidaemia, obesity etc. All of these are also risk factors for AD and vascular dementia. Additional non-vascular risk factors are often present, such as anaemia, hyperparathyroidism, aluminium, polypharmacy, sleep disturbances and depression. Reflective of that, sub-clinical vascular WMH on MRI are found in many CKD patients. However, a direct association between vascular factors and cognitive decline in CKD patients has not been established. Apparently, the cognitive deficits in CKD are primarily caused by several factors independent of vascular co-morbidity, which rather have a synergistic effect on clinical outcome measures.

Clinically, patients with CKD may present with deficits in the domains of memory and executive function consistent with a subcortical pattern of cognitive dysfunction. The impairment in terms of memory

Clinical manifestations:

- During/after dialysis procedure: (*'dialysis disequilibrium syndrome'*)
  - Acute cognitive impairment – a delirium state
    - ° (Sub)acute disturbance of consciousness
    - ° Disorientation, attention deficit
    - ° Impaired cognitive function

In general, the symptoms resolve completely. Recurrent delirium state in haemodialysis patients is associated with higher risk of chronic cognitive decline and dementia.

- Chronic stage
  - Progressive neurological symptoms associated with dementia
    - ° Dysarthria-apraxia
    - ° Mutism
    - ° Myoclonus
    - ° Personality change, mood disorders
    - ° Seizures

Compared to patients undergoing classical haemodialysis, the clinical manifestations are less severe in patients undergoing peritoneal dialysis.

Treatment:

- Improve the renal organ function and optimise the dialysis procedure.
  - Renal transplantation
  - Reconstitution of metabolic changes

Neuroimaging strategy

CT shows a low sensitivity in the detection of abnormalities associated with CKD. Neuroimaging should include a multi-sequence MRI protocol, including FLAIR (WM changes and atrophy). T2 (infratentorial WM changes and thalamic lesions). T2* (microbleeds) and 3D-T1 with multi-planar reconstructions to evaluate the hippocampus. Optional: Diffusion-weighted imaging (DWI). The imaging findings are summarized below:

Neuroimaging findings in CKD

- No consistent and specific imaging findings have been reported concerning dementia associated with CKD and/or dialysis.
- High incidence of cardiovascular/cerebrovascular co-morbidity in CKD patients
  - Moderate to severe vascular white matter changes (in 50% of CKD patients)
  - Silent subcortical (lacunar) infarctions can be frequently observed
  - Neurodegeneration, reflected by global cortical atrophy
  - During haemodialysis cerebral oedema can be observed

The amount of the vascular white matter damage seems related to the degree of renal failure and is more pronounced in patients undergoing haemodialysis

capacity may resemble AD. In most of the cases, however, the deficits in memory functions reflects a synergistic pattern of Alzheimer pathology and vascular cerebral compromise.

Differential Diagnosis

Alzheimer's disease, vascular dementia, normal ageing brain, metabolic/toxic encephalopathy, reversible posterior leukoencephalopathy syndrome (mimicking vasoenic oedema occurring during dialysis procedure).

7.5.4 Carbon Monoxide Poisoning

Synonyms, Abbreviations
Delayed encephalopathy, delayed neuropsychiatric syndrome

History

Carbon monoxide (CO) was identified by Cruikshank in 1800 and toxic effects on dogs were investigated by Claude Bernard around 1846.

Aetiology and Histopathology

CO is a colourless, tasteless, inodorous, and non-irritant toxic gas produced by incomplete combustion of hydrocarbons. As a result, CO poisoning is easily missed. Sources of CO poisoning include exhaust fumes, impaired heating systems, inhaled smoke and methylene chloride (a frequent component of solvents such as paint remover). CO is easily absorbed through the lungs, and binds to haemoglobin 200–250 times more effectively than oxygen, thus blocking its capacity to carry oxygen. In addition to cellular hypoxia, CO-mediated toxicity includes reoxygenation injury (particularly in the CNS) and peroxygenation of lipids leading to (transient) demyelination. CO also affects the mitochondrial respiratory enzyme chain by binding to cytochrome oxidase. This leads to a substantial and prolonged impairment of the intracellular oxidative metabolism.

CNS damage in the delayed stage includes bilateral demyelination and necrosis of the white matter, globus pallidus, cerebellum, hippocampus and the cerebral cortex.

Clinical Presentation, Epidemiology and Treatment

CO poisoning is one of the most frequent causes of injury and death due to poisoning world wide. In the USA, more than 40,000 people per year are admitted to medical institutions for CO poisoning. The clinical symptoms of acute CO poisoning are non-specific (flu-like symptoms) and dominated by systemic reactions of cellular hypoxia such as tachycardia and tachypnea. As a result of cellular hypoxia and vasodilatation in the CNS, neurological symptoms including seizures and syncopes frequently occur. Chronic exposure to relatively low levels of CO can lead to persistent symptoms such as headaches, lightheadedness, depression, attention deficits, memory decline, nausea and vomiting. Up to 30% of patients may develop a delayed onset of neuropsychiatric symptoms (*delayed neuropsychiatric syndrome, delayed encephalopathy*) 3–240 days after the initial exposure – see Sect. 7.5.5.

Initial treatment in the acute setting of poisoning is the removal from the exposure. Immediate administration of oxygen significantly reduces the half life of CO. Hyperbaric oxygen therapy is used in order to support the dissociation of CO from HbCO and cytochrome oxidase. Further treatment options include the symptomatic treatment of seizures, cardiac complications, hypotension, pulmonary oedema and metabolic acidosis.

Neuroimaging findings in CO poisoning:

Necrotic and ischaemic lesions in the deep grey matter structures
* Globus pallidus, to a lesser extent putamen and caudate nucleus
  - Hippocampal atrophy may develop
* T2-hyperintense lesions surrounded by hypointense rim (haemosiderin)
  - CT hypodense
* Can show disruption of the BBB in terms of contrast enhancement

Less frequent the cerebral and cerebellar cortex may be involved:
* Diffuse hyperintensities on T2-weighted images particularly in the temporal lobe.

White matter demyelination in the subcortical white matter
* Bright on T2/FLAIR, sparing U-fibres
* Confluent aspect, predominantly located in the periventricular WM and centrum semi-ovale
* In delayed stages WM damage may extend to corpus callosum, internal/external capsule and subcortical U-fibres
* Might improve slightly during follow-up of several months.

In the acute stage of carbon monoxide poisoning, diffusion-weighted images show:
* Restricted diffusion (high signal on DWI, low ADC values)
* In the affected areas of the white and grey matter indicating cytotoxic oedema

In the chronic stage, the ADC values may gradually increase.

MR spectroscopy shows decreased levels of NAA (axonal damage) and increased lactate concentrations (hypoxia). SPECT shows areas of cerebral hypoperfusion, particularly in the temporal cortex.

Differential diagnosis:
Clinically, CO is one of the 'great mimickers' due to the diverse and non-specific presentation. The radiological DD includes:
* Metabolic diseases (Wilson disease)
* Japanese encephalitis
* Vascular lesions (small vessel disease)
* Creutzfeldt–Jackob disease (CJD)
* Leigh syndrome

Neuroimaging Strategy and Findings

Rapid initial evaluation involves CT in the acute setting. However, CT is less sensitive in detecting changes associated with carbon monoxide poisoning (Figs. 7.27). Further neuroimaging should include a more sensitive multi-sequence MRI protocol, including FLAIR/T2-weighted images, T1-weighted images (pre- and postcontrast administration) and DWI (Fig. 7.28). Optional: MR spectroscopy, MR perfusion, SPECT to evaluate cerebral perfusion status (see box above).

Fig. 7.27 A 21-year-old female fire victim who was found in a comatose state in her office. At the time of arrival the police was alerted by CO-alarm. The oxygen saturation at the emergency department was 65% and the percentage of HbCO was 32%. CT shows symmetric hypodense lesions in the globus pallidus (*arrows*) due to hypoxic grey matter damage

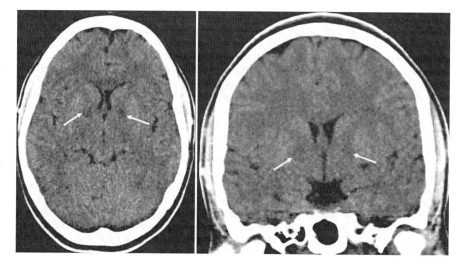

Fig. 7.28 A 51-year-old woman with drowsiness and instability that was diagnosed as accidental carbon monoxide poisoning caused by faulty domestic heating. (**a**) Axial T2-weighted image shows increased signal intensity of bilateral globus pallidi. (**b**) Corresponding diffusion-weighted image shows high signal due to restricted diffusion in bilateral globus pallidi. (Reprinted with permission from Pranshu Sharma, *Am J Roentgenol* 2009; 193: 879–886)

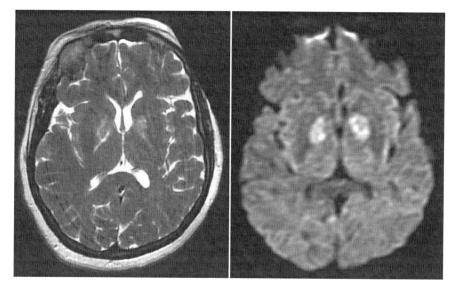

7.5.5 Delayed Post-hypoxic Demyelination

Synonyms

Delayed post-hypoxic leukoencephalopathy (DPHL), delayed anoxic leukoencephalopathy

Aetiology, Histopathology, Clinical Presentation and Treatment

Delayed post-hypoxic leukoencephalopathy (DPHL) develops typically 2–40 days after the initial hypoxic/

anoxic event. Clinically, DPHL is characterised by cognitive deficits which are typically subcortical with problems of attention and speed of processing, frontal executive deficits are common – memory may appear to be affected but this may partly be due to the frontal and attentional problems. In addition, gait disturbance and hypo/akinetic motor syndromes are characteristic but psychotic syndromes, mutism, tremor, incontinence and speech disturbances can be observed. DPHL is connected to carbon oxide (CO) poisoning (see Sect. 7.5.4) since ~3% of patients with CO intoxication develop DPHL. DPHL is also frequently seen as a complication of drug overdose (heroin, benzodiazepine), surgical anaesthesia and cardiac arrest, etc.

The exact pathophysiologic mechanism is not completely understood. The reduced oxygenation of blood and oxygen delivery to the brain tissue probably results in a breakdown of ATP-dependent enzymatic pathways leading to demyelination. This is reflected in confluent areas of diffuse demyelination, particularly located in periventricular and subcortical deep white matter. The subcortical U-fibres and the infratentorial white matter are relatively spared.

Interestingly, only a subset of patients with the above risk factors will actually develop DPHL. The degree of hypoxia does not completely correspond to the risk of developing DPHL. Obviously, other markers must influence the clinical course and the severity of symptoms. Deficiency in arylsulphatase-A, an enzyme involved in the sulphatide metabolism, has been described to be a predisposing factor. A complete deficiency of arylsulphatase-A results in a metachromatic leukodystrophy, which is characterised by demyelination of the central and peripheral nervous system. An association between a partial deficiency of arylsulphatase-A and DPHL has been observed. However, a substantial number of patients developing DPHL have no impairment of arylsulphatase-A.

No proven treatment options exist. Recovery rates up to 75% have been reported but depend on the severity of the insult (Fig. 7.29).

Neuroimaging strategy and differential diagnosis in DPHL:
CT may be helpful as an initial imaging approach. MRI has a higher sensitivity and is the image modality of choice – no contrast material administration is necessary. Optional: ^{1}H-MRS, SPECT
CT: Confluent hypodense areas in periventricular and deep white matter In later stages, global cortical atrophy with ventricular enlargement is seen.

MRI findings:
* T2/FLAIR: diffuse/confluent hyperintensity in periventricular and deep white matter
   – Representing diffuse demyelination
* Subcortical U-fibres and white matter in the posterior fossa are spared
* No cavitation can be observed
* Generalised atrophy is a late finding
* DWI: restricted diffusion (up to 30 days after the anoxic/hypoxic event)
   – Much longer than in ischaemic stroke
[1]H-MR spectroscopy: Low NAA and increased Cho in affected white matter; lactate mostly normal

---

Differential diagnosis:
* Toxic encephalopathy (heroin inhalation, inhalation of organic solvents)
* Chronic inflammatory disease (e.g. multiple sclerosis, PML)
* Metabolic diseases associated with demyelination (e.g. leukodystrophies)
* Radiation/chemotherapy-induced encephalopathy

7.5.6 Organic Solvent Inhalation

Synonyms
Organic solvents dementia, House Painter's dementia, toluene misuse

History

Organic solvents were introduced in the second half of the nineteenth century. The rapid increase of their use led to reports dealing with toxicity, in general, and neurotoxicity, in particular, in the 1920s. Organic solvents are a heterogeneous group of agents. Only a limited number of agents have been sufficiently tested for neurotoxic effects. Toluene (methyl-benzene) is one of the best investigated solvents in terms of toxicity and cognitive impairment.

Epidemiology, Clinical Presentation and Treatment

Organic solvent inhalation is a common form of substance abuse especially in children and young adults because of the low costs and easy availability. Besides substance abuse, chronic organic solvent inhalation may also result from occupational exposure in dry cleaning, aviation and chemical industries.

Organic solvents can be found in spray paint, paint thinner, pharmaceuticals, rubber, gasoline, varnishes and glue. Among the lipophilic organic solvents that might cause long-term damage, toluene is a frequently used industrial solvent which is a lipid soluble clear liquid aromatic hydrocarbon.

Acute intoxication states due to inhalation are characterised by transient symptoms including behavioural changes, euphoria, headache and ataxia.

Fig. 7.29 Delayed post-hypoxic demyelination. A 40-year-old woman presented with a history of neurobehavioural disturbance beginning with confusion and decreased attention and distractability, progressing to include urinary and faecal incontinence. She had recently been resuscitated after an inadvertent overdose of morphine. On admission, there was massive swelling and T2 hyperintensity of the supratentorial white matter with sparing of the basal ganglia (*upper row*). She developed severe global cognitive impairment, with marked deficits in memory and verbal fluency. A second MRI after 3 weeks showed high signal on DWI due to restricted diffusion on the ADC map (*middle row*). Clinical improvement occurred with supportive care only. A third scan after 6 months showed almost complete normalisation of WM abnormalities (*bottom row*) without atrophy. This was mirrored by clinical improvement and she returned to work. (Reprinted with permission from Sophie Molley, *Am J Neuroradiol* 2006;1763–1765)

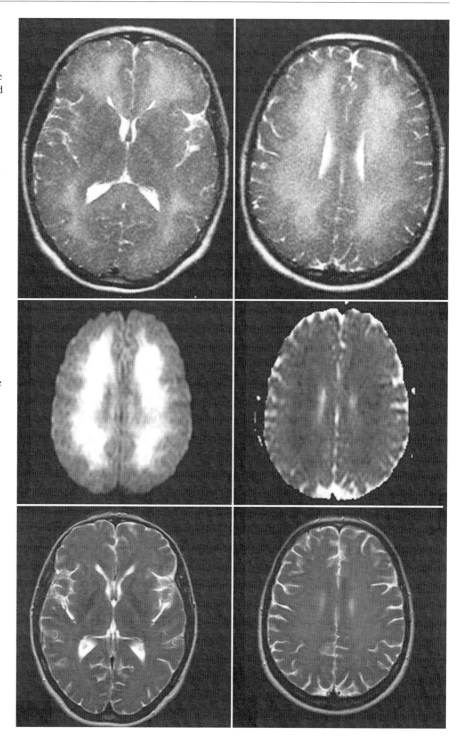

Persistent symptoms are usually related to chronic exposure. Cognitive deficits occur in 65–79% of patients with a history of chronic exposition to organic solvents. Patients may present with substantial cognitive impairment across many domains including learning and memory, visuo-spatial, attention/concentration, working memory, deficits in speed of information processing and executive abilities. Some patients meet the criteria of dementia. Children or young adults show a significant decrease in IQ.

There probably is a relationship between the duration of solvent (toluene) abuse and degree of neuropsychological impairment as well as imaging findings in terms of white matter abnormalities (axonal damage and increased glial cell activity) measured by ¹H-MR spectroscopy. Clinical improvement may occur in patients with long-term abstinence. No other treatment options are available.

Aetiology and Histopathology

The histopathological mechanism of toluene-mediated damage to the CNS is not completely understood. Toluene enters the blood easily through the respiratory tract and reaches the brain tissue by passing the blood-brain barrier. Due its lipophilic nature, toluene has a high affinity to lipid-rich tissue components and therefore accumulates in the white matter, particularly in myelin. It is assumed that the storage of toluene in the white matter subsequently leads to activation of glial cells, demyelination and axonal damage. Of note, there is evidence that axonal pathology occurs independently from demyelination and that demyelination might occur secondary to axonal damage and loss. Neuropathologically, toluene-mediated brain damage is reflected by demyelination and gliosis mostly localised in the periventricular, subcortical and cerebellar white matter. Further neurodegenerative changes including a thinning of the corpus callosum, widening of the ventricles and atrophy of the brain and cerebellum can be observed in long-term toluene users.

Neuroimaging Strategy and Findings

CT may be an initial approach, but multi-sequence MRI (including T2-weighted sequences and DWI) is the image modality of choice (Fig. 7.30). Optional investigations include ¹H-MR spectroscopy, SPECT (see box on the next page). Neuroradiological changes are only observed in patients with chronic exposure to organic solvents. Neuroimaging studies in acute intoxications fail to show abnormalities.

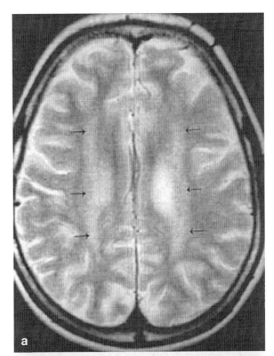
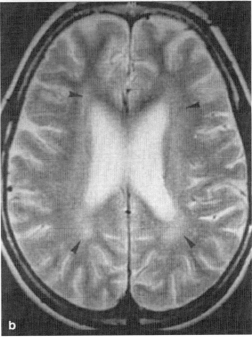

Fig. 7.30 Axial T2-weighted images in a 16-year old who had inhaled toluene for 6 years. (**a**) High signal intensity is seen in the centrum semi-ovale (*arrows*) on both sides. The peripheral cerebral white matter and grey matter–white matter differentiation are preserved. (**b**) High-signal-intensity changes involve the frontal and parietal periventricular white matter (*arrowheads*). Note that the lateral ventricles and cerebral sulci are enlarged. (With permission from Aydin et al., *AJNR* 2002)

Neuroimaging findings and differential diagnosis in Organic Solvent Disorders

CT may reveal global atrophy or areas of focal atrophy in the brain
- Particularly of the cerebellum, and brainstem
- Sulcal widening and ventricular dilatation
- Thinning of the corpus callosum (better seen on sagittal MRI)

MRI may reveal atrophy (as for CT) and white matter changes:
- Deep white matter, but particularly in the periventricular white matter
- Around the deep grey matter structures
- Decreased differentiation between the white matter–grey matter interface

Structural MRI abnormalities can be found in ~20–50% of patients with chronic toluene inhalation.

[1]H-MR spectroscopy in the deep white matter, and cerebellum (but not in the thalamus):
- Low concentrations of NAA
- High concentrations of myo-inositol

indicating a combination of axonal damage and probable gliosis in the white matter

SPECT: Areas of low perfusion can be frequently (in up to 70% of cases) found in prefrontal, parietal and temporal brain regions.

Differential diagnosis:
Includes toxic, metabolic and infectious diseases with diffuse demyelination
- Delayed post-hypoxic demyelination
- Post-treatment demyelination (chemotherapy, radiation therapy)
- Osmotic demyelination syndrome
- CO-intoxication
- Post-hypoglycaemic encephalopathy

7.5.7 Heroin Vapour Inhalation

Synonyms

Heroin vapour inhalation-induced spongiform leukoencephalopathy, heroin-induced toxic leukoencephalopathy, 'chasing the dragon', 'chinesing', 'Chinese blowing'.

History

Inhalation of heated heroin or 'Chasing the dragon' gained popularity around 1950 in Hong Kong and more recently worldwide. Spongiform leukoencephalopathy related to heroin vapour inhalation was first described in 1982 in the Netherlands, when Wolters reported 47 patients with spongiform leukoencephalopathy due to inhalation of heroin pyrolysate. Since then, many similar reports and case series have been published.

Aetiology, Clinical Presentation and Treatment

Heroin powder (heroin pyrolysate) is usually heated on aluminium foil over a flame and the vapour is inhaled through a straw or pipe ('chasing the dragon'). The inhaled heroin is absorbed rapidly with high bioavailability and immediate central nervous effects. What is unclear is how and why damage to CNS structures (in particular oligodendrocytes and myelin) occurs in some cases. It has been suggested that toxicity may be related to additives/contaminants of heroin or to tin in the aluminium foil. However, no specific toxin has been definitively identified.

The clinical course is poorly defined. A latent period between exposure and manifestations of clinical symptoms ranges from several days to months. In general, clinical manifestations can be subdivided into three stages (see Table 7.4). The clinical course is variable and patients may remain in the initial or intermediate clinical stage. However, approximately 25% of the patients will enter the terminal stage, which is characterised by a mortality rate of up to 23%. There is no established treatment; and management is supportive aiming at preventing complications (e.g. aspiration pneumonia). In some case series experimental treatment with antioxidant therapy including coenzyme Q was associated with substantial improvement of symptoms.

Table 7.4 Clinical manifestations of heroin vapour inhalation-induced spongiform leukoencephalopathy

Initial stage	Speech problems (pseudobulbar), cerebellar ataxia, motor restlessness, apathy, bradyphrenia
Intermediate stage	Pyramidal signs (spastic paraparesis), extrapyramidal signs (tremor, choreathetoid movements, myoclonic jerks).
Final stage	Stretching spasms, hypotonic paresis, akinetic mutism, central pyrexia, death

Histopathology

Histopathologically, heroin-leukoencephalopathy involves spongiform degeneration of the white matter with vacuolisation of the myelin lamellae at the intraperiodic lines on electron microscopy. In addition, swelling of the mitochondria and a distension of the endoplasmatic reticulum is seen in oligodendrocytes.

Neuroimaging Strategy and Findings

In general, the imaging findings in patients with spongiform toxic leukoencephalopathy due to heroin vapour inhalation correlate well with the neuropathological findings. Hence, the pattern of involvement on structural neuroimaging is fairly specific for this leukoencephalopathy (Box 7.3). Imaging therefore may usefully point to the aetiology (heroin inhalation), which may not have been available (or not revealed) on history taking. CT is often used in the acute setting but may be normal in patients with mild manifestations. Multi-sequence MRI (including T2-weighted and /or FLAIR sequences) provides a higher sensitivity and represents the image modality of choice (Figs. 7.31 and 7.32). Optional image modalities: Diffusion-weighted imaging (DWI), ^{1}H-MR spectroscopy, SPECT, PET. Typical imaging findings and differential diagnosis and presented in Box 7.3.

7.6 Cognitive Dysfunction Associated with Cancer Therapy

Mike P. Wattjes

> **Synonyms, Abbreviations**
> Chemobrain, chemo fog syndrome, radiation-induced leukoencephalopathy, radiation necrosis

History

Compared to the effects of chemotherapy, the clinical consequences of radiotherapy for the CNS have been described several decades ago. Sheiline first described in 1977 a classification of different stages of radiation-induced encephalopathy which is still being used. Advances in cancer treatment, in particular the combination of different treatment modalities (chemo-,

> ## Box 7.3 Neuroimaging findings and differential diagnosis in Heroin encephalopathy
>
> CT and MRI findings
> - CT may be normal in patients with only mild manifestations
> - Confluent hypodense white matter changes suggestive of diffuse demyelination can be observed.
>   - Infratentorial changes are located in the cerebellum, brain stem and cerebellar peduncles.
>   - Supratentorial brain white matter confluent hypodense white matter lesions.
>   ○ Posterior WM typically involving the posterior limb of the internal capsules.
>   ○ The anterior limbs are classically spared.
> - Deep GM structures, cortex and subcortical U-fibres are classically spared.
> - Symmetric involvement of corticospinal tract, medial lemniscus and tractus solitarius (MRI).
> - No disruption of the blood–brain barrier in terms of contrast enhancement can be observed.
> - Findings on DWI may include increased diffusion (suggestive of vasogen oedema), but also decreased diffusion (due to vacuolisation and intracellular oedema).
>
> Advanced neuroimaging
> - ^{1}H – MR spectroscopy
>   - Low NAA and choline in the affected areas indicating axonal damage
>   - Increased levels of lactate reflecting abnormal energy metabolism
> - SPECT: Low levels of perfusion can be observed bilaterally in the parietal and occipital cortex
> - ^{18}F-FDG PET decreased uptake in the described affected white matter areas on MRI
>
> Differential diagnosis
> - Toxic encephalopathy due to other agents (e.g. inhalation of organic solvents)
> - Chronic inflammatory disease (e.g. multiple sclerosis, PML)
> - Metabolic diseases associated with demyelination (e.g. forms of leukodystrophy)
> - Radiation-induced encephalopathy
> - Chemotherapy-induced encephalopathy
> - Post-hypoxic demyelination (e.g. due to carbon monoxide poisoning)
> - Reversible posterior leukoencephalopathy syndrome (RPLS)

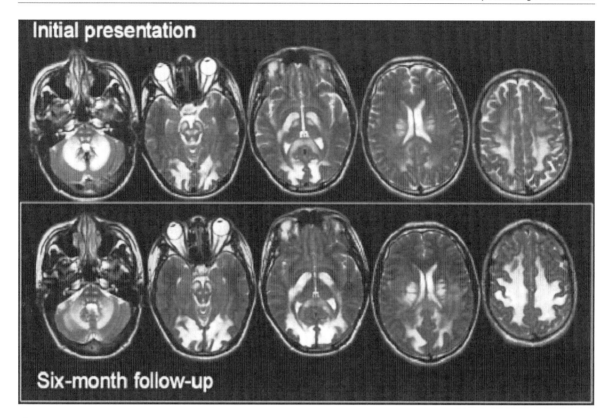

Fig. 7.31 Toxic leukoencephalopathy in 'chasing the dragon'. This 23-year-old HIV-negative male was admitted after a 2–3 week history of bilateral weakness, ataxia and dysarthria that progressed to an inability to speak. The patient later admitted to inhaling heroin vapour daily for the past 3 years. T2-weighted MR images at initial presentation (*above*) and at 6-month follow-up (*below*) show increased signal within the cerebellar white matter with some sparing of the dentate nuclei; involvement of the anterior horns (motor) of the brainstem grey matter and the corticospinal tracts; anterior thalami; posterior limb of the internal capsule, with sparing of the anterior limb; splenium of the corpus callosum; posterior corona radiata; and the white matter of the occipital, parietal, posterior-temporal and posterior-frontal lobes, with sparing of the sub-cortical white matter. (Reprinted with kind permission from *Br J Radiol* 2005;78:997–1004; Sophy Molloy, *AJNR* 2006)

hormone-, and radiotherapy), have led to further appreciation of the necessity to understand and investigate their accompanying toxic effects and clinical consequences.

Aetiology and Histophysiology

Radiotherapy

A combination and interaction of effects on different cell types (e.g. mature neurons, neural progenitor cells, oligodendrocytes and endothelial cells) is responsible for radiation-induced CNS damage, and may lead to cognitive decline. The degree and pattern of radiation-induced effects strongly depends on the time and

dosage. The effect on oligodendroglia is characterised histopathologically by a time-dependent breakdown of myelin, apoptosis of oligodendroglial precursor cells and, finally, a loss of mature oligodendrocytes and myelin several months after the radiation exposure.

Radiation can have a dose-dependent direct effect on the function of mature neurons; more pronounced, however, are the effects on neural progenitor cells. These self-renewing cells are crucial for the neurogenesis (generation of neurons, astroglia, oligodendroglia, lineage-restricted precursor cells) in the post-natal and adult brain. This effect becomes most obvious in the hippocampus, where radiation can substantially decrease the production of neurons (Fig. 7.33). The decline in the neurogenic cell population after radiotherapy is probably the result of an

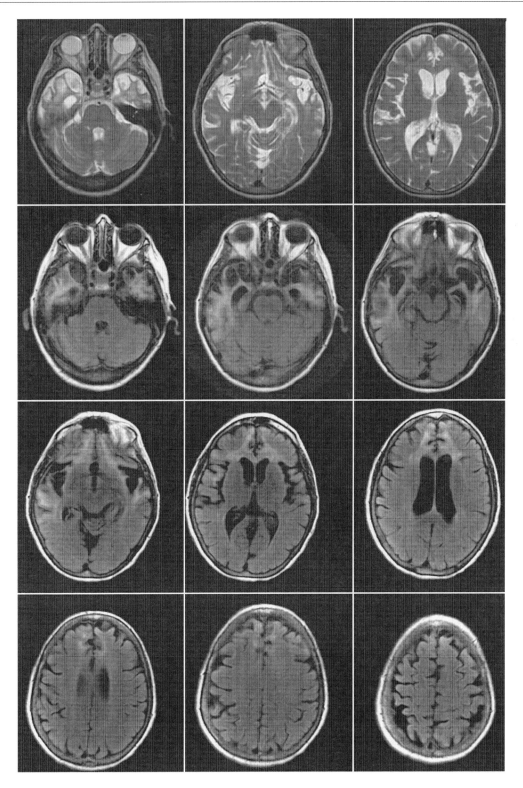

Fig. 7.32 Heroin intoxication. This 55-year-old lady with a vacuolising myelinopathy of the temporal lobes presented with dementia over the course of the last 4 years following extensive heroin inhalation approximately 10 years ago. She smoked heroin for more than 10 years on a regular basis. Images kindly provided by Timo Krings. Note high signal on T2-weighted (*top*) FLAIR (bottom three rows) images in the frontal and temporal lobes, with end-stage atrophy of the temporal lobes bilaterally, similar to FTLD

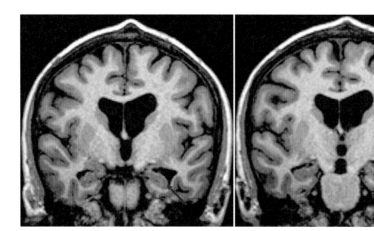

Fig. 7.33 Hippocampal atrophy after radiation therapy. This 31-year-old woman developed acute lymphoblastic leukaemia at the age of 14. She was treated with combination whole-brain radiation and chemotherapy (including vincristine, prednisone, l-asparaginase, 6-mercaptopurine methotrexate). She developed late neuropsychological deficits, including impairment of working memory, attention deficits, visual spatial deficits and impaired cognitive flexibility and eye-hand coordination. Her total IQ was 63 at time of MRI. Coronal T1-weighted images shows severe atrophy of especially the left hippocampus (*red arrows*), probably reflecting damage to neural progenitor cell population

impairment of proliferative activity in combination with direct cell damage (and death). The inhibition of neurogenesis is enhanced by changes in the neuronal microenvironment: radiation-induced microglial inflammatory reaction secondarily leading to neurovascular changes.

Time-dependent changes in blood-brain barrier permeability play a role in the pathophysiology of radiation-induced damage in the CNS. The early (transient) phase of blood-CNS barrier disruption is strongly associated with apoptosis of endothelial cells, which recovers within a few weeks. During the regeneration process endothelial proliferation occurs. The late (second) phase of blood-CNS permeability is associated with tissue necrosis, inflammatory reactions, vasogenic oedema and tissue hypoxia leading to cell damage and loss. This process is also referred to as radiation necrosis and occurs between 6 months and 10 years after therapy (see also Chapter 8.4.4).

Chemotherapy

The current pathophysiological concept of chemotherapy-induced damage is based on a variety of factors including direct vascular injury, oxidative damage, inflammation, autoimmune reactions, toxic neuronal injury, chemotherapy-induced systemic factors such as anaemia, and genetic factors (e.g. APOE-ε4 allele).

The degree of neurotoxicity strongly depends on the type, dosage and combination of various chemotherapeutic agents. Due to toxic effects on vascular endothelial cells and release of free radicals, vascular damage occurs leading to alterations of cerebral blood flow, perfusion and metabolism. Chemotherapy-induced inflammatory responses are driven by cytokine-mediated cascades (including IL-1, IL-6, TNF-α) inducing glial cell activation with secondary axonal damage. Typical manifestations of this inflammatory process are vasculitis and allergic hypersensitivity responses. The direct neuronal injury is further characterised by an imbalance of neurotransmitters (particularly acetylcholine).

Chemotherapeutic agents may not only directly lead to neurotoxicity, but also indirectly by promoting conditions associated with cognitive decline, such as metabolic and hormonal abnormalities, anaemia, liver and renal dysfunction etc.

Combined Therapy

Combined therapeutic strategies including both chemotherapy and radiotherapy do have higher neurotoxic effects than single modality treatment (Fig. 7.34). The single pathological mechanisms are acting in a synergistic way but are also likely to interact with each other.

Clinical Presentation, Epidemiology and Treatment

Chemotherapy

The incidence of chemotherapy-induced cognitive dysfunction can reach up to 70%. The occurrence and severity of cognitive manifestations related to chemotherapy depends on different variables including the treatment protocol, patient's age, integrity of the blood-brain barrier, and the level of cognitive functioning before treatment. In general, chemotherapy broadly affects various domains of neuropsychological functioning including verbal and visual memory, mental flexibility, information processing, attention and concentration, visuo-spatial ability, language and motor function. Regarding memory function, patients may display impaired learning efficiency and problems with memory recall, due to damage in frontal and subcortical white matter networks (Fig. 7.35).

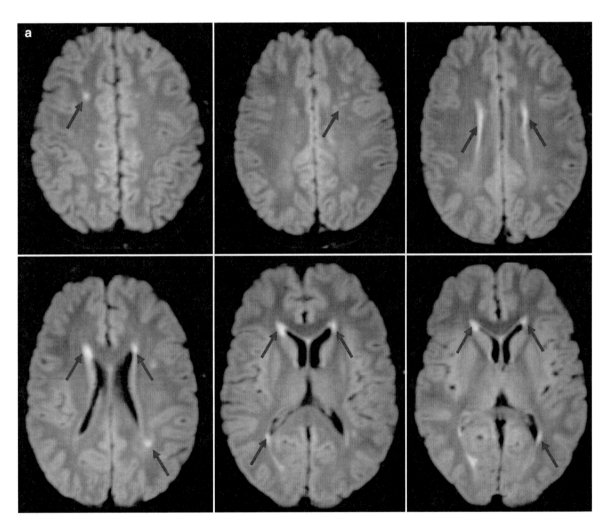

Fig. 7.34 Combined radio- and chemotherapy effects. This 31-year-old woman developed acute lymphoblastic leukaemia at the age of 13. She was treated with combination therapy of brain radiation and systemic and intrathecal chemotherapy (including vincristine, prednisone, l-asparaginase, 6-mercaptopurine methotrexate). She developed severe attention deficits (sustained attention) and impairment of cognitive flexibility. Axial FLAIR images (**a**) show focal WMH in the deep white matter and diffuse white matter changes in the periventricular white matter indicating treatment-induced leukoencephalopathy. Axial T2-weighted gradient-echo images (**b**) reveal microhaemorrhages (*arrows*) in the deep grey matter, neocortex, thalamus and white matter–grey matter junction, suggestive of radiation-induced vasculopathy (*continued on next page*)

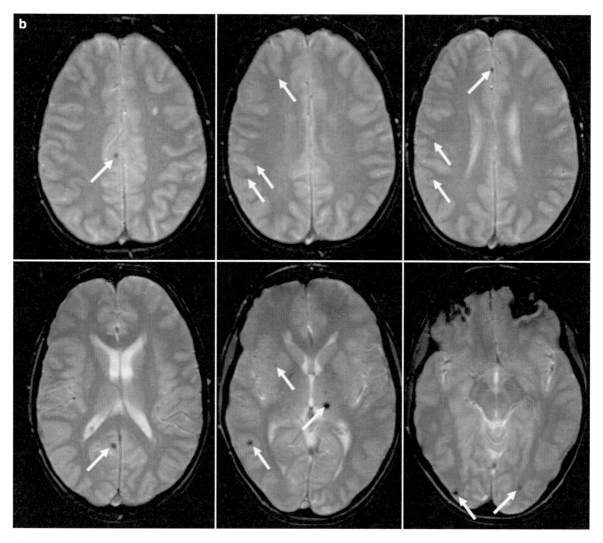

Fig. 7.34 (continued)

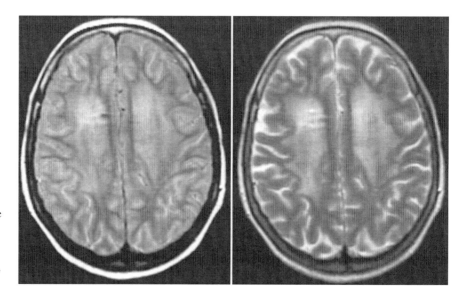

Fig. 7.35 Leukoence-phalopathy after intrathecal methotrexate treatment. Note diffuse mild signal increase in white matter on proton-density-weighted (*left*) and heavily T2-weighted images, with sparing of the U-fibres

Due to the complex mechanisms and interactions of different chemotherapeutic agents no specific treatment exists. Current treatment strategies focus on antioxidant regiments such as vitamin E and the herb *Ginkgo biloba* in order to reduce the formation of free radicals. Other pharmacological interventions include recombinant human erythropoietin to increase the haemoglobin levels or use of the psychostimulant dexmethylphenidate to improve cognitive performance.

Radiotherapy

Cognitive impairment is one of the most common sequelae of radiotherapy in children and adults. The incidence can reach 86% and strongly depends on patient's age, dose per fraction, cumulative dose, target volume (Fig. 7.36), time interval, additional use of chemotherapy and co-morbidity (anaemia, renal failure, diabetes etc.). Radiation encephalopathy is traditionally subdivided into three different clinical stages: acute, early-delayed and late-delayed reaction:

1. During the *early stage* (first weeks) patients may remain asymptomatic or may develop self-limiting (sub) acute focal neurological deficits.
2. The *early-delayed stage* is characterised by somnolence, fatigue and symptoms of cognitive impairment suggestive of frontal network malfunction, with slowed information processing, decline in memory retrieval, attention deficits and decline in executive functioning.
3. The symptoms of *late-delayed* stage occur months to years after treatment and are linked to the hippocampus function. They are probably resulting from damage to neural progenitor cells. The main symptoms are impairment of memory function, attention deficits and visual motor processing. Most of the symptoms are irreversible or even progress over time.

Management of radiation encephalopathy (Fig. 7.37) includes treatment of oedema and increased intracranial pressure. No specific treatment of radiation-induced cognitive dysfunction exists. Several experimental in vivo studies focus on indometacin in order to repair the network of neurogenesis or try to use anti-inflammatory drugs and antioxidants.

Differential Diagnosis

The differential diagnosis should include the whole gamut of transient and/or persistent white matter changes associated with demyelination and gliosis such as post-hypoxic demyelination, metabolic disturbances, hereditary and inflammatory white matter diseases.

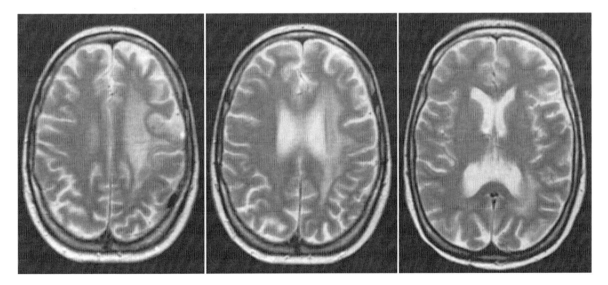

Fig. 7.36 Localised effects of radiotherapy. This patient was treated with radiotherapy applied to the left hemisphere only, resulting in unilateral diffuse leukoencephalopathy of the left hemispheric white matter, with sparing of the U-fibres

Neuroimaging Strategy and Findings after Chemotherapy and Radiotherapy
Neuroimaging should include a multi-sequence MRI protocol:

- Axial FLAIR: assessment of white matter changes and global and focal cortical atrophy
- Axial T2: assessment of infratentorial white matter changes and thalamic lesions
- Axial T2*: detection of vasculopathy in terms of microhaemorrhages/microbleeds
- Coronal 3D T1 with multi-planar reconstructions: evaluation of hippocampal structures
- Contrast-enhanced T1-weighted images: detection of blood–brain barrier disruption
- Optional: Diffusion-weighted imaging, MR perfusion, MR spectroscopy

Stage	*Chemotherapy*	*Radiotherapy*
Acute	Symmetric and focal or diffuse Deep and periventricular WMH Sparing the U-fibres, corpus Callosum, anterior commissure	Asymmetric, focal or diffuse Deep and periventricular WMH Sparing of U-fibres Initial phase: interstitial oedema, acute demyelination, transient disruption of the blood–brain barrier (contrast enhancement)
Early/late subacute	Contrast enhancement in necrotic areas ¹H-MRS: ↓ *tNAA* and ↑ *Choline*	Contrast enhancement and swelling (radionecrosis); Vasculopathy: microbleeds
Chronic	Global atrophy	Global *and* local atrophy (hippocampus) Secondary tumours (e.g. meningeoma) Cavernomas

Combined therapy

The combination of chemotherapy and radiotherapy can result in more severe manifestations in terms of acute changes (oedema) but also chronic white matter changes, vasculopathy, axonal damage and secondary neurodegeneration.

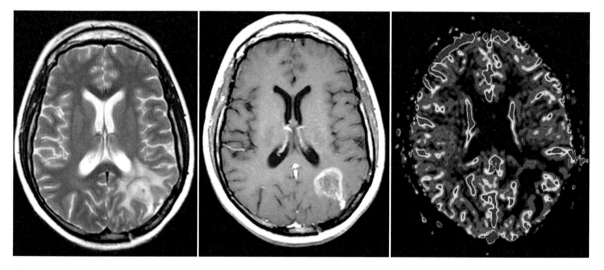

Fig. 7.37 Radiation necrosis. On the T2-weighted image (*left*), an area of oedema is seen, with central, partly patchy, enhancement after gadolinium administration (*middle*). Perfusion-weighted image (*right*) shows low CBV, consistent with the biopsy finding of radiation necrosis

7.7 Trauma

7.7.1 Dementia Associated with Traumatic Brain Injury

Mike P. Wattjes

Synonyms, Abbreviations
Traumatic brain injury (TBI), diffuse axonal injury (DAI)

History

The pathophysiological mechanism of traumatic brain injuries (TBI) has first been described in 1943 by Holbourn. Using 2D gelatine moulds he showed that rotational forces rather than linear and/or translational forces are causing shear injury in the brain, in particular at locations with substantial differences in tissue density.

Aetiology

TBI is caused by an outside force manifesting as a direct or indirect impact (coup or contre-coup) and/or acceleration leading to primary and secondary injury of the brain tissue. Diffuse axonal injury (DAI) in patients with TBI is a result of rapid rotational acceleration/deceleration of the brain due to unrestricted head movement. This results in shearing forces particularly occurring in areas of the brain containing tissues of different density such as the grey–white matter interface. The degree of injury depends on the distance from the centre of rotation, the arc of rotation and duration of the force.

Histophysiology

Following a traumatic event, acute neuronal pathology develops over a time interval of hours to days. The primary brain injury occurs in the acute trauma setting by stretching, compression and tearing the brain tissue and vascular structures. The subsequent axonal pathology is characterised by a damage of sodium channels in the cell membrane resulting in a pathological influx of sodium and subsequently increased intracellular calcium concentration leading to an activation of proteolytic activity and intracellular oedema. Neuronal/axonal swelling leads to elongated varicosities and bulb formation. Histopathologically, this results in 'retraction ball' formation at the terminal end of the axon followed by Wallerian degeneration and reactive gliotic changes and restructuring.

The concept of secondary brain injury is based on a complex interaction of inflammation, release of free radicals, excitotoxicity, oedema and changes in cerebral blood perfusion. It has been postulated that TBI may induce a molecular cascade of inflammation that leads to the formation of amyloid plaque formation. APOE-ε4, a protein which is involved in the transportation of produced β-amyloid peptide, seems to play a key role in this process. In fact, patients can show features of neurodegeneration and Alzheimer pathology after TBI in terms of an increase of *tau* and A-β-amyloid precursor protein levels in the CSF and amyloid-β plaque deposition in the temporal cortex. Another possibility is that TBI exhausts reserve capacities and thus lower the threshold for subsequent Alzheimer pathology to become manifest (stepping-stone theory), which could explain the association between brain trauma and AD.

Clinical Presentation, Epidemiology and Treatment

Traumatic Brain Injury

Cognitive impairment is one the most important and frequent long-term consequences of traumatic brain injury. The clinical outcome shows a relationship with the trauma severity and mechanism. In the acute trauma setting, patients can present with a wide range of symptoms ranging from focal cognitive deficits, amnesia to a vegetative state. At one year follow-up, patients with severe TBI frequently show slower information processing, decline of anterograde memory. However, also patients with mild to moderate TBI frequently present with cognitive complaints such as lack of concentration and memory deficits usually immediate after the traumatic event which usually recovers completely during short-term follow up. The manifestation of TBI is very heterogeneous. Depending on the localisation and severity of the traumatic lesions in the CNS (Fig. 7.38), different patterns of clinical manifestations

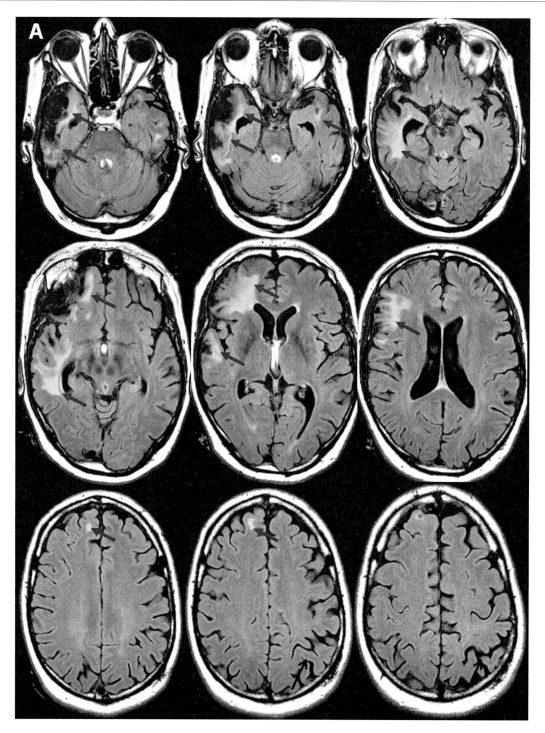

Fig. 7.38 Multifocal TBI and DAI. MRI images of a 57-year-old man who was involved in a car accident leading to a severe brain contusion. Subsequently he developed short term memory complaints and concentration deficits. The transverse FLAIR images (**a**) shows post-contusional changes in the right frontal en temporal lobe (including the medial temporal lobe) with focal loss of brain tissue (grey and white matter) and perifocal gliotic changes (*red arrows*). Please note also the slight asymmetry of the ventricles due to ex vacuo dilatation of the right lateral ventricle. This patient already presents with neurodegenerative changes of the brain suggestive of Alzheimer pathology on the coronal T1-weighted images (**b**) – disproportionate bilateral atrophy of the hippocampus (*white arrows*). Please not also the right temporal post-traumatic changes and the ex vacuo dilatation of the temporal horn. The susceptibility-weighted images (**c**) show haemosiderin deposits in the contusion area of the right frontal lobe but also microbleeds in the deep and juxtacortical white matter of the left parietal lobe (*red arrows*) indicating axonal injury grade 1 according to Adams classification (*continued on next page*)

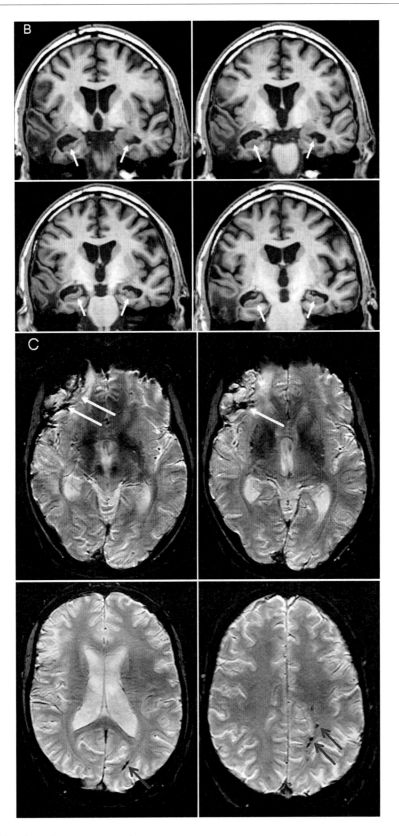

Fig. 7.38 continued (*see legend on previous page*)

can occur: *memory function* characterised by working memory impairment, *language* problems including fluent and non-fluent aphasia and high order language alterations and *executive functions*.

TBI in children can have even more consequences in term of clinical outcome and is therefore an important socio-economic issue (Fig. 7.39). TBI is a frequent cause of developmental disability involving intellectual, behavioural, attention, language and memory functions possibly leading to long-term cognitive impairment.

An association between chronic TBI and dementia showing neuropathological features of Alzheimer's disease referred to as the term *Dementia pugilistica*

(see Sect. 7.7.2) has been reported. The crucial question whether TBI might increase the risk of developing dementia is still a matter of debate. Several retrospective and prospective studies have shown a positive association between a history of TBI and dementia (particularly AD) and have reported that TBI might be related to an earlier onset of dementia. However, the data of all performed studies are not completely conclusive since some other studies and meta-analysis failed to reproduce these findings. In addition, an effect of the APOE-ε4 allele on the clinical outcome in patients with TBI could not conclusively be demonstrated.

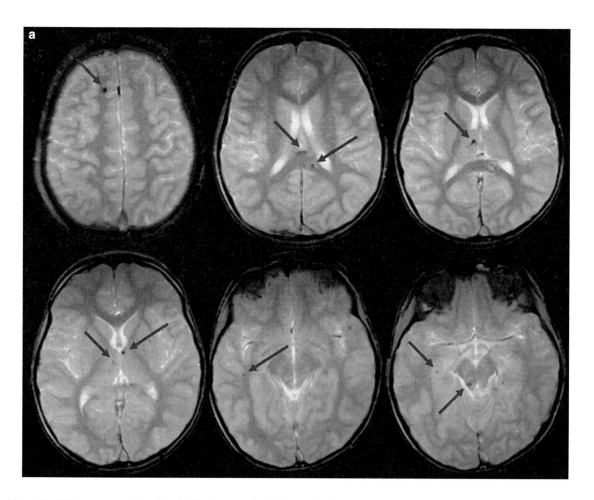

Fig. 7.39 Diffuse axonal injury (DAI). This 5 year-old child was involved in a car accident presenting with prolonged post-traumatic unconsciousness and a low Glasgow coma scale score. (**a**) Transverse T2* images show multiple focal hemorrhagic lesions (microbleeds) associated with DAI in typical anatomic locations such as in the grey–white matter junction of the supratentorial brain, the corpus callosum, the basal ganglia and the brain stem (*red arrows*). The distribution is consistent with grade 3 severity according to Adams classification. (**b**) Transverse T2-weighted turbo spin echo (*left*), DWI (*middle*) and corresponding AIDS dementia complex (ADC) map (*right*) demonstrate non-haemorrhagic lesions, with restricted diffusion due to intracellular oedema (*arrows*)

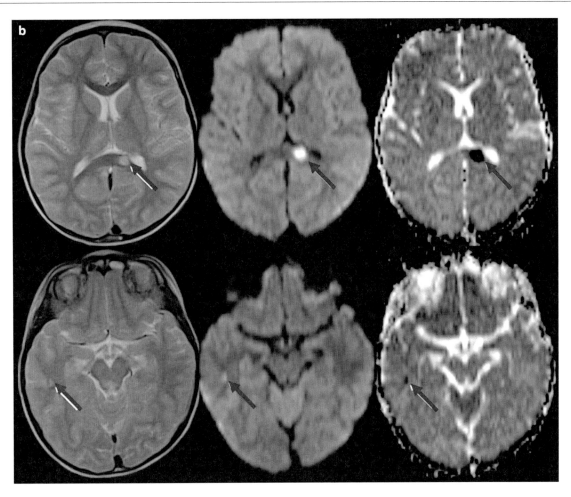

Fig. 7.39 continued (*see previous page for legend*)

Diffuse Axonal Injury (DAI)

A more severe manifestation of TBI is DAI, which is associated with a rather poor clinical outcome. DAI is one of the most frequent manifestations of rotational deceleration, classically resulting from high-speed motor vehicle accidents. Patients presenting with DAI can also show other traumatic changes of the central nervous system including epidural and sub-dural haematoma, traumatic subarachnoid haemorrhage and haemorrhagic brain contusion. However, DAI can occur independently from other traumatic lesions and be the sole traumatic manifestation in the brain.

In contrast to focal traumatic changes (e.g. epidural haematoma or cortical contusion) DAI does not always present clinically with focal neurological deficits or a lucid interval. Prolonged post-traumatic unconsciousness and coma are the most common clinical manifestation in patients with DAI. While rarely resulting in death, depending on the degree of involvement, DAI can result in a persistent vegetative state. Patients with TBI associated with DAI usually pass through an interval of global cognitive disturbance (post-traumatic amnesia) which potentially can recover. The clinical manifestation and prognosis correlates with the extent of involvement assessed by neuroimaging.

Anatomically, DAI most frequently occurs bilaterally in the supratentorial brain particularly at the grey-white matter junction, but also involving the corpus callosum, basal ganglia region (caudate nucleus thalamus, internal capsule) and the brain stem (mesencephalon) – see Fig. 7.39. The Adams classification recognises – according to the anatomic location – the following three stages which has been used to classify DAI as mild, moderate or severe:

Stage 1. Supratentorial brain damage including the parasagittal regions of frontal lobes, periventricular temporal region, parietal and occipital lobes, internal capsule and cerebellum

Stage 2. In addition to stage 1, involvement of the corpus callosum

Stage 3. In addition to stage 2, involvement of the brain stem

The treatment of both common TBI and DAI includes established treatment schemes for TBI (stabilising the patient, treatment of increased intracranial pressure, neurosurgical treatment). However, TBI in general and DAI in particular currently lack specific treatment options.

Neuroimaging strategy and findings in TBI:

Acute setting
- CT: the image modality of choice
  - Spiral CT with multi-planar reconstructions (bone and soft tissue kernel)
  - Demonstrate haematoma and mass-effect
  - Cerebral contusions demarcate only after days
- MRI: if CT fails to explain clinical presentation
  - Imaging protocol: include FLAIR, T2*/SWI, DWI, T1-weighted images
  - T2*-weighted (microbleeds) and DWI (cytotoxic oedema) may reveal DAI
  - FLAIR and T2-weighted imaging for cortical contusion
  - T1-weighted imaging: detection of early subacute haemorrhage

- Cerebral hypoperfusion
  - HMPAO-SPECT and FDG PET: reduced perfusion/metabolism

Chronic setting
- Ventricular widening and communicating hydrocephalus (CT or MR)

- Focal cortical contusions with gliotic changes (esp. temporal and baso-frontal)
  - MRI (FLAIR) more sensitive than CT
  - MRI (T2*/SWI): detection of haemosiderine deposition
- Wallerian degeneration
  - Signal changes in the corticospinal tract
  - Hypertrophic degeneration of the olivary nuclei
- Neurodegenerative changes
  - Global cortical atrophy
  - Focal atrophy (particularly hippocampus) on coronal T1-weighted MRI
- DAI
  - Cerebral microbleeds only seen using T2*-weighted MRI
    - Differential diagnosis: CAA, cavernomata, vasculitis

7.7.2 Dementia Pugilistica

Synonyms

Chronic traumatic brain injury; post-traumatic Alzheimer's disease; chronic traumatic encephalopathy; chronic boxer's encephalopathy; traumatic boxer's encephalopathy; boxer's dementia; 'punch-drunk' syndrome

History and Histopathology

The repeated head trauma experienced by boxers can lead to the development of dementia pugilistica. The clinical symptom complex of this disorder was first described by Martland in 1928, and the neuropathology was first described in a classic report by Corsellis et al. in 1973. They describe the presence of numerous neurofibrillary tangles in the absence of plaques, in contrast to the profusion of tangles and plaques seen in AD. Later investigations found that there were diffuse plaques instead of neuritic plaques. Epidemiological studies have shown that head injury and ApoE 4 genotype are risk factors for AD. It is probable that dementia pugulistica and AD share common pathogenic mechanisms leading to tangle and plaque formation.

Clinical Presentation, Epidemiology and Treatment

Dementia pugulistica associated with boxing occurs in approximately 20% of professional boxers who suffered multiple concussions. Typical manifestations include problems with memory, or even dementia and Parkinsonism, or tremors and lack of coordination. Risk factors associated with chronic traumatic brain injury include increased exposure (i.e. duration of career, age of retirement, total number of bouts), poor performance, increased sparring and APOE genotype, notably the epsilon 4 allele. A recent report also demonstrated that dementia in retired boxers may be caused and/or exacerbated by other etiologic factors, such as multiple cerebral infarcts and Wernicke–Korsakoff syndrome. Clinically, boxers exhibiting dementia pugilistica will present with varying degrees of motor, cognitive and/or behavioural impairments. The diagnosis is dependent upon documenting a progressive neurological condition attributable to brain trauma and unexplainable by an alternative pathophysiological process. There is no specific therapy.

Neuroimaging Strategy and Findings

The association between the finding of a cavum septi pellucidi and dementia in old boxers was first described by Ferguson and Mawdsley in 1965. Since then it has commonly been listed as one of the features of this condition ('born to box'). Fenestration of the septum pellucidum, with formation of a cavum, however, does not appear to correlate with neurological or physiological evidence of brain damage and should probably be classified as a chance finding, as documented in a large Swedish study. A number of imaging techniques have been used to investigate changes produced in the brain by boxing. The most commonly reported finding is diffuse cerebral atrophy with a non-specific pattern. Most morphological studies have failed to show significant correlations between putative abnormalities on imaging and clinical evidence of brain damage. Serial studies on large groups may be more informative but are not available to date. The link with AD, as denoted above, may lead to findings of bilateral hippocampal shrinkage.

Differential Diagnosis

Given that retired boxers are usually middle-aged and older, a comprehensive clinical and neuroimaging evaluation is necessary to exclude other causes of dementia. Actually, as the MRI findings are non-specific, a number of alternative diagnoses may be considered: Alzheimer's disease, Parkinson's disease with dementia, frontotemporal dementia being the most prevalent.

Suggested Reading

Subacute Sclerosing Panencephalitis

Achim CL, Wiley CA (1998) Virus-mediated dementias. In: Markesbery WR (ed) Neuropathology of dementing disorders. Edward Arnold, London, pp 312–339
Everall IP, Chong WK, Wilkinson ID, Paley MN, Chinn RJ, Hall-Craggs MA, Scaravilli F, Lantos PL, Luthert PJ,

Harrison MJ (1997) Correlation of MRI and neuropathology in AIDS. J Neurol Neurosurg Psychiatry 62:92–95
Garg RK (2002) Subacute sclerosing panencephalitis. Postgrad Med J 78:63–70

Bacterial Infections: Whipple's Disease

Fenollar F, Puechal X, Raoult D (2007) Whipple's disease. N Engl J Med 356:55–66
Panegyres PK (2008) Diagnosis and management of Whipple's disease of the brain. Pract Neurol 8:311–317

Neurosyphilis

Hildenbrand P, Craven DE, Jones R, Nemeskal P (2009) Lyme neuroborreliosis: manifestations of a rapidly emerging zoonosis. AJNR Am J Neuroradiol 30(6):1079–87
Peng F, Hu X, Zhong X, Wei Q, Jiang Y, Bao J, Wu A, Pei Z (April 2008) CT and MR findings in HIV-negative neurosyphilis. Eur J Radiol 66(1):1–6
http://emedicine.medscape.com/article/1169231-overview
van Eijsden P, Veldink JH, Linn FH, Scheltens P, Biessels GJ (2008 Feb) Progressive dementia and mesiotemporal atrophy on brain MRI: neurosyphilis mimicking pre-senile Alzheimer's disease? Eur J Neurol 15(2):e14–5

Multiple Sclerosis and ADEM

Amato MP, Zipoli V, Portaccio E (2006) Multiple sclerosis-related cognitive changes: a review of cross-sectional and longitudinal studies. J Neurol Sci 245:41–46
Calabrese M, Rinaldi F, Mattisi I et al (2010) Widespread cortical thinning characterizes patients with MS with mild cognitive impairment. Neurology 74:321–328
Geurts JJ, Barkhof F (2008) Grey matter pathology in multiple sclerosis. Lancet Neurol 7:841–851
Kutzelnigg A, Lassmann H (2008) Neuropathology of cognitive dysfunction in multiple sclerosis, Handbook of clinical neurology, demyelinating diseases, vol 89 (3rd series). Elsevier, New York, pp 719–723
Staff NP, Lucchinetti C, Keegan BM (2009) Multiple Sclerosis With Predominant, Severe Cognitive Impairment. Arch Neurol 66:1139–1143
Zarei M, Chandran S, Compston A, Hodges J (2003) Cognitive presentation of multiple sclerosis: evidence for a cortical variant. J Neurol Neurosurg Psychiatry 74:872–877

Coeliac Disease

Hu WT, Murray JA, Greenaway MC, Parisi JE, Josephs KA (2006) Cognitive Impairment and Celiac Disease. Arch Neurol 63:1440–1446

Fragile X-Associated Tremor/Ataxia Syndrome (FXTAS)

Amiri K, Hagerman RJ, Hagerman PJ (2008) Fragile X-associated tremor/ataxia syndrome: an aging face of the fragile X gene. Arch Neurol 65:19–25

Brunberg JA, Jacquemont S, Hagerman RJ et al (2002) Fragile X premutation carriers: characteristic MR imaging findings of adult male patients with progressive cerebellar and cognitive dysfunction. AJNR 23:1757–66

www.nfxf.org

Mitochondrial Dementia

Finsterer J (2008) Cognitive decline as a manifestation of mitochondrial disorders (mitochondrial dementia). J Neurol Sci 272:20–33

Knaap MS and Valk J. (2005) Mitochondria and mitochondrial disorders. In: Magnetic resonance of myelination and myelin disorders (3rd ed). Springer, Heidelberg, pp 195–203

Diffuse Neurofibrillary Tangles with Calcification (DNTC)

Ito Y, Kato T, Suzuki T, Yokokawa Y, Aiba I, Arahata Y et al (2003) Neuroradiological and clinical abnormalities in dementia of diffuse neurofibrillary tangles with calcification (Kosaka-Shibayama disease). J Neurol Sci 209:105–109

Kosaka K (1994) Diffuse neurofibrillary tangles with calcification: a new presenile dementia. J Neurol Neurosurg Psychiatry 57:594–596

Modrego PJ, Mojonero J, Serrano M, Fayed N (2005) Fahr's syndrome presenting with pure and progressive presenile dementia. J Neurol Sci 26:367–369

Cerebrotendinous Xanthomatosis

Barkhof F, Verrips A, Wesseling P, Van der Knaap MS, Van Engelen BGM, Gabriels FJM, Keyser A, Wevers RA, Valk J (2000) Cerebrotendinous xanthomatosis: the spectrum of imaging findings and the correlation with neuropathologic findings. Radiology 217:869–876

Verrips A, Hoefsloot LH, Steenbergen GC, Theelen JP, Wevers RA, Gabriels FJ, Van Engelen BG, Van den Heuvel LP (2000) Clinical and molecular genetic characteristics of patients with cerebrotendinous xanthomatosis. Brain 123:908–919

Adult Polyglucosan Body Disease

Rifai A, Klitzke M et al (1994) Dementia of adult polyglucosan body disease, evidence of cortical and subcortical dysfunction, Arch. Neurol 51:90–94

Savage G, Ray F et al (2007) Stable neuropsychological deficits in adult polyglucosan body disease. J Clin Neurosci 14:473–477

Van der Knaap M, Valk J (2005) Magnetic resonance of myelination and myelin disorders, 3rd edn. Springer, New York, Chapter 16: Adult polyglucosan body disease

Neuroaxonal Leukodystrophy (NAL)

Freeman SH, Hyman BT, Sims KB, Hedley-Whyte ET, Vossough A, Frosch MP, Schmahmann JD (2009) Adult onset leukodystrophy with neuroaxonal spheroids: clinical, neuroimaging and neuropathologic observations. Brain Pathol 19(1):39–47

http://www.ulf.org/types/NeuroaxonalLeukoencephalopathy.html

Marotti JD, Tobias S, Fratkin JD, Powers JM, Rhodes CH (2004) Adult onset leukodystrophy with neuroaxonal spheroids and pigmented glia: report of a family, historical perspective, and review of the literature. Acta Neuropathol 107(6):481–8

Mascalchi M, Gavazzi C, Morbin M, Giaccone G, Arnetoli G, Zappoli R, Bugiani O (2000) CT and MR imaging of neuroaxonal leukodystrophy presenting as early-onset frontal dementia. AJNR Am J Neuroradiol 27(5):1037–9

Late-Onset Metachromatic Leukodystrophy

Eichlera F, Groddd W, Grant E, Sessa M, Biffi A, Bley A, Kohlschuetter A, Loes DJ, Kraegeloh-Mann I (2009) Metachromatic Leukodystrophy: A Scoring System for Brain MR Imaging Observations. AJNR 30:1893–97

Van der Knaap MS, Valk J (2005) Magnetic resonance of myelin, myelination and myelin disorders, 3rd edn. Springer, New York. Chapter 6 - Metachromatic leukodystophy

Adreno-Myelo-Neuropathy

Edwin D, Speedie LJ, Kohler W, Naidu S, Kruse B, Moser HW (1996) Cognitive and brain MRI findings in adrenomyeloneuropathy. Ann Neurol 40:675–8

Powers JM, DeCiero DP, Ito M, Moser AB, Moser HW (2000) Adrenomyeloneuropathy: a neuropathologic review featuring its noninflammatory myelopathy. Neuropathol Exp Neurol 59:89–102

Vanishing White Matter Disease

Gascon-Bayarri J, Campdelacreu J, Sánchez-Castañeda C et al (2009) Leukoencephalopathy with vanishing white matter presenting with presenile dementia. J Neurol Neurosurg Psychiatry 80(7):810–1

van der Knaap MS, Pronk JC, Scheper GC (2006) Vanishing white matter disease. Lancet Neurol 5(5):413–23

Globoid Cell Leukodystrophy (GLD)

Loes DJ, Peters C, Krivit W (1999) Globoid cell leukodystrophy: distinguishing early-onset from late-onset disease using
a brain MR imaging scoring method. AJNR Am J Neuroradiol
20:316–323

Wang C, Melberg A, Weis J, Mansson J-E, Raininko R (2007)
The earliest MR imaging and proton MR spectroscopy
abnormalities in adult-onset Krabbe's disease. Acta Neurol
Scand 116:268–272

Cognitive Dysfunction and Vitamin B Deficiencies

Sakurai K et al (2009) Wernicke's encephalopathy with cortical
abnormalities: clinicoradiological features: report of 3 new
cases and review of the literature. Eur Neurol 62:274–280

Sechi G, Serra A (2007) Wernicke's encephalopathy: new clinical settings and recent advances in diagnosis and management. Lancet Neurol 6:442–455

Weidauer S, Nichtweiss M, Lanfermann H, Zanella FE (2003)
Wernicke encephalopathy: MR findings and clinical presentation. Eur Radiol 13(5):1001–1009

Zuccoli G, Pipitone N (2009) Neuroimaging findings in acute
Wernicke's encephalopathy: review of the literature. AJR
Am J Roentgenol 192:501–508

Dementia Associated with Chronic Kidney Disease and Dialysis

Buchman AS, Tanne D, Boyle PA et al (2009) Kidney function
is associated with rate of cognitive decline in elderly.
Neurology 73:920–927

Krishnan AV, Kiernan MC (2009) Neurological complications
of chronic kidney disease. Nat Rev Neurol 5:542–551

Madero M, Gul A, Sarnak MJ (2008) Cognitive function in
chronic kidney disease. Semin Dial 21:29–37

Miu AC, Benga O (2006) Aluminium and Alzheimer's disease:
a new look. J Alzheimers Dis 10:179–201

Muhteşem Ağıldere A et al (2001) MRI of neurologic complications in end-stage renal failure patients on hemodialysis:
pictorial review. Eur Radiol 11:1063–1069

Murray AM (2008) Cognitive impairment in aging dialysis and
chronic kidney disease populations: an occult burden. Adv
Chronic Kidney Dis 15:123–132

Pereira AA, Weiner DE, Scott T et al (2005) Cognitive function
in dialysis patients. Am J Kidney Dis 24:448–462

Cognitive Dysfunction Related to Carbon Monoxide Poisoning

Kondziella D et al (2009) 1H MR spectroscopy of gray and
white matter in carbon monoxide poisoning. J Neurol
256:970–979

Lin WC et al (2009) White matter damage in carbon monoxide
intoxication assessed in vivo using diffusion tensor MR
imaging. AJNR Am J Neuroradiol 30:1248–1255

Lo CP et al (2007) Brain injury after carbon monoxide poisoning: early and late complications. AJR Am J Roentgenol
189:W205–211

Sharma P et al (2009) Toxic and acquired metabolic encephalopathies: MRI appearance. AJR Am J Roentgenol
193:879–886

Tapeantong T, Poungvarin N (2009) Delayed encephalopathy
and cognitive sequelae after acute carbon monoxide poisoning: report of a case and review of the literature. J Med Assoc
Thai 92:1374–1379

Weaver LK (2009) Carbon monoxide poisoning. N Engl J Med
360:1217–1225

Cognitive Decline Related to Delayed Posthypoxic Demyelination

Gottfried JA et al (1997) Delayed posthypoxic demyelination:
association with arylsulfatase-A deficiency and lactate
acidosis on proton MR spectroscopy. Neurology 49:
1400–1404

Lee HB, Lyketsos CG (2001) Delayed post-hypoxic leukoencephalopathy. Psychosomatics 42:530–533

Molloy S et al (2006) Reversible delayed posthypoxic leukoencephalopathy. AJNR Am J Neuroradiol 27(8):1763–1765

Sharma VK, Tan JH (2007) Post-hypoxic early selective putaminal necrosis followed by delayed extensive sub-cortical
demyelination. Brain Inj 21:1199–1202

Shprecher DR et al (2008) Clinical and diagnostic features of
delayed hypoxic leukoencephalopathy. J Neuropsychiatry
Clin Neurosci 20:473–477

Cognitive Dysfunction Related to Organic Solvent Inhalation

Aydin K et al (2002) Cranial MR findings in chronic toluene
abuse by inhalation. AJNR Am J Neuroradiol 23:1173–1179

Borne J et al (2005) Neuroimaging in drug and substance abuse
part II: opioids and solvents. Top Magn Reson Imaging
16:239–245

Yücel M et al (2008) Toluene misuse and long-term harms:
a systematic review of the neuropsychological and neuroimaging lieterature. Neurosci Biobehav Rev 32:910–926

Cognitive Decline Related to Heroin Vapour Inhalation

Bartlett E, Mikulis DJ (2005) Chasing "chasing the dragon"
with MRI: leukoencephalopathy in drug abuse. Br J Radiol
78(935):997–1004

Blasel S et al (2010) Toxic encephalopathy after heroin abuse
without heroin vapor inhalation. Clin Neuroradiol 20:48–53

Hagel J et al (2005) "Chasing the dragon"- Imaging of heroin inhalation leukoencephalopathy. Can Assoc Radiol 56:199–203

Jee RC et al (2009) Heroin vapor inhalation-induced spongiform leukoencephalopathy. J Fromos Med Assoc 108:518–522

Keogh CF et al (2003) Neuroimaging features of heroin inhalation toxicity. Am J Roentgenol 180:847–850

Offfiah C, Hall E (2008) Heroin-induced leukoencephalopathy: characterization using MRI, diffusion weighted imaging, and MR spectroscopy. Clin Radiol 63:146–152

Wolters EC et al (1982) Leukoencephalopathy after inhaling "heroin pyrolysate". Lancet 2:123–1237

Cognitive Dysfunction Associated with Cancer Therapy

Dietrich J et al (2008) Clinical patterns and biological correlates of cognitive dysfunction associated with cancer therapy. Oncologist 13:1285–1295

Nelson CJ et al (2007) Chemotherapy and cognitive deficits: mechanisms, findings, and potential interventions. Palliat Support Care 5:273–280

Reddik WE et al (2007) Quantitative morphologic evaluation of magnetic resonance imaging during and after treatment of childhood leukemia. Neuroradiology 49:889–904

Ricard D et al (2009) Brain damage from anticancer treatments in adults. Curr Opin Oncol 21:559–565

Soussain C et al (2009) CNS complications of radiotherapy and chemotherapy. Lancet 374:1639–1651

Verheyde J, Benotmane MA (2007) Unraveling the fundamental molecular mechanisms of morphological and cognitive defects in the irradiated brain. Brain Res Rev 53:312–320

Dementia Associated with Traumatic Brain Injury

Azouvi P et al (2009) Cognitive deficits after traumatic coma. Prog Brain Res 177:89–110

Hurley RA et al (2004) Traumatic axonal injury: novel insights into evolution and identification. J Neuropsychiatry Clin Neurosci 16:1–7

Li XY, Feng DF (2009) Diffuse axonal injury: novel insights into detection and treatment. J Clin Neurosci 16:614–619

Nandoe RDS et al (2002) Head trauma and Alzheimer's disease. J Alzheimers Dis 4:303–308

Starkstein SE, Jorge R (2005) Dementia after traumatic brain injury. Int Psychogeriatr 162:2086–2093

Suskauer SJ, Huisman TAGM (2009) Neuroimaging in pediatric traumatic brain injury: current and future predictors of functional outcome. Dev Disabi Rev 15:117–123

Dementia Pugilistica

Bodensteiner JB, Schaefer GB (1997) Dementia pugilistica and cavum septi pellucidi: born to box? Sports Med 24(6):361–5

Jordan BD, Relkin NR, Ravdin LD, Jacobs AR, Bennett A, Gandy S (1997) Apolipoprotein E epsilon4 associated with chronic traumatic brain injury in boxing. JAMA 278(2):136–140

Nandoe RD, Scheltens P, Eikelenboom P (2002) Head trauma and Alzheimer's disease. J Alzheimers Dis 4(4):303–308

Nowak LA, Smith GG, Reyes PF (2009) Dementia in a retired world boxing champion: case report and literature review. Clin Neuropathol 28:275–280

Dementias with Associated 'Brain Swelling'

8

Contents

8.1 Introduction

In addition to volume loss and signal change, structural MRI allows an assessment of 'brain swelling' – we use this here as an umbrella term to cover a range of conditions with apparent increases in volume (or pressure) of a part of the brain or a cerebrospinal fluid (CSF) compartment – the brain may not be literally 'swollen' in all of these. This appearance however is an important pointer in a number of different disorders and have been considered here together. Brain swelling and/or volume shifts may be thought of as affecting the brain either globally, locally or compartmentally (Table 8.1). This chapter focuses on disorders that have brain swelling as their most striking feature, for example hydrocephalus or glioblastoma, but there may be disorders that have other lead symptoms, and have swelling as a secondary feature, for example congophilic amyloid angiopathy (CAA), which will be discussed elsewhere under the heading of their leading feature (in case of CAA – vascular).

8.2 Giant Virchow–Robin Spaces (VRS)

The perivascular Virchow–Robin Spaces (VRS) may be mildly dilated in normal ageing, as discussed in Chap. 4, with diameters of individual VRS not exceeding 1–2 mm. In certain subjects, however, so-called giant VRS may occur, with diameters of individual VRS reaching 5–10 mm or even more. Giant VRS may occur in a localised or general pattern (Fig. 8.1), and are generally considered to be a benign derangement of CSF drainage, with limited clinical significance. The diagnostic finding is that of multiple

Table 8.1 Differential diagnosis in global and local brain swelling

Global brain swelling may occur in:
- Giant Virchow–Robin Spaces (VRS)
- Traumatic swelling (acute phase)
- Reversible posterior leukoencephalopathy (RPLS)
- Venous congestion due to dural AV fistula
- Hydrocephalus, either obstructive or communicating
- Leptomeningeal disease, with or without hydrocephalus
- Malignant disease, for example gliomatosis, intravascular malignancies
- Infectious and inflammatory disease
- Other causes of generalised cerebral oedema

Local brain swelling may occur in:
- Giant Virchow–Robin Spaces (VRS)
- Infectious disease, for example HSV infection and PML or tuberculosis
- Radiation necrosis
- Malignant disease, for example glioma, lymphoma
- Paraneoplastic disease
- Arterio-venous malformation (AVM)
- Congophilic amyloid angiopathy (CAA)
- Infarction – temporarily in subacute phase
- Local inflammation – from a range of causes for example vasogenic oedema secondary to immunotherapy in AD

sharply delineated structures with CSF-like signal on all sequences within the brain parenchyma. In case of giant VRS, the mass effect often leads to compression of the surrounding tissue, with sometimes a slight gliotic appearance with high signal on for example fluid attenuation inversion recovery (FLAIR) images. Despite their frightening appearance, giant VRS are a 'don't touch' lesion, and most probably not the explanation for the cognitive symptoms. The differential diagnosis encompasses epidermoid cyst, certainly when located in the midline and the basal cisterns – in that case diffusion weighted imaging (DWI) is mandatory to show restricted diffusion caused by the keratin lamellae (whereas in VRS the will be increased diffusion).

8.3 Infections

Many infections will be accompanied with either local of global brain swelling. Here, we restrict ourselves to those conditions that may be accompanied by cognitive impairment and/or dementia. Conversely, there are infections that may lead to cognitive decline or dementia, without accompanied brain swelling, which

are discussed under the heading of their key MRI feature (e.g. white matter changes) in Chap. 7, for example HIV encephalitis.

8.3.1 Herpes Simplex Encephalitis

8.3.1.1 Aetiology and Histopathology

Herpes simplex encephalitis (HSE) is the most common form of acute viral encephalitis. The usual cause in adults is herpes simplex virus type-1 (HSV-1), while HSV-2 is more common in neonates and immunologically compromised patients. The typical location of HSV encephalitis is in the medial temporal and ventral frontal lobes. This has led to the suggestion that spread of virus along the cranial nerves serves as the Porte d'entrée. Although the clinical presentation is insidious and variable, early diagnosis is important, since the effectiveness of antiviral treatment is strongly dependent on its early initiation.

At gross inspection, a typical distribution of lesions is found, affecting the medial temporal and fronto-basal areas. The lesions may appear necrotic, and are frequently haemorrhagic. HSE is also known as inclusion-body encephalitis, referring to the microscopic appearance of eosinophilic intranuclear inclusions. These inclusions identify infected neurons and glial cells. Astrogliosis and microglial activation are found, as well as oedema and necrosis.

8.3.1.2 Clinical Presentation and Treatment

The symptoms of HSE develop over the course of several days, with fever, headache, seizures and drowsiness. The drowsiness can evolve into stupor and coma, a poor prognostic sign. Features related to the location of the lesion in the temporal lobes and ventral parts of the frontal lobe include anosmia, gustatory hallucinations, personality change and psychosis. The EEG shows slow-wave complexes at a frequency of 2–3 Hz, and lumbar puncture reveals increased CSF pressure with a mild pleocytosis (usually lymphocytes, but occasionally also neutrophils and erythrocytes), and increased protein.

In addition to typical symptoms and EEG findings, the role of MRI in the acute stage can be pivotal. A

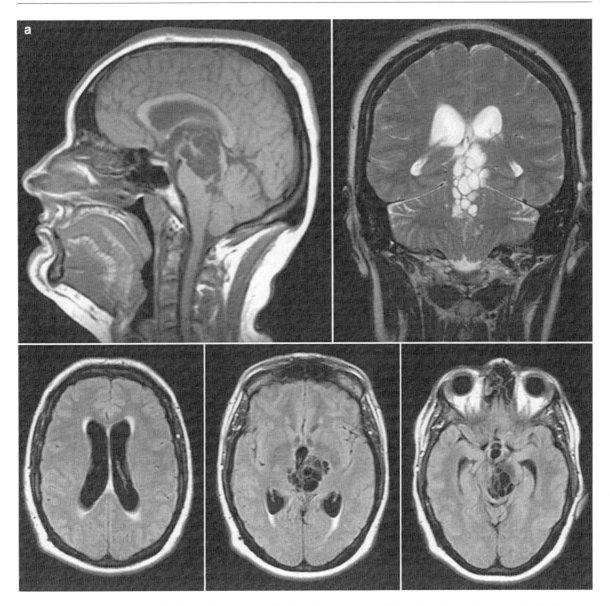

Fig. 8.1 (**a**) Localised giant VRS. Note a cluster of giant VRS at the mesencephalic level, leading to obstruction in CSF flow and hydrocephalus with pressure mantles (high signal on FLAIR) (*continued on next page*)

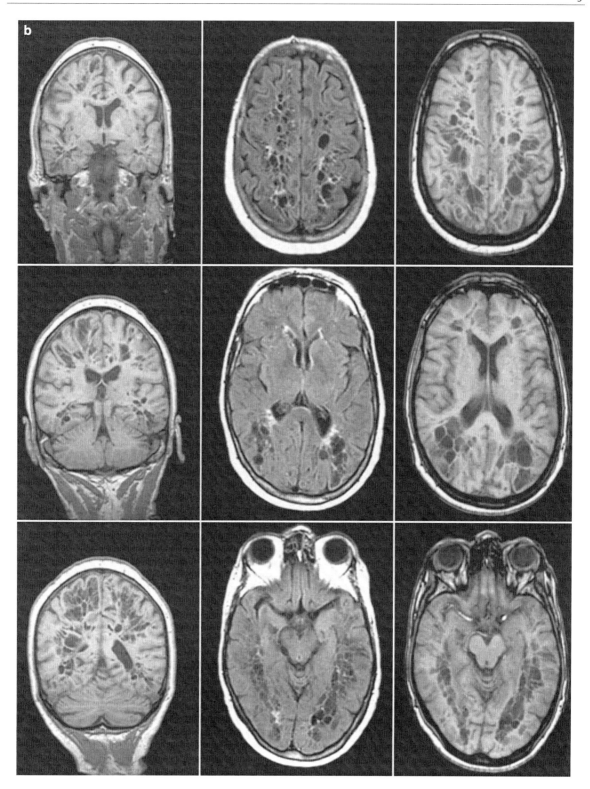

Fig. 8.1 (continued) (**b**) Generalised giant VRS. Note extremely wide VRS scattered throughout the white matter with a signal mostly similar to CSF, but with occasional zones of high signal intermingled. Symptomatology can be surprisingly mild despite frightening appearance

definite diagnosis is made using the polymerase chain reaction (PCR) analysis of the CSF for HSV. Despite its being available usually with 24 h, by the time the results of this test become available, an important therapeutic window has already been missed and here MRI plays an important role. The combination of PCR and MRI is completely diagnostic. Treatment with antiviral drugs can influence prognosis and survival when instituted early. In patients who recover, serious neurological deficits are common: including amnesia, frontal behavioural problems, dementia and seizures.

8.3.1.3 Neuroimaging Strategy and Key Findings

CT is useful to rule out gross pathology such as a glioma, but particularly suffers from artefacts in the medial temporal lobes, and thus may miss early pathology in these regions. MRI is therefore the preferred imaging modality and reflective of the more or less pathognomonic histopathological findings (Fig. 8.2).

Neuroimaging strategy and differential diagnosis

- MRI preferred over CT

  - Hyperintense signal on axial and coronal T2/FLAIR
  - DWI more sensitive in early phase
  - May be hyperintense on T1 (haemorrhage)
  - Variable gadolinium enhancement

- Main finding: temporal lobe lesions

  - Bilateral in typical cases, but may initially be unilateral/asymmetric
  - Medial temporal lobe frequently involved
  - Extension to either the insular region or lateral temporal lobe
  - More extensive disease: temporal lobe, basal frontal areas

Differential diagnosis

- Other causes of limbic encephalitis: other infections, paraneoplastic and auto-immune encephalitis
- Low-grade tumour or dysplasia
- Limbic glioma (can be bilateral)

Differential Diagnosis

High signal lesions in the medial temporal lobe may be caused by a variety of disorders, including neoplasms such as glioma and cortical dysplasia. In combination with the clinical picture with drowsiness and fever, the differential diagnosis is very limited. Causes of a limbic encephalitis (other infections, paraneoplastic or auto-antibody mediated – for example voltage-gated potassium channel antibodies) need to be considered as a cause of the MRI appearance. However, the combination of a positive PCR for HSV and a typical MRI is virtually diagnostic.

8.3.2 Progressive Multifocal Leukoencephalopathy (PML)

8.3.2.1 Aetiology and Histopathology

Before the occurrence of AIDS, progressive multifocal leukoencephalopathy (PML) was an extremely rare condition that occurred in patients with immunosuppression, mostly (or usually) in those with haematological malignancies. Other important immunosuppressive settings include chemotherapy for other malignancies; transplant patients; and more recently patients receiving immunotherapy with humanised monoclonal antibodies (e.g. natalizumab for the treatment of multiple sclerosis). PML is an opportunistic infection that occurs in up to 5% of AIDS patients, and has a very poor prognosis, even with treatment. PML is caused by the JC papovavirus, a ubiquitous DNA virus that (almost exclusively in immunocompromised patients) infects oligodendrocytes and leads to massive demyelination. HIV infection and PML infection produce synergistic damage, perhaps explaining why PML in AIDS causes death in a matter of months, whereas other patients may survive up to a year or longer. While the final diagnosis depends on histopathology, a combination of clinical findings, CSF analysis and characteristic findings on MRI will generally suffice.

Antibodies against JC virus are found in 80–90% of normal healthy adults. Latent virus can be detected in many organs, and PML probably results from infiltration of the brain by infected lymphocytes after reinfection. Once the virus has crossed the blood–brain barrier, astrocytes, and then oligodendrocytes become infected.

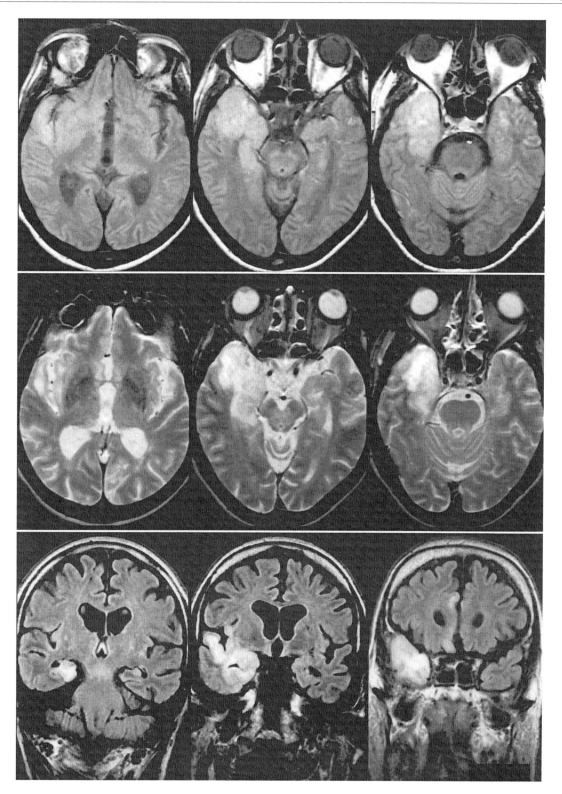

Fig. 8.2 MRI findings in acute HSV. This 65-year-old man presented with drowsiness, disorientation and epilepsy. Axial proton density (*top row*), heavily T2-weighted (*middle row*) and coronal FLAIR (*bottom row*) images show signal increase and swelling of the left temporal lobe, extending far back into the hippocampus and to a certain degree also in the insula and cingulum. The findings are suggestive HSV encephalitis, which was confirmed by serology

This leads to the cardinal macroscopic finding, massive demyelination producing a greyish discoloration of the white matter. Histopathologically, a triad of multifocal demyelination, hyperchromatic enlarged oligodendroglial nuclei, and enlarged bizarre astrocytes with lobulated hyperchromatic nuclei, is found. In rare cases without underlying HIV infection, inflammatory infiltrates can be found.

8.3.2.2 Clinical and Laboratory Findings

The most common findings include weakness and speech disturbances, but a range of cognitive abnormalities, headache and visual impairment also occur. The progress of the disease and deterioration is rapid, with a median time to death of only 6 months. The diagnosis is based on imaging findings (see below) and CSF analysis; PCR analysis of the CSF is positive in the majority of cases.

8.3.2.3 Neuroimaging Strategy and Findings

CT can be useful in demonstrating a multifocal disorder with swelling and marked hypodensity with little contrast enhancement. MRI is more sensitive and the protocol should include gadolinium administration. The typical MR findings in PML consist of multiple focal areas of markedly increased signal on T2-weighted and markedly hypointense on T1-weighted MR images, located in the subcortical white matter, with gyral swelling – the appearances can be very characteristic with the signal change in white matter 'pushing against' but sparing the cortical ribbon – so called 'scalloping out' of the grey-white border (Fig. 8.3). On plain

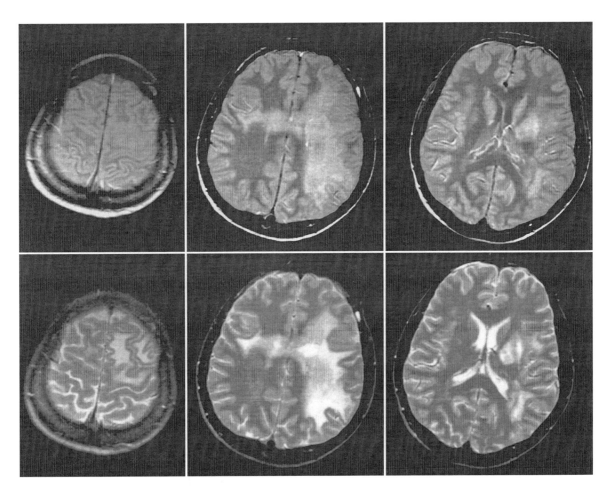

Fig. 8.3 PML in a young man with AIDS. Note the signal increase on the proton density (*top row*) and heavily T2-weighted (*lower row*) images, with gyral swelling and involvement of the U-fibres (*continued on next page*)

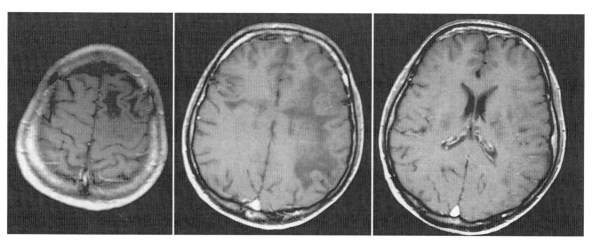

Fig. 8.3 (continued) After gadolinium injection, no enhancement is seen on the T1-weighted images (*bottom row*); note that the lesions are severely hypointense, consistent with the finding of complete demyelination in these patients

T1-weighted images, the signal intensity is very low, and severely reduced MT ratios are found. Both findings relate to complete demyelination, and provide a means to differentiate these lesions from HIVE, in which they are nearly isointense on T1-weighted images (Box 8.1). They also correlate with the finding of increased choline on MR spectroscopy. Follow-up studies in patients with clinical progression can show progressive decrease of the signal on T1-weighted images, which in the absence of growth of the lesions indicates progressive demyelination. In the rare occurrence of PML in a HIV-negative patient, the likelihood of enhancement is much higher, consistent with the histopathological finding of inflammation. There is no established role for SPECT/PET in the work-up of PML.

Differential diagnosis of PML and distinguishing MRI features
- Acute disseminated encephalomyelitis (ADEM) – more frequent gadolinium enhancement (Sect. 7.3.2)
- Lymphoma – homogeneous enhancement (Sect. 8.4.3)
- HIV encephalitis – less well delineated, not T1-hypointense (Sect 7.2.2.1)
- Multifocal glioma – more diffuse borders (Sect 8.4.1)
- Immune reconstitution inflammation syndrome (IRIS) – patchy enhancement
- RPLS/PRES – parasagittal distribution (Sect. 8.8)

Box 8.1 Neuroimaging findings in PML

MRI preferred over CT, showing:
- Multifocal bilateral hemispheric lesions
  - Mostly subcortical, with limited cortical involvement
    Basal ganglia and thalamus may also be involved
  - T2 hyperintense, T1 strongly hypointense
  - Limited or absent contrast enhancement
    More likely in HIV negative individuals and treated patients
- Frequent infratentorial involvement (e.g. cerebellar peduncles)

8.4 Neoplastic Disease and Radiation Necrosis

8.4.1 Eloquent Areas and Common Tumours

Brain tumours, when situated in eloquent areas, may present with neuro-psychological and behavioural disorders and mimic a neurodegenerative disease. In slow-growing tumours, these symptoms may even for years be the only signs of cerebral pathology. Critical locations are the temporal lobes (Fig. 8.4), basal

Fig. 8.4 Temporal lobe glioma presenting with memory impairment. This 70-year-old woman developed memory disturbances over the course of 2 years. She presented with language problems, difficulty in reading and calculation. MRI showed an infiltrative mass lesion in the left temporal lobe, which was proven to be a low-grade glioma

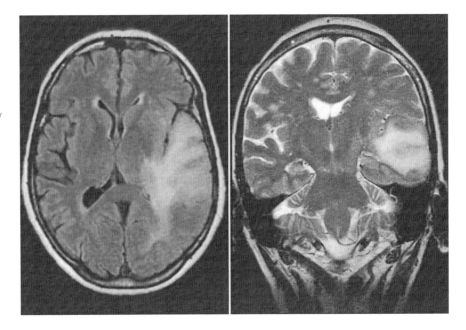

frontal areas, bilateral thalamic lesions, involvement of the cingulum and/or corpus callosum (Fig. 8.5) and the limbic system. Modern imaging modalities can easily clarify whether such a mass lesion is present. Ruling out space-occupying lesions as a possible cause of dementia should be an early part of the investigation of patients with neurocognitive or neurobehavioral changes of unknown cause, as discussed in Chap. 2.

There are however, two malignant lesions which may be difficult diagnostically: gliomatosis diffusa and intravascular lymphomatosis. These diffuse widespread malignant proliferative lesions of the brain will be discussed separately. Gliomatosis cerebri deserves particular consideration since it often presents with psychiatric and cognitive symptoms, occasionally leading to a diagnostic delay of several years. The MR pattern, in addition, is highly characteristic. In the case of intravascular lymphomatosis one is dealing with a rare but treatable disorder, often misdiagnosed for a long time, with MR characteristics that may suggest the diagnosis.

8.4.2 Gliomatosis Cerebri Diffusa

Synonyms

Gliomatosis cerebri; Diffuse glioma of the brain; Gliomatous hypertrophy, Blastomatous type of diffuse sclerosis; Central diffuse schwannosis

8.4.2.1 History and Classification

In 1938 Nevin described three cases of what he called 'gliomatosis cerebri', characterised histopathologically by a diffuse cellular overgrowth throughout wide areas of the cerebral hemispheres. The disease is now recognised as a separate entity in the WHO classification of brain tumours. Diagnosis requires an infiltrative primary brain tumour involving at least two, usually three lobes of the brain, in patients where MR shows a diffusely infiltrating process, involving grey and white matter, with initially only minor mass effect and without necrosis and with, in the beginning, relatively minor neurological and cognitive changes. Ultimately, all brain areas, both supra- and infratentorially, may become diffusely involved.

8.4.2.2 Clinical Presentation and Therapy

The clinical presentation is variable. Headache is present in many cases, and may be accompanied by focal seizures, ataxia and dysphagia as well as a variety of cognitive and behavioural presentations including dysphasia and memory problems. The clinical picture suggests a multifocal progressive disorder. In general, neurosurgery is not an option; radiotherapy and chemotherapy results are mostly disappointing.

Fig. 8.5 Corpus callosum involvement in cerebral lymphoma. This 78-year-old man developed severe memory impairment and confusional states in the preceding half year. He was considered in the first instance to be suffering from a rapidly progressive dementing illness. The images show a relatively well-preserved brain with small ventricles and a normal-appearing cortex. Around the ventricles and in the corpus callosum, with extension in the left frontal and right peritrigonal white matter, there is a lesion with high signal intensity on the T2-weighted images (*upper row*), and strong and homogeneous enhancement after contrast injection (*bottom row*). The pattern is typical of cerebral lymphoma

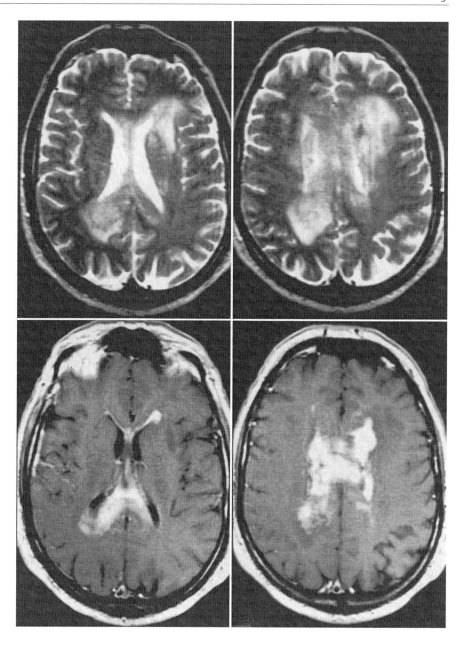

8.4.2.3 Neuropathology and Genetics

Macroscopically, there is diffuse swelling of the involved structures. Although all parts of the brain may be involved, including the brainstem and cerebellar hemispheres, there is a rather strong preference for the central, periventricular areas and mesolimbic parts of the temporal lobe. Moderately pleomorphic glial cells which infiltrate pre-existing structures without significant destruction are the microscopic hallmark of gliomatosis diffusa. Usually, these cells are of astrocytic type and react positively with glial fibrillary acid protein (GFAP). No familial cases of gliomatosis diffusa have been reported; various karotype abnormalities have been reported.

8.4.2.4 Neuroimaging Strategy and Findings

MRI is the preferred imaging modality. Proton density, T2-weighted and FLAIR sequences will show increased signal intensity of the involved areas, with unsharp

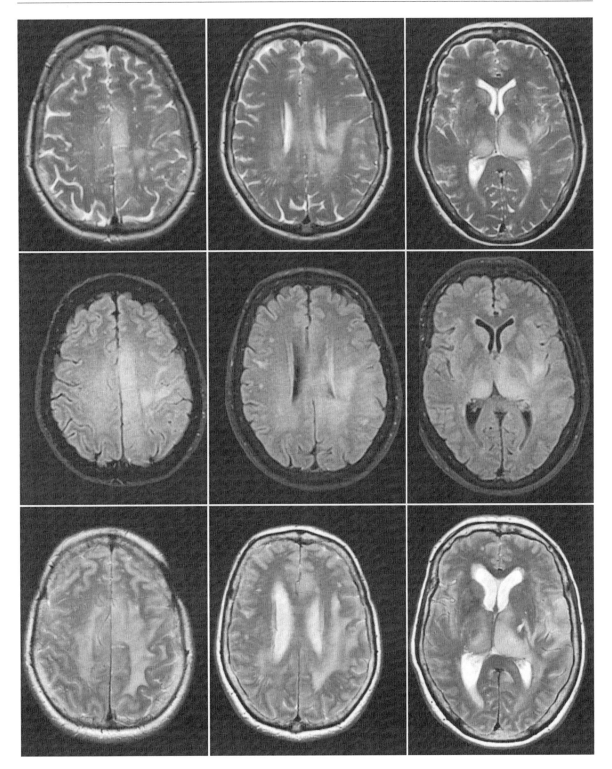

Fig. 8.6 Gliomatosis cerebri. This 59-year-old woman initially presented with motor symptoms and subsequently developed cognitive impairment at presentation, T2-weighted (*upper row*) and FLAIR (*middle row*) images show multifocal involvement of the cingulate gyrus and white matter of centrum semiovale on the left side, and thalamus on both sides. At follow-up (*bottom row*), there is diffuse spreading of the disease with patchy white matter involvement of both hemispheres

demarcation from non-involved tissue (Fig. 8.6). There may be striking loss of demarcation of grey and white matter and mild swelling. When the periventricular structures are involved, the ventricles will be locally narrowed with an irregular border. In these centrally located cases, the corpus callosum is nearly always involved and can be swollen; the anterior commissure is also frequently involved. There may be symmetry of the abnormalities but this may disappear when the disease progresses. In most cases, there is no enhancement of the involved structures after contrast injection. The MR pattern of the disease is mostly diagnostic. In cases with central location, with symmetrical involvement of the thalamus, central venous thrombosis has to be excluded.

8.4.3 Intravascular Lymphomatosis

Synonyms

Malignant angioendotheliomatosis; Angioendotheliomatosis proliferans systematica; Proliferating endotheliosis; Intravascular histiocytosis; Angioendotheliomatosis proliferans; Neoplastic endotheliosis; Neoplastic endotheliomatosis; Angiotrophic (intravascular) large-cell lymphoma; Intravascular endothelioma; Reactive inflammatory systematised angioendotheliomatosis; Reactive angioendotheliomatosis; Proliferating systematised endotheliosis

8.4.3.1 Aetiology and Clinical Presentation

Intravascular lymphomatosis (IVL) is a rare non-Hodgkin lymphoma, characterised by proliferation of lymphoma cells within the vascular (arterial and venous) lumen, without involvement of adjacent parenchymal tissue. IVL with cerebral involvement is predominantly of B-cell lineage. The reasons for including this rare disorder here are twofold: first, MR imaging can contribute to the diagnosis; second, IVL is a treatable condition and an early diagnosis may prevent irreparable damage. Patients with cerebral IVL have been described in the age range of 30–80 years. Clinically, the cerebral form of IVL has a variable presentation. Often there are stroke-like episodes or repetitive transient ischaemic attacks. Focal cortical symptoms, confusion, disorientation, dementia and seizures may follow. A meningo-encephalitis-like presentation has also been described. The prognosis is fatal in 80% of patients within a year; treatment consists of poly-chemotherapy.

8.4.3.2 Neuroimaging Strategy and Findings

Cerebral manifestations on MR include infarct-like lesions dispersed throughout the brain, venous occlusion of the large sinus or cortical veins with venous infarctions, and enhancement of dura or arachnoid after contrast injection. Diffusion weighted imaging may help to indicate the nature of the cerebral lesions and identify them as infarctions. MR angiography may show abnormalities of arteries but is not diagnostic.

In cases with venous involvement, CT/MR venography is useful to show the obstructed veins. Manifestations in other organs (skin or lungs for example), anaemia and a high erythrocyte sedimentation rate may help in arriving at the correct diagnosis (Fig. 8.7).

8.4.4 Radiation Necrosis

8.4.4.1 Aetiology, Histopathology and Clinical Findings

Radiation necrosis is a late complication of radiotherapy, occurring between 6 months and more than 10 years after therapy – often a diagnostic challenge. Firstly, to determine whether or not cognitive and imaging findings are new, in other words, truly progressing, or date from the original insult. Secondly, whether this represents tumour recurrence or the late effects of radiation. Symptoms and signs are determined by the location of the damage. Cognitive and memory impairment after radiation is seen especially frequently in patients irradiated for naso- and oropharyngeal tumours. In these patients bilateral involvement of the temporal lobes can be seen (Fig. 8.8).

In late delayed leukoencephalopathy, the changes consist of demyelination, astrogliosis, multifocal coagulative necrosis and cavitation. The lesions may be symmetrically distributed over the hemispheres. Focal radiation damage will occur in and around the area receiving the highest radiation dose. They are markedly vascular lesions, with hyalinisation, fibrosis and necrosis of vessel walls with vascular thrombosis. Endothelial proliferation and varying degrees of

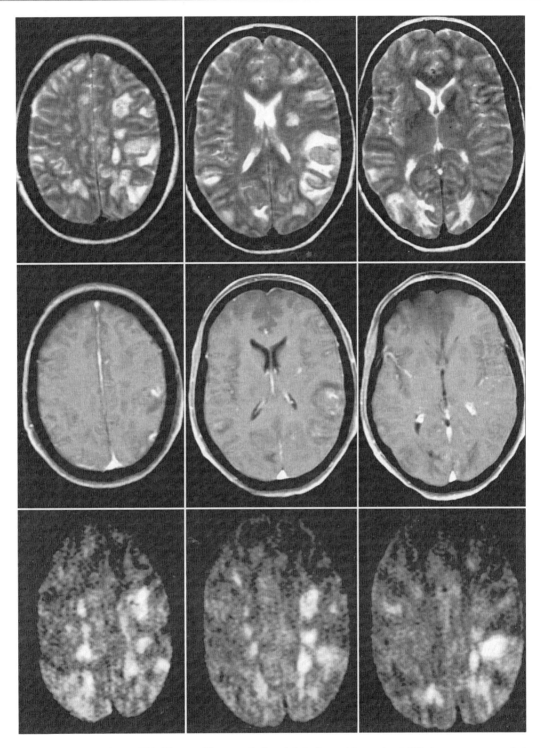

Fig. 8.7 Intravascular lymphoma. This 43-year-old woman had a series of transient ischaemic attacks, all referring to different neurological locations and presented with cognitive disturbances and lowered consciousness. The T2-weighted images (*top row*) show multiple hyperintense foci involving both cortical layers and underlying white matter. T1-weighted images (*middle row*) show patchy enhancement in the left temporoparietal region. Diffusion trace images (*bottom row*) show multiple lesions with restricted diffusion, some arranged along the vascular watershed zones. The combination of border zone infarctions and partial widespread infarction must be a combination of thrombo-embolic lesions and diminished perfusion. This pattern may, for example, be seen after resuscitation after cardiac arrest and in coagulative disorders during sepsis. In this case the diagnosis of intravascular lymphomatosis was confirmed by biopsy

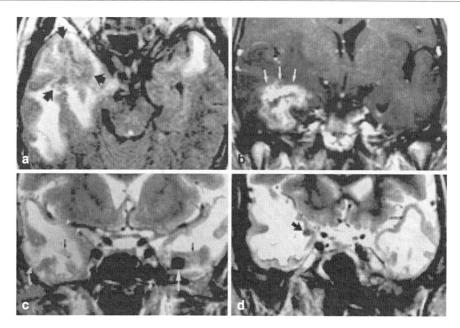

Fig. 8.8 Radiation necrosis of the temporal lobes. MRI scans of patients who underwent radiation therapy for a nasopharyngeal tumour between 5 months and several years previously. The FLAIR image in (**a**) shows bilateral temporal lobe hyperintensity, more pronounced on the right side. Within these areas cystic regions occur (*arrows*) with lower signal intensity. Coronal T2-weighted images in two other patients (**c** and **d**) show extensive involvement of both temporal lobes; both cases presented with severe cognitive and behavioural consequences. Contrast enhancement (**b**) and central necrosis may occur in these patients. (Reprinted with permission from Radiology 1999;213:800–807)

adventitial fibroblast proliferation occur. In the areas of most severe vascular changes cavitation ensues.

8.4.4.2 Neuroimaging Strategy and Findings

The protocol for post-radiotherapy/chemotherapy MR studies should include T1-weighted images (with fat suppression if the skull base is involved), without and with contrast. Both grey and white matter can be involved, the swollen white matter showing a homogeneously high signal intensity on the T2-weighted and FLAIR images, while the overlying cortex is thin. Contrast enhancement may occur both in tumour recurrence and in radiation necrosis. Perfusion MRI and SPECT/PET (thallium/FDG/methionine) may be useful in distinguishing high-grade vital tumour from necrosis by showing low perfusion/metabolism in the latter. MR spectroscopy may also help in this differentiation: the typical tumour spectrum will show decreased N-acetylaspartate, lactate and high choline concentrations – in radiation necrosis the MR spectrum is nearly devoid of metabolites, while lactate may still be present.

8.5 Autoimmune Limbic Encephalitis

Jonathan Schott

Synonyms

Limbic encephalitis (LE); Paraneoplastic encephalomyelitis

8.5.1 History and Nosology

Concepts of non-infectious limbic encephalitis have changed considerably over the past few years. Initially the term was used exclusively to describe patients presenting with a subacute onset of memory impairment and confusion, focal, usually medial temporal lobe seizures, and pathological changes within the temporal lobe occurring in the context of a distant malignancy. The isolation of first a range of causative paraneoplastic antibodies, and latterly antibodies targeting cell-surface proteins has strengthened the view that this syndrome occurs on an autoimmune basis, and thus may be non-neoplastic as well.

8.5.2 Clinical Presentation, Pathology and Differential Diagnosis

From a practical perspective, autoimmune LE is a clinical diagnosis, based on the subacute (usually days to a number of weeks) onset of severe episodic memory impairment, usually in association with a range of behavioural and psychiatric symptoms including irritability, confusion, hallucinations, confabulation or depression. Patients often have complex partial seizures with or without secondary generalisation. Autonomic dysfunction and sleep disturbance are likely to reflect diencephalic involvement, and there may be overlap with a more generalised 'encephalomyelitis' with associated brainstem signs, or movement disorders.

In the absence of an identifiable infectious cause, patients with LE are likely to have an underlying autoimmune aetiology. While 'traditional' autoimmune disorders including SLE, Behcet's and relapsing polychondritis can present in this way, the majority of such patients will have either (1) a paraneoplastic syndrome, where the antibody is directed against an intraneuronal protein; or more frequently (2) an antibody directed against a cell-surface antigen. Of these, antibodies against components of the voltage-gated potassium channel (VGKC) or N-methyl d-aspartate (NMDA) receptor are emerging as distinct, important and potentially treatable conditions. Other antibodies directed against cell-surface antigens, including GABA-B and Glycine receptors, have recently been described, usually in the context of an autoimmune LE syndrome.

The differential diagnosis at presentation is wide and includes infections such as HSE and human herpes viruses 6 and 7, especially in immunocompromised patients (e.g. bone marrow transplant). Imaging, ideally with MRI, including coronal FLAIR sequences, will typically show asymmetric temporal lobe swelling and signal change in the case of HSE (see Sect. 8.3.1), and often very symmetric isolated hippocampal signal change in HHV 6 and 7. Examination of the CSF with PCR is the gold standard for diagnosing these conditions.

8.5.3 Paraneoplastic Limbic Encephalitis

Conceptually, the paraneoplastic syndromes are thought to represent an immune response directed against an underlying neoplasm, with cross-reaction against neural structures expressing similar protein epitopes. In keeping with this hypothesis, antibodies produced by tumours can bind to neural tissue; and there is also evidence that the immune response directed against the underlying primary may slow tumour growth or even cause tumour regression. In LE, the most common responsible antibodies all against intraneuronal antigens are shown in Table 8.2.

In cases of 'pure' paraneoplastic LE, post-mortem examination reveals involvement of structures of the limbic system only, with severe neuronal loss, gliosis and microglial infiltration. There is commonly relatively mild perivascular and leptomeningeal inflammation. In more extensive disease and generalised encephalomyelitis, other brain structures including the cerebral hemispheres and cerebellum may also be involved.

Diagnosis of paraneoplastic LE may be straightforward in patients with a typical phenotype, known

Table 8.2 Paraneoplastic limbic encephalitis: causes and presentation

Antibody	Tumour	Suggestive Clinical Features
Hu	Small cell lung	Smoking history Sensory/autonomic neuropathy
Ma2	Testicular	Young male Hypothalamic dysfunction
CRMP5	Small cell lung Malignant thymoma	Smoking history Sensory/autonomic neuropathy Chorea Uveitis
Amphiphysin	Breast Small cell lung	Stiff person syndrome Sensory/autonomic neuropathy
AMPAR	Lung Breast Thymus	Female predominance Seizures CSF pleocytosis
Ri	Neuroblastoma Breast Ovarian	Opsoclonus-myoclonus Cerebellar degeneration Brainstem encephalitis
VGKC	Thymoma NSCLC (may also be non-neoplastic)	Neuromyotonia, myoclonus, hyponatraemia Rapid progression, dd CJD
NMDA receptor	Ovarian teratoma (may also be non-neoplastic)	Young woman Psychiatric symptoms and/or memory loss

malignancy and positive antibodies. However, in practice this is rarely the case, not least as the neurological syndrome often predates the detection of the underlying tumour, and antibodies hard to detect. Once infection is excluded, investigations are required both to establish the diagnosis of LE, and to look for an underlying malignancy. The former requires blood analysis, EEG, CSF examination and MRI. Imaging findings and their differential are presented in Box 8.2, Fig. 8.9.

Once paraneoplastic LE is suspected, an occult malignancy should be sought. Depending on local availability and expertise, this may be achieved by imaging chest, abdomen and pelvis with CT, supplemented by mammography and testicular ultrasound as appropriate. Whole body FDG-PET where available may detect small tumours not seen using conventional imaging, with some authorities advocating that it may be most cost-effective to image first with FDG-PET when a paraneoplastic syndrome is strongly considered.

Response to treatment in 'classical' paraneoplastic LE – where the responsible antigen is intraneuronal – is variable, and often poor. Treating the underlying tumour if possible is associated with a better outcome than immunosuppression.

Box 8.2 Neuroimaging Strategy in Paraneoplastic Limbic Encephalitis

MRI preferred imaging modality

- Axial FLAIR, T2, T1, DWI
  - Add coronal images to show limbic lobe involvement
- Post-gadolinium T1

MRI findings

- Abnormalities in medial temporal lobes
  - Can be very circumscribed and remarkably symmetrical
- May be asymmetric and more widespread
  - Amygdala, cingulate, fornix, hypothalamus, basal frontal cortex
- On T2-weighted and FLAIR images the involved regions are hyperintense
- Some lesions may enhance after gadolinium injection
- MRI lesions may regress without clinical improvement

Differential diagnosis of Paraneoplastic Limbic Encephalitis

- MRI is unreliable in distinguishing different forms of paraneoplastic LE
    - Ma-2 antibodies patients often show widespread MRI abnormalities in the hypothalamus, thalamus, midbrain and medial temporal lobes
- Anti-NMDA antibodies – often normal imaging
- Infections: HSE, HHV6 and 7
- Non-neoplastic causes of autoimmune LE
- Central venous thrombosis
    - Bilateral involvement of the basal ganglia
    - Can be excluded by MR or CT venography

8.5.4 Non-neoplastic Autoimmune Encephalitis

In addition to the neoplastic causes of autoimmune LE, there is a growing number of non-neoplastic causes of autoimmune LE, which encompasses non-neoplastic antibodies against potassium channels or the NMDA-receptor, or involve antibodies that are gluten-sensitive or thyroid related (Hashimoto encephalitis), and anti-

glutamic acid decarboxylase antibody syndrome. In some cases, auto-immune encephalitis can be demonstrated by binding to rat-brain tissue, without the antigen being determinable. The encephalopathy seen in SLE and sarcoid is also partly caused by autoimmune antibody mechanisms.

8.5.4.1 Voltage-Gated Potassium Channelopathies (VGKC)

VGKCs are expressed widely throughout the nervous system, and are responsible for repolarisation of neurons after the passage of an action potential. Antibodies against components of the potassium channel (here termed VGKC antibodies) can be detected by radioimmunoprecipitation assay using labelled α-dendrotoxin, and are associated with neuromyotonia and Morvan's syndrome (the combination of neuromyotonia and CNS features including hallucinations, memory and sleep impairment, and autonomic hyperactivity). The similarity between Morvan's syndrome and LE led to the identification of VGKC antibodies in patients with 'pure' LE.

A fairly consistent core phenotype has been established: patients are usually middle aged with a male predominance, and have a classical LE presentation with profound episodic memory disturbance, confusion, complex partial seizures and often sleep/wake

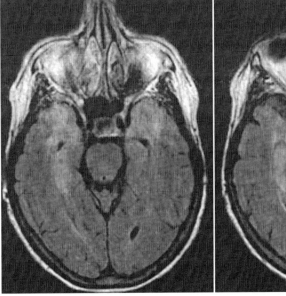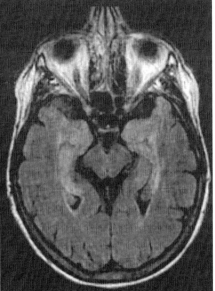

Fig. 8.9 Paraneoplastic LE. FLAIR images of a 74-year-old woman with progressive cognitive deterioration, memory disorders, changes in behaviour and confusional states. MRI shows involvement of both temporal lobes, in particular the temporal polar areas, the hippocampus and parahippocampal gyrus. This was diagnosed as limbic encephalitis. Post-mortem examination revealed a small cell lung carcinoma

disturbance. Persistent hyponatraemia, resistant to treatment, is often present, and appears to be a strong pointer towards the diagnosis. Some patients may have very frequent stereotyped motor seizures. The CSF examination is usually unremarkable, occasionally showing a slightly elevated protein and cell count. Elevated VGKC antibody titres establish the diagnosis. MRI is usually abnormal, showing focal, bilateral or unilateral T2 or FLAIR hyperintensity within medial temporal lobe structures, best seen in the coronal plane (Fig. 8.10). Early in the disease, the hippocampi may appear swollen, but as the disease progresses and the signal change resolves, profound hippocampal atrophy may be seen, representing irreversible tissue loss.

Most cases of VGKC-LE appear not to be associated with an underlying malignancy, but as associated tumours (notable thymoma) have been described, these should be excluded by appropriate imaging. Some cases of VGKC-LE may resolve spontaneously, but given the profound atrophy, persistent cognitive deficits and ongoing seizures that are often seen in untreated cases, once the diagnosis has been made by demonstration of significantly elevated VGKC antibody titres in serum (and often CSF), treatment with anticonvulsants and immune suppression should not be delayed. Short of randomised controlled trials, best evidence suggests treatment with plasma exchange or intravenous immunoglobulin, followed by oral steroids, using VGKC antibody titre, neuropsychometry and brain imaging to monitor response.

8.5.4.2 Anti-NMDA-Receptor Encephalitis

A characteristic neuropsychiatric syndrome with antibodies against NR1–NR2 heteromers was first described in 2007. Patients present with psychosis, memory impairment and seizures, before developing dyskinaesias, autonomic disturbance and hypoventilation (often requiring intensive care admission). The median age is in the 20s, with a significant proportion occurring in children. The disease is more common in women, and may be the cause of new onset focal seizures in adults. The CSF typically shows a moderate pleocytosis. MRI is often normal, but may show hippocampal or white matter abnormalities. A proportion of affected women has an ovarian teratoma, and may respond rapidly to tumour removal. In non-tumour-related cases, rapid reduction of antibody titre with immunosuppression is associated with better outcomes (Fig. 8.11).

8.5.4.3 Hashimoto Encephalitis

Hashimoto encephalitis is a rare encephalopathy associated with elevated antithyroperoxidase or antithyroglobulin antibodies. It occurs five times more frequently in women, with or without associated thyroid dysfunction. The presentation can be fluctuating with stroke-like episodes or have an insidious onset of cognitive dysfunction with seizures. Other symptoms include tremor, ataxia, sleep disturbance, headache, psychosis,

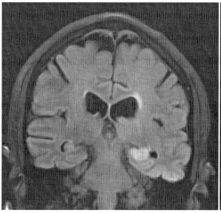 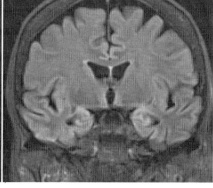

Fig. 8.10 Coronal FLAIR images in VGKC limbic encephalitis. *Left*: Coronal FLAIR imaging in a 68 year old man presenting with a 1 month history of amnesia, confusion and focal motor seizures. High signal and swelling is observed in the hippocampus. *Right*: Coronal FLAIR imaging in a 76 year old woman with a 1 year history of relapsing VGKC LE shows bilateral hippocampal atrophy with some residual high signal is seen in both hippocampi

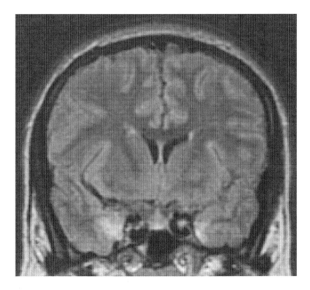

Fig. 8.11 FLAIR findings in anti-NMDA receptor encephalitis. FLAIR imaging in a 22-year-old woman with a 1 month history of psychiatric features, amnesia, dyskinaesias and autonomic dysfunction shows bilateral high signal in the medial temporal lobes on both sides. NMDA antibodies were positive in serum, without evidence for an underlying neoplasm

visual hallucinations and myoclonus. The disorder is steroid responsive and as such falls into the all-important category of potentially reversible causes of cognitive decline. However, it is worth adding that Hashimoto encephalitis remains a controversial entity with some authors questioning its existence and suggesting that the elevated anti-thyroid antibodies may reflect cross-reaction in the context of other auto-immune/paraneoplastic targets – the implication of this point of view is that elevated antithyroperoxidase or antithyroglobulin antibodies should be interpreted cautiously (Fig. 8.12).

8.6 Normal Pressure Hydrocephalus (NPH)

8.6.1 History and Nosology

The term normal pressure hydrocephalus (NPH) was coined by Adams et al. in 1965, to describe a condition of hydrocephalus presenting with a clinical triad of symptoms (mental deterioration, gait instability, urinary incontinence) with normal CSF pressure.

The syndrome attracts significant attention, since it is one of the few potentially treatable dementias; however, as estimated by Vanneste in 1992, treatable (i.e. reversible) NPH is extremely rare, with an estimated incidence of 2.2 cases per million per year. The prevalence of NPH varies hugely between countries and between different clinical centres with some centres claiming they never see a case while others suggest it is the most common treatable (by shunting) cause of dementia. This disparity reflects uncertainty and differences in opinion about the nature (and existence) as well as diagnostic criteria and procedures of this disorder.

8.6.2 Clinical Presentation

To make a presumed clinical diagnosis of NPH the clinical triad of Hakim should be present:

1. *An unstable gait*, with unsteadiness upon turning, small steps and, in advanced stages a magnetic phenomenon with 'gluing' of feet to the ground.
2. *Mental deterioration*, consisting of cognitive slowing and impaired recall, in the absence of clear episodic memory deficits.
3. *Increased urinary urgency*, which can range from slightly increased frequency to frank incontinence of the frontal type.

8.6.3 Neuropathology and Aetiological Considerations

Pathologically the brain shows relatively little cortical atrophy with clear dilatation of the ventricular system, disruption of the ependymal lining, mild reactive subependymal gliosis and demyelination. Whether more extensive areas of demyelination (as in subcortical small vessel disease) can be associated with NPH is still a matter of debate. In rare cases, leptomeningeal fibrosis can be found as a cause for reduced CSF absorption; cases with aqueductal stenosis have also been reported. In biopsied cases of clinical NPH, a high prevalence of Alzheimer pathology has been found.

NPH is thought to be due to a disturbance of CSF dynamics, caused by decreased absorption of CSF.

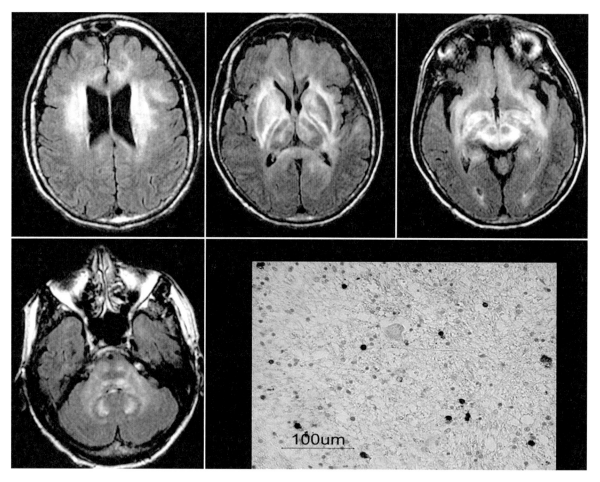

Fig. 8.12 Hashimoto encephalitis. This 62-year-old man presented with ataxia and progressive cognitive decline, and was suspected to have CJD. FLAIR images revealed a diffuse leukoencephalopathy with symmetric and extensive involvement of white matter but also deep grey matter nuclei. Histological examination revealed neuronal destruction with perivascular lymphocytic infiltrates with cytotoxic CD8-positive (*dark brown*) T-cells (*lower right*). While this raises the suspicion of Hashimoto encephalitis, antibodies against VGKC rather than thyroid were found – illustrating the difficulties surrounding the diagnosis of auto-immune encephalitis (although treatment options with immune modulation are similar). (Case kindly provided by Jan Braakhekke)

In most cases, the hydrocephalus is of the communicating type, with the site of reduced CSF absorption at the convexity of the brain. The cause of CSF absorption most frequently is subarachnoid haemorrhage, meningitis or trauma; however, in many cases the origin of NPH is 'idiopathic'. By definition, the CSF pressure is 'normal' in NPH, although temporary increases can be found (and may perhaps be present in all cases in the very early stage); during CSF pressure monitoring, so-called B-waves (shortly-lived pressure elevations at a frequency of 0.5–2/min) can be found.

Alternative routes of CSF absorption (e.g. via the ependyma and perivascular spaces) may account for the normalised pressure. This normal pressure is accompanied by reduced cerebral compliance and supposedly a 'water-hammer' effect with increased CSF pulsation causing damage to the periventricular structures such as the corona radiata and fronto-caudate tracts. Other theories for white matter damage include disturbance of cerebral blood flow autoregulation, leading to ischeemic damage of the periventricular white matter.

8.6.4 Neuroimaging Strategy and Findings

CT can be used to demonstrate hydrocephalus, although MRI is preferred. The cardinal findings in NPH include ventricular widening (including upward bowing of the corpus callosum and dilatation of the temporal horns); usually in the absence of sulcal widening and/or cortical atrophy (Box 8.3). On coronal images, there is a small CC angle (Fig. 8.14). The hydrocephalus is mostly of the communicating type, but cases with aqueductal stenosis have also been reported.

On MRI, NPH should not be considered when there is associated sulcal widening (.i.e. cortical atrophy) in conjunction with ventricular widening – this constellation fits with simple atrophy (e.g. due to Alzheimer's disease). No signs of increased CSF pressure should present, although small bands of transependymal CSF flow may be present despite the 'normal' pressure. Such a pressure band consists of a relatively thin, slightly blurry zone of increased signal best seen on moderately T2-weighted or FLAIR images, extending homogeneously around the whole ventricular system, including the inferior (temporal) horns. Pressure bands should not be confused with periventricular caps and bands that occur with normal aging, which are more sharply delineated, sometimes with irregular extension into the adjacent white matter; such bands certainly do not extend into the temporal lobes. More deeply located periventricular white matter abnormalities are not a prerequisite for the diagnosis, but their presence may not necessarily indicate poor responsiveness to shunting.

There is considerable debate about the value of flow imaging in NPH. Increased CSF flow-void on T2-weighed spin-echo MRI in the aqueduct has been reported, but is not very predictive of shunt responsiveness. CSF flow can be quantified using phase-contrast (PC) MR imaging; its diagnostic/predictive role in NPH remains uncertain.

8.6.5 Differential Diagnosis

In fact all dementias may be accompanied by the clinical symptoms of NPH. Therefore, all should be considered in the differential diagnosis and should be treated accordingly. Notably AD and VAD/ischaemic subcortical white matter changes should be considered. Parkinson's disease with incipient dementia may mimic NPH and so does PSP. Both NPH and AD may present with enlarged temporal horns, but in NPH the hippocampal volumes should be normal; in AD the reduced hippocampal volume (if present) will be accompanied by widening of the parahippocampal sulci, while in NPH these sulci will not be widened, despite enlargement of the inferior (temporal) horn of the lateral ventricle.

In VAD or ischaemic white matter changes, white matter damage leads to secondary ventricular enlargement, and the same pathway may be damaged by white matter tract disruption, explaining why the clinical signs may also overlap with those of NPH.

In addition microbleeds are more often seen in VAD than NPH (Figs. 8.13 and 8.14).

Box 8.3 Imaging criteria for NPH

Exclusion of mass lesion, obstruction or primary neurodegenerative disease
   Communicating hydrocephalus without sulcal widening

- Evans (frontal horns to inner table skull) index >0.3
- May have small zone of transependymal CSF flow
- Associated white matter hyperintensities deemed irrelevant

Upward bowing of the corpus callosum (sagittal) or small CC angle (coronal)
   Increased CSF flow void may be indicative of success for shunting
   NB: Cisternography, SPECT/PET are of uncertain value

8.6.6 Shunt Response

There is considerable debate as to what constitutes the ideal patient to shunt. Clinically, absence of cognitive deficits favours response, and so does prominent gait disorder. Radiologically, absence of white matter changes, Evan's index >0.4 and absence of cortical atrophy have all been associated with good response.

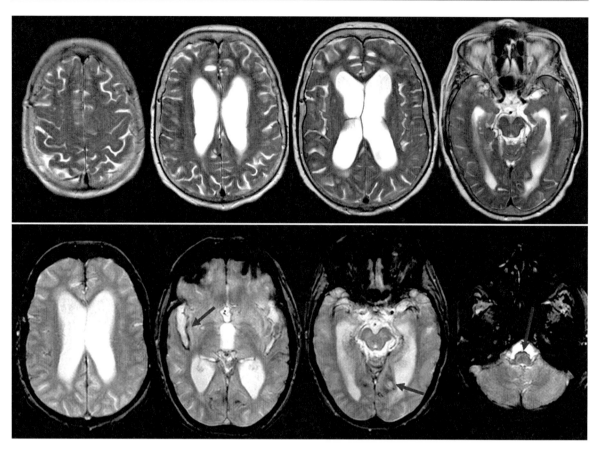

Fig. 8.13 MRI findings in normal pressure hydrocephalus (NPH). This 85-year-old man developed memory impairment and walking difficulties over the cause for several months. Note the hydrocephalus with narrow sulci at the convexity. Lumbar puncture revealed a normal pressure of 12 cm H$_2$0 and led to an improvement in the MMSE score from 12–23 and improvement in walking. The cause of NPH in this patient was previous subarachnoid haemorrhage as evidence by leptomeningeal haemosiderin deposits on T2*-weighted images (*red arrows – lower row*). He was treated with a lumboperitoneal drain that led to marked improvement, but was subsequently complicated by infections, requiring multiple drain revisions

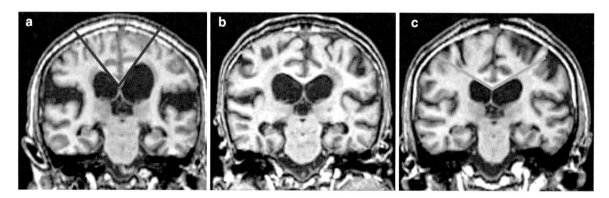

Fig. 8.14 Callosal angles in idiopathic NPH, AD and healthy man. The 64-year-old male NPH patient in (**a**) had an MMSE score of 27, Evans index of 0.30 and a callosal angle of 69°. The 67-year-old male AD patient in (**b**) had an MMSE score of 26, an Evans index of 0.29 and a callosal angle of 102°. The 70-year-old healthy man in (**c**) had an MMSE score of 30, an Evans index of 0.24, and a callosal angle of 118°. (Adapted with permission from Eur Radiol (2008) 18:2678–2683)

Enlargement of temporal horns has been reported as a poor prognostic sign – possibly reflecting the presence of hippocampal atrophy and hence AD pathology. CSF tap and MRI flow void are not predictive.

8.7 Arteriovenous Malformations (AVM) and Dural Arteriovenous Fistula (dAVF)

Arterio-venous malformations (AVM) typically present with haemorrhage or epilepsy at a relatively young age. Occasionally, the initial symptoms are cognitive in nature, especially when the AVM is strategically located, for example in the limbic system (Fig. 8.15). The diagnostic clue is an increased number of flow voids on T2-weighted images, with or without accompanying signal increase on T2/FLAIR – or signs of past bleeding as evidenced by haemosiderin deposits (i.e. low signal on T2*-weighted images).

Brain swelling may be due to increased venous pressure, for example venous thrombosis. For example,

thrombosis of the deep venous sinus may result in marked oedema of the thalamus bilaterally and reduced consciousness. Brain swelling may also occur in patients with dural arterio-venous fistula (dAVF), certainly when there is cortical venous reflux. The associated area of venous engorgement can be lobar, or even engage the complete brain (Fig. 8.16a). The clinical picture may be of a global dementia and may even mimic CJD in terms of a rapid and widespread decline.

MR imaging findings in dAVF include:
- Brain swelling
- WM signal increase (oedema)
- Increased numbers of leptomeningeal venous flow voids

Although a very rare cause of dementia, dAVF should be recognised, since the condition is fully reversible upon intravascular treatment (by injecting glue into the external carotid artery branches where they enter the dural sinus through a fistula point). The diagnosis is confirmed by catheter angiography (Fig. 8.16b).

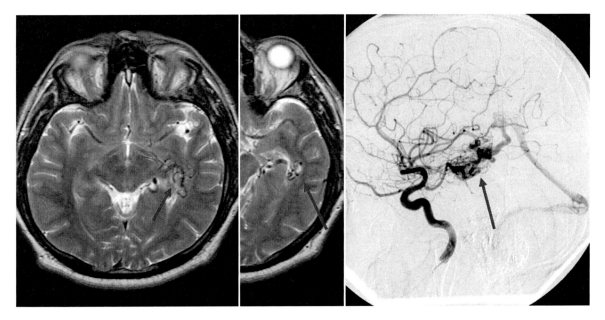

Fig. 8.15 Medial temporal lobe AVM. This 44-year-old man presented with headache, character change, language and memory problems. MRI revealed a cluster of abnormal vessels in the left medial temporal lobe on T2-weighted images. The diagnosis of AVM was confirmed by catheter angiography showing a nidus of abnormal arteriovenous connections fed by the posterior choroidal artery, with abnormal early drainage to the vein of Galen and strait sinus

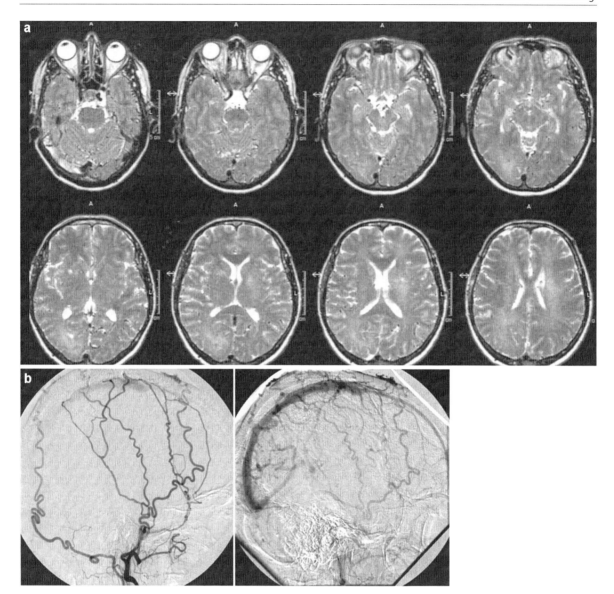

Fig. 8.16 (**a**) dAVF in patient with rapid dementia. Note diffuse high signal in white matter and increased number of leptomeningeal vessels with dark flow voids. (**b**) Catheter angiography revealed hypertrophic external carotid braches with abnormally early filling of intracranial sinus and cortical veins. Subsequent endovascular treatment led to complete clinical and radiological recovery

8.8 Reversible Posterior Leukoencephalopathy (RPLS)

Synonyms

Posterior reversible encephalopathy syndrome (PRES)

Brain swelling may be caused by white matter oedema (with or without accompanying myelin damage) in subjects treated with a variety of underlying disorders, including (malignant) hypertension, chemotherapy and others. In the so-called reversible posterior leuko-encephalopathy syndrome (RPLS), the pattern will consist of lesions located predominantly in a posterior and paramedian or parasagittal fashion. In most cases,

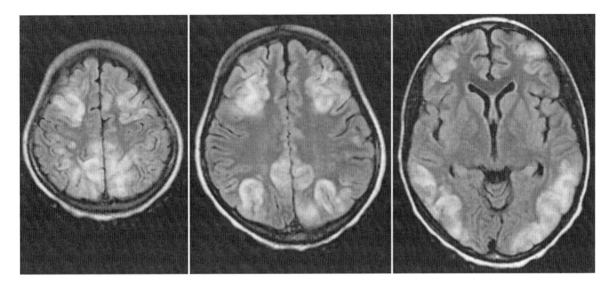

Fig. 8.17 MRI findings in RPLS/PRES. Note the bilateral areas of high signal on FLAIR in the cortex and adjacent white matter with mild swelling. The parasagittal distribution, often with a posterior preponderance is highly suggestive of RPLS/ PRES and should prompt the search for an underlying cause, such as hypertension. Diagnosis is confirmed by complete resolution of the MRI findings upon treatment of the alleviation of the cause

the abnormalities are fully reversible upon discontinuation of chemotherapy and treatment of hypertension, thus confirming the diagnosis of RPLS. However, it is worth noting that the terms PRES and RPLS have been criticised as being misnomers since the condition does necessarily involve posterior regions (or may involve many lobes); may not be reversible and is not strictly an encephalopathy (Fig. 8.17).

8.9 Amyloid-Targeted Therapy

The pathophysiology of AD is partly understood only, but the prevailing theory regarding the amyloid cascade has lead to the development of therapies directed at the removal of amyloid in the form of plaques (see Sect. 5.2). Current strategies include both active and passive immunisation. While the clinical benefit of these strategies remains uncertain, it has become clear that those therapies do indeed target cerebral amyloid and may lead to complete removal of plaques from the human brain.

However, removal of amyloid from the brain may lead to serious side effects and even death. Although clearly dose related, oedematous changes may occur on FLAIR images even at low doses in the subcortical region with associated brain swelling, usually without any clinical sequelae. This phenomenon is sometimes referred to as vasogenic oedema (or oedema – and hence the abbreviation of VE) and resembles the naturally occurring inflammatory response in sporadic CAA (see Sect. 6.4.3.2) with some similarities being drawn to RPLS. In addition to the above, pure sulcal hyperintense changes may occur as well as more widespread punctiform changes. Finally, cerebral microbleeds may occur, either at the same location or distant from it (Fig. 8.18).

An important characteristic of these signal changes and associated swelling is the transient nature. Upon cessation of therapy, the high signal changes usually disappear and after reinstalling the therapy do not re-occur usually, but may re-appear at another location. Risk factors for the occurrence of these side effects include amyloid angiopathy, APOE 4 carriership as well as the presence of age-related white matter changes and microbleeds (with a correlation with the location of subsequent VE).

There is no consensus regarding the aetiology of these changes associated with anti-amyloid therapy. Among the theories put forward is that amyloid, being removed from the parenchyma is accumulating in the perivascular drainage system leading to vessel fragility and obstruction, with either secondary microbleed formation or oedema or inflammation, or combinations of these. These observation link parenchymal to vascular amyloid and bridge the gap between vascular dementia and AD – despite the former being

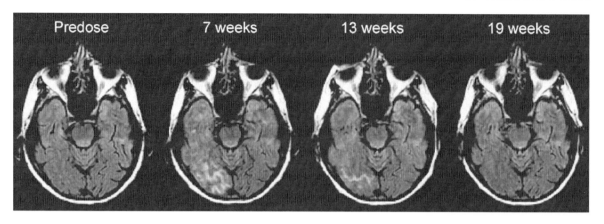

Fig. 8.18 Vasogenic oedema (VE) after anti-amyloid therapy. This 69-year-old woman is an APOE-4 homozygote who was treated with bapineuzumab 1.0 mg/kg IV. She remained asymptomatic despite the appearance of multiple areas of VE evident on the MRI. The VE was apparent on MRI by 7 weeks after her first infusion and resolved by 19 weeks. The patient was redosed at 0.5 mg/kg of bapineuzumab IV and followed for over 2 years without recurrence of VE. Reproduced with permission from Neurology 2009;73:2061–2070

associated with $A\beta$1-40 and the latter to $A\beta$1-42. See also Chap. 6.7 for a discussion of the overlap between these disorders.

Suggested Reading

Gliomatosis Cerebri Diffusa

Desclée P, Rommel D, Hernalsteen D, Godfraind C, de Coene B, Cosnard G (2010) Gliomatosis cerebri, imaging findings of 12 cases. J Neuroradiol 37(3):148–58, Epub ahead of print

Radiation Necrosis

Yu A, Li K, Li H (2006) Value of diagnosis and differential diagnosis of MRI and MR spectroscopy in gliomatosis cerebri. Eur J Radiol 59(2):216–21
Chan YL, Leung SF, King AD, Choi PH, Metreweli C (1999) Late radiation injury to the temporal lobes: morphologic evaluation at MR imaging. Radiology 213:800–807

Non-neoplastic Autoimmune Encephalitis

Dalmau J, Gleichman AJ, Hughes EG, Rossi JE, Peng X, Lai M et al (2008) Anti-NMDA-receptor encephalitis: case series and analysis of the effects of antibodies. Lancet Neurol 7:1091–8
Gultekin SH, Rosenfeld MR, Voltz R et al (2000) Paraneoplastic limbic encephalitis: neurological symptoms, immunological findings and tumour association in 50 patients. Brain 123:1481–94
Irani SR, Bera K, Waters P et al (2010) N-methyl-D-aspartate antibody encephalitis: temporal progression of clinical and paraclinical observations in a predominantly non-paraneoplastic disorder of both sexes. Brain 133:1655–1667
Rosenbloom MH, Smith S, Akdal G, Geschwind MD (2009) Immunologically mediated dementias. Curr Neurol Neurosci Rep 9(5):359–67
Vincent A, Buckley C, Schott JM et al (2004) Potassium channel antibody-associated encephalopathy: a potentially immunotherapy-responsive form of limbic encephalitis. Brain 127:701–12
Voltz R (2002) Paraneoplastic neurological syndromes: an update on diagnosis, pathogenesis, and therapy. Lancet Neurol 1:294–305

Normal Pressure Hydrocephalus (NPH)

Relkin N, Marmarou A, Klinge P, Bergsneider M, Black PM (2005) Diagnosing idiopathic normal-pressure hydrocephalus. Neurosurgery 57(3 Suppl):S4–16

Index

F. Barkhof et al., *Neuroimaging in Dementia*,
DOI: 10.1007/978-3-642-00818-4, © Springer-Verlag Berlin Heidelberg 2011